Algebra 2
and Pre-Calculus
(Volume I)

Lesson/Practice Workbook
for Self-Study and Test Preparation

Build Your Self-Confidence and Enjoyment of Math!

Comprehensive Solutions Manual Sold Separately

Aejeong Kang

MathRadar

Send all inquiries to:

MathRadar, LLC
5705 Spring Hill Dr.
Mckinney, Texas 75072

Visit www.mathradar.com for more information and a sneak preview of the MathRadar series of math books.

Send inquires via email: info@mathradar.com

Algebra 2 and Pre-Calculus (Volume I): Lesson/Practice Workbook for Self-Study and Test Preparation

ISBN-13: 978-0-9893689-5-7

ISBN-10: 0989368955

Printed in the United States of America.

Preface

I wrote these books because I am a mother and I have a strong academic background in mathematics. I have a BS degree in Mathematics and Master's degree in Mathematics as well. I have completed Ph.D. program in Biostatistics.

After receiving the big blessing of our first child, a daughter, I decided to forgo my personal career goals to become a full-time mother. When our daughter entered 7th grade, that meant lots of help with her study of math-my passion. However, I struggled to find good math books that would help her understand difficult concepts both clearly and quickly. About two years ago, I talked with my husband and my kids (now I have 2 children 8th grader, Nichole and 1st grader, Richard) about an idea that it would be better to write math books myself at least for my kids because I really want my kids study math with best books. After the conversation, I decided that the best way to help my children was by writing math books for them myself. They wholeheartedly agreed.

That's why I've been able to pour all my knowledge, energy, and soul into these books. Because I'm a mom, I would do anything for my children. Thanks to my family's endless support, I wrote them eight books, designed for use in junior high and high-school mathematics.

And that would have been the end of my journey, but my husband and children insisted that I share my work outside of our family. They encouraged me to make my work available to other parents looking, as I was, for well-written, great mathematics books for their children.

So I finally decided to publish these books. I do so with the hope that they will help your children find success and confidence in learning and studying mathematics.

But I would never have begun or finished this project without the support of my family. Kyungwan, Nichole, and Richard, you are my world. Thank you.

Introduction

☑ *After reading several pages of explanation/description about a certain mathematical concept, you still don't get it.*

☑ *You have worked on many related problems to understand mathematical concepts, but you still feel completely lost in the mathematical jungle.*

☑ *You bought a math book with good reviews, but it only offers short answers without detailed solutions. You feel confused and frustrated.*

☑ *You've tried multiple learning math books, but you've still not getting good grades in math. It seems like math is just not for you.*

If any one of these situation sound familiar, the MathRadar series will help you escape!

Everyone has different learning abilities and academic skill. MathRadar series is written and organized with emphasis on helping each individual study mathematics at his/her own pace.

Each book consists of clean and concise summaries, callouts, additional supporting explanations, quick reminders and/or shortcuts to facilitate better understanding. Each concept is thoroughly explained with step-by-step instruction and detailed proofs.

With the numerous examples and exercises, students can check their comprehension levels with both basic and more advanced problems.

Carry the MathRadar series with you!

Work on them anytime and anywhere!

Finally, you can start to enjoy mathematics!

Whether you are struggling or advanced in your math skills, the MathRadar series books will build your self-confidence and enjoyment of math.

I hope Math Radar is what you need and will be a great tool for your hard work.

Your comments or suggestions are greatly appreciated.

Please visit my website at www. mathradar.com or email me at ae-jeong@mathradar.com

Thank you very much. And remember, math can be fun!

Aejeong Kang

Algebra 2 and Pre-Calculus

Algebra 2 and Pre-Calculus

Volumes I and II have  , respectively.

The solutions manuals make it possible for students to study difficult concepts on their own. With the solutions manuals, students will be able to better understand how to solve problems through step-by-step for each problem.

TABLE OF CONTENTS

Chapter 1. The Number System

Chapter 2. Polynomials

Chapter 3.   Equations and Inequalities

Chapter 4.   Elements of Coordinate Geometry and Transformations

Chapter 5. Functions

Chapter 6.   Exponential and Logarithmic Functions

Answer Key

Index

Volume II

Chapter 1   Trigonometric Functions

Chapter 2   Matrices and Determinants

Chapter 3   Sequences and Series

Chapter 4   Probability and Statistics

Chapter 5   Conic Sections

Chapter 6   Vectors

Solutions Manual for Volume I

Solutions Manual for Volume II

Chapter 1. The Number System

1-1 Real Numbers

1. Introduction

(1) Number Lines

In a line, the counting numbers are arranged from left to right. We call them *positive integers*.

Arranging the number 0 and negative integers (positive integers with negative signs in front) on the line from right to left, we have a *number line*.

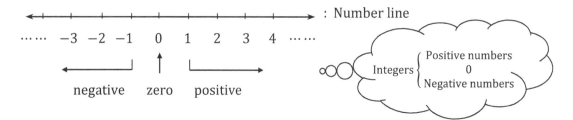

To fill a number line, we also include *rational numbers* that can be written as integers divided by other non-zero integers.

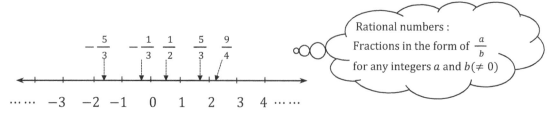

There are also many other numbers which cannot be expressed as fractions. We call them *irrational numbers*.

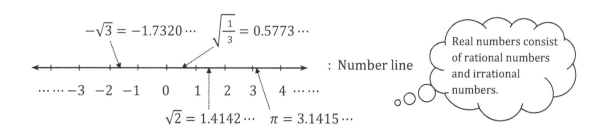

Every number has a point assigned to it on a number line.

All numbers that can be attached to points on a number line are called *real numbers*.

Note : One-to-One Correspondence

   ① *Every number corresponds to exactly one point on a real number line.*

   ② *Every point on a real number line corresponds to exactly one real number.*

(2) Classification of Real Numbers

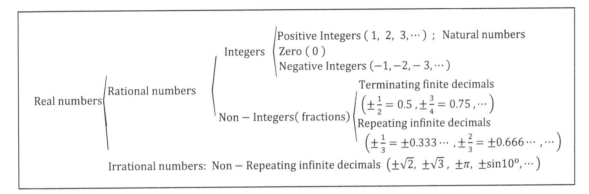

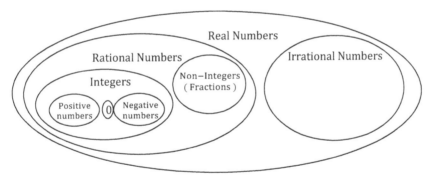

2. Exponents

(1) Positive Integral Exponents

1) Definition

For any real number a and positive integer n, the nth *power of a* is written as :

$$a^n = \underbrace{a \times a \times \cdots\cdots \times a}_{n \text{ factors}}$$: product of n repeated factors of a.

In the expression a^n, n is called the *exponent* (or *power*) of a and a is called the *base*.

a^n is read "a to the nth power."

2) Properties of Exponents

For any real numbers a, b, and positive integers m, n,

> ① $a^m \times a^n = a^{m+n}$
>
> ② $(a^m)^n = a^{mn}$
>
> ③ $(ab)^m = a^m b^m$
>
> ④ $\dfrac{a^m}{a^n} = a^{m-n}$

(2) Negative Integral and Zero Exponents

1) Definition

Consider $\dfrac{a^2}{a^5} = a^{2-5} = a^{-3}$

Since $\dfrac{a^2}{a^5} = \dfrac{1}{a^3}$, $a^{-3} = \dfrac{1}{a^3}$

In general, for any real number a $(a \neq 0)$ and positive integer n,

> $a^{-n} = \dfrac{1}{a^n}$; $a^0 = 1$

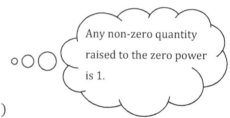

Any non-zero quantity raised to the zero power is 1.

For any non-zero number a, a^0 is equal to 1.

($\because$ Since $\dfrac{a^2}{a^2} = a^{2-2} = a^0$ and $\dfrac{a^2}{a^2} = 1$, $a^0 = 1$)

For example, $2^0 = 1$ and $(-2)^0 = 1$

2) Properties of Exponents

For any real numbers $a(a \neq 0)$, $b(b \neq 0)$, and positive integers m, n,

> ① $a^m \times a^n = a^{m+n}$
>
> ② $(a^m)^n = a^{mn}$
>
> ③ $(ab)^m = a^m b^m$

3. Radicals

(1) Square Roots

1) Definition

① A *perfect square* is an integer that is equal to the product of which two identical numbers.

Example

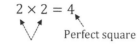

$$2 \times 2 = 4 \qquad 3 \times 3 = 9$$

Perfect square   Perfect square

Two identical numbers   Two identical numbers

② The *square root* of a perfect square is equal to the number which is multiplied twice to get the perfect square and is represented by a radical sign "$\sqrt{}$".

Note : $\sqrt{a}$ means the positive square root of a number, a.

$-\sqrt{a}$ means the negative square root of a number, a.

Example

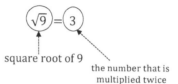

$$\boxed{\sqrt{4}} = \boxed{2} \qquad \boxed{\sqrt{9}} = \boxed{3}$$

square root of 4   square root of 9

the number that is multiplied twice   the number that is multiplied twice

Note : ① Since $3 \times 3 = 9$ and $(-3) \times (-3) = 9$, square roots of 9 are 3 and -3.

② Since $\sqrt{a}$ represents the positive square root of a number, a, $\sqrt{9} = 3$.

But $\sqrt{9} = -3$ is incorrect.

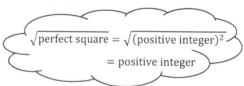

$$\sqrt{\text{perfect square}} = \sqrt{(\text{positive integer})^2}$$
$$= \text{positive integer}$$

③ Let a and b be real numbers and let $n \geq 2$ be a positive integer.

If $a = b^n$, then b is the nth root of a.

Specially, if $n = 2$, then the root is a square root. If $n = 3$, then the root is a cube root.

In the symbol $b = \sqrt[n]{a}$, $\sqrt[n]{a}$ is called a *radical*, n is called an *index*, and a is called a *radicand*.

The index 2 of the square root is usually omitted. That is, $\sqrt[2]{a} = \sqrt{a}$.

Example

$5^2 = 5 \cdot 5 = 25$

Since 5 is one of the two equal factors of 25, 5 is a square root of 25.

Since $(-2)^3 = (-2) \cdot (-2) \cdot (-2) = -8$, -2 is a cube root of -8.

Since $3^4 = 3 \cdot 3 \cdot 3 \cdot 3 = 81$, 3 is a fourth root of 81.

④ For any number a,

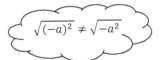

$$\text{if } a \geq 0 \Rightarrow \begin{cases} (\sqrt{a})^2 = a \;\; ; \;\; (-\sqrt{a})^2 = (\sqrt{a})^2 = a \\ \sqrt{a^2} = a \;\; ; \;\; \sqrt{(-a)^2} = \sqrt{a^2} = a \end{cases}$$

$$\text{if } a < 0 \Rightarrow \sqrt{a^2} = -a \;\; (\text{ Positive integer})$$

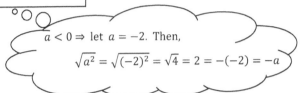

Note : For any number a,

$$\sqrt{a^2} = |a| = \begin{cases} a, \; a \geq 0 \\ -a, \; a < 0 \end{cases} \text{ where } |a| \text{ denotes the absolute value of a.}$$

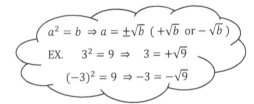

Example

$a = 3 \Rightarrow \sqrt{a^2} = \sqrt{3^2} = \sqrt{9} = 3 = a$ (Positive number)

$a = -3 \Rightarrow \sqrt{a^2} = \sqrt{(-3)^2} = \sqrt{9} = 3 = -(-3) = -a$ (Positive number)

Therefore, $\sqrt{a^2} = |a| \geq 0$

2) Magnitude of Square Roots

For any positive numbers a and b,

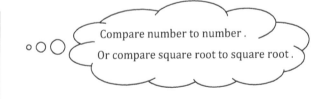

$$① \;\; a < b \;\; \Rightarrow \;\; \sqrt{a} < \sqrt{b}$$

$$② \;\; \sqrt{a} < \sqrt{b} \;\; \Rightarrow \;\; a < b$$

$$③ \;\; \sqrt{a} < \sqrt{b} \;\; \Rightarrow \;\; -\sqrt{b} < -\sqrt{a}$$

Note : ① $0 < a < \sqrt{b} < c \;\; \Rightarrow \;\; a^2 < b < c^2$

② *To compare a and $\sqrt{b}$, remove the radical sign squaring both comparing numbers.*

That is, compare $(a)^2 = a^2$ and $(\sqrt{b})^2 = b$

Or convert a number into a radical. That is, compare $a = \sqrt{a^2}$ and $\sqrt{b}$.

Example

Compare 3 to $\sqrt{5}$

Since $3^2 = 9$ and $(\sqrt{5})^2 = 5$, $3^2 > (\sqrt{5})^2$ $\therefore 3 > \sqrt{5}$

OR since $3 = \sqrt{9}$ and $9 > 5$, $\sqrt{9} > \sqrt{5}$ $\therefore 3 > \sqrt{5}$

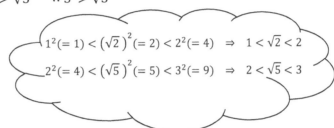

$$1^2(= 1) < \left(\sqrt{2}\right)^2(= 2) < 2^2(= 4) \Rightarrow 1 < \sqrt{2} < 2$$
$$2^2(= 4) < \left(\sqrt{5}\right)^2(= 5) < 3^2(= 9) \Rightarrow 2 < \sqrt{5} < 3$$

(2) Properties of Radicals

Note : ① $\sqrt[n]{a} = a^{\frac{1}{n}}$ *: the nth root of a.*

② *The square root of a :* $\sqrt{a} = \sqrt[2]{a} = a^{\frac{1}{2}}$

③ *The cube root of a :* $\sqrt[3]{a} = a^{\frac{1}{3}}$

For example, $\sqrt{4} = \sqrt[2]{4} = \sqrt[2]{2^2} = (2^2)^{\frac{1}{2}} = 2^1 = 2$,

$\sqrt[3]{8} = \sqrt[3]{2^3} = (2^3)^{\frac{1}{3}} = 2^1 = 2$, *and* $\sqrt[4]{16} = \sqrt[4]{2^4} = (2^4)^{\frac{1}{4}} = 2^1 = 2$

1) For any non-negative real numbers a and b,

① $\sqrt{a^2} = a$

② $\sqrt{ab} = \sqrt{a}\sqrt{b}$

③ $\sqrt{\dfrac{a}{b}} = \dfrac{\sqrt{a}}{\sqrt{b}}$, $b \neq 0$

④ $\sqrt[n]{a} = a^{\frac{1}{n}}$

⑤ $\sqrt[n]{a^m} = (a^m)^{\frac{1}{n}} = a^{\frac{m}{n}}$

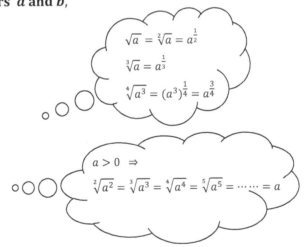

$$\sqrt{a} = \sqrt[2]{a} = a^{\frac{1}{2}}$$
$$\sqrt[3]{a} = a^{\frac{1}{3}}$$
$$\sqrt[4]{a^3} = (a^3)^{\frac{1}{4}} = a^{\frac{3}{4}}$$

$$a > 0 \Rightarrow$$
$$\sqrt[2]{a^2} = \sqrt[3]{a^3} = \sqrt[4]{a^4} = \sqrt[5]{a^5} = \cdots\cdots = a$$

2) For any real numbers a and b,

For a positive real number a, consider $\sqrt[3]{-a}$.

Since there is a negative real number $-b$ $(b > 0)$ such that $(-b)^3 = -a$,

$-b = \sqrt[3]{-a} = (-a)^{\frac{1}{3}}$.

① $\sqrt[n]{a^n} = \begin{cases} |a|, & \text{if } n \text{ is even} \\ a, & \text{if } n \text{ is odd} \end{cases}$

Example

$$a = -4 \Rightarrow \begin{cases} \sqrt[2]{a^2} = \sqrt[2]{(-4)^2} = \sqrt[2]{16} = \sqrt[2]{4^2} = (4^2)^{\frac{1}{2}} = 4^1 = 4 = |-4| = |a| \\ \sqrt[3]{a^3} = \sqrt[3]{(-4)^3} = (-4)^{3 \cdot \frac{1}{3}} = (-4)^1 = -4 = a \end{cases}$$

② $\sqrt{a^2} = |a|$

③ $\sqrt{a^3} = a\sqrt{a}$

n is odd $\Rightarrow \sqrt[n]{-a} = -\sqrt[n]{a}$

Ex. $\sqrt[3]{-8} = \sqrt[3]{-2 \cdot -2 \cdot -2} = \sqrt[3]{(-2)^3} = -2$

$-\sqrt[3]{8} = -\sqrt[3]{2^3} = -2$

$\therefore \sqrt[3]{-8} = -\sqrt[3]{8}$

Example $\sqrt{2^3} = \sqrt{8} = \sqrt{2 \cdot 4} = \sqrt{2}\sqrt{4} = \sqrt{2} \cdot 2 = 2\sqrt{2}$

4. Irrational Numbers

(1) Definition

An irrational number is a real number that cannot be expressed as a fraction.

It is expressed as the square root of a number that is not a perfect square.

Example $\sqrt{2}, \sqrt{3}, \sqrt{5}, -\sqrt{2}, -\sqrt{3}, -\sqrt{5}, \cdots\cdots$ are irrational numbers.

Note : Using a calculator, you can find the square roots of numbers.

$\sqrt{1} = 1$ $\sqrt{2} = 1.4142\cdots$ $\sqrt{3} = 1.7320\cdots$ $\sqrt{4} = 2$ $\sqrt{5} = 2.2360\cdots$ $\sqrt{6} = 2.4494\cdots$

$\sqrt{7} = 2.6457\cdots$ $\sqrt{8} = 2.8284\cdots$ $\sqrt{9} = 3$ $\sqrt{10} = 3.1622\cdots$ $\sqrt{11} = 3.3166\cdots$

Note :

$$\text{Decimal} \begin{cases} \text{Finite} \Rightarrow \text{Terminating Decimal} \Rightarrow \text{Rational Number} \\ \text{Infinite} \begin{cases} \text{Repeating Decimal} \Rightarrow \text{Rational Number} \\ \text{Non} - \text{Repeating Decimal} \Rightarrow \text{Irrational Number} \end{cases} \end{cases}$$

(2) Operations of Irrational Numbers

1) Multiplying and Dividing Square Roots

For any $a > 0$, $b > 0$,

$$\sqrt{a} \cdot \sqrt{b} = \sqrt{ab} \qquad m\sqrt{a} \cdot n = mn\sqrt{a} \qquad m\sqrt{a} \cdot n\sqrt{b} = mn\sqrt{ab}$$

$$\frac{\sqrt{a}}{\sqrt{b}} = \sqrt{\frac{a}{b}} \qquad m \div \sqrt{a} = \frac{m}{\sqrt{a}} \qquad m\sqrt{a} \div n\sqrt{b} = \frac{m}{n}\sqrt{\frac{a}{b}}$$

A. If the indexes are the same,

⇒ Apply the properties of roots.

① $\sqrt[n]{a} \cdot \sqrt[n]{b} = \sqrt[n]{ab}$

② $\dfrac{\sqrt[n]{a}}{\sqrt[n]{b}} = \sqrt[n]{\dfrac{a}{b}},\ b \neq 0$

Example

$\sqrt{3} \cdot \sqrt{5} = \sqrt{3 \cdot 5} = \sqrt{15}$ $3\sqrt{2} \cdot 5 = 15\sqrt{2}$ $3\sqrt{2} \cdot 4\sqrt{3} = 12\sqrt{6}$

$\dfrac{\sqrt{3}}{\sqrt{5}} = \sqrt{\dfrac{3}{5}}$ $5 \div \sqrt{2} = \dfrac{5}{\sqrt{2}}$ $3\sqrt{2} \div 4\sqrt{3} = \dfrac{3}{4}\sqrt{\dfrac{2}{3}}$

$\sqrt[3]{2} \cdot \sqrt[3]{5} = \sqrt[3]{10}$ $\dfrac{\sqrt[3]{2}}{\sqrt[3]{5}} = \sqrt[3]{\dfrac{2}{5}}$

B. If the indexes are different,

⇒ Rewrite each radical as an exponential expression and apply the properties of exponents.

① $\sqrt[m]{a} \cdot \sqrt[n]{a} = a^{\frac{1}{m}} \cdot a^{\frac{1}{n}} = a^{\frac{1}{m} + \frac{1}{n}}$

② $\dfrac{\sqrt[m]{a}}{\sqrt[n]{a}} = \dfrac{a^{\frac{1}{m}}}{a^{\frac{1}{n}}} = a^{\frac{1}{m} - \frac{1}{n}}$

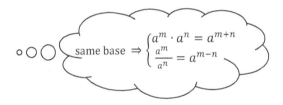

same base $\Rightarrow \begin{cases} a^m \cdot a^n = a^{m+n} \\ \dfrac{a^m}{a^n} = a^{m-n} \end{cases}$

Example

$\sqrt[3]{5} \cdot \sqrt{5} = 5^{\frac{1}{3}} \cdot 5^{\frac{1}{2}} = 5^{\frac{1}{3} + \frac{1}{2}} = 5^{\frac{5}{6}}$

$\dfrac{\sqrt[3]{5}}{\sqrt{5}} = \dfrac{5^{\frac{1}{3}}}{5^{\frac{1}{2}}} = 5^{\frac{1}{3} - \frac{1}{2}} = 5^{-\frac{1}{6}} = \dfrac{1}{5^{\frac{1}{6}}}$

$a^m \cdot a^{-m} = a^{m+(-m)} = a^0 = 1$

$\therefore a^{-m} = \dfrac{1}{a^m}$

$\sqrt{4} \cdot \sqrt[3]{8} = \sqrt[2]{2^2} \cdot \sqrt[3]{2^3} = (2^2)^{\frac{1}{2}} \cdot (2^3)^{\frac{1}{3}} = 2^1 \cdot 2^1 = 2^{1+1} = 2^2 = 4$

$\dfrac{\sqrt{4}}{\sqrt[3]{8}} = \dfrac{\sqrt[2]{2^2}}{\sqrt[3]{2^3}} = \dfrac{(2^2)^{\frac{1}{2}}}{(2^3)^{\frac{1}{3}}} = \dfrac{2^1}{2^1} = 2^{1-1} = 2^0 = 1$

2) Using the Distributive Property for Square Roots

For any $a > 0,\ b > 0,\ c > 0,$

① $\sqrt{a} \cdot (\sqrt{b} + \sqrt{c}) = \sqrt{ab} + \sqrt{ac}$ $(\sqrt{a} + \sqrt{b}) \cdot \sqrt{c} = \sqrt{ac} + \sqrt{bc}$

② $\sqrt{a} \cdot (\sqrt{b} - \sqrt{c}) = \sqrt{ab} - \sqrt{ac}$ $(\sqrt{a} - \sqrt{b}) \cdot \sqrt{c} = \sqrt{ac} - \sqrt{bc}$

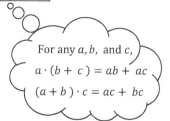

For any $a, b,$ and $c,$

$a \cdot (b + c) = ab + ac$

$(a + b) \cdot c = ac + bc$

Example

$$\sqrt{2} \cdot (\sqrt{3} + \sqrt{5}) = \sqrt{2} \cdot \sqrt{3} + \sqrt{2} \cdot \sqrt{5} = \sqrt{6} + \sqrt{10}$$

$$(\sqrt{6} - \sqrt{2}) \cdot \sqrt{3} = \sqrt{6} \cdot \sqrt{3} - \sqrt{2} \cdot \sqrt{3} = \sqrt{18} - \sqrt{6} = 3\sqrt{2} - \sqrt{6}$$

3) Simplifying Square Roots

A. Moving inside of the radical

For any $a > 0,\ b > 0,$

① $\sqrt{a^2 b} = \sqrt{a^2} \cdot \sqrt{b} = a\sqrt{b}$

② $\sqrt{\dfrac{a}{b^2}} = \dfrac{\sqrt{a}}{\sqrt{b^2}} = \dfrac{\sqrt{a}}{b}$

$a < 0,\ b > 0$

$\Rightarrow \sqrt{a^2 b} = -a\sqrt{b}$

Example

$$\sqrt{12} = \sqrt{2^2 \cdot 3} = \sqrt{2^2} \cdot \sqrt{3} = 2\sqrt{3} \qquad \sqrt{\dfrac{2}{9}} = \sqrt{\dfrac{2}{3^2}} = \dfrac{\sqrt{2}}{\sqrt{3^2}} = \dfrac{\sqrt{2}}{3}$$

B. Moving outside of the radical

For any $a > 0,\ b > 0,$

① $a\sqrt{b} = \sqrt{a^2} \cdot \sqrt{b} = \sqrt{a^2 b}$

② $\dfrac{\sqrt{a}}{b} = \dfrac{\sqrt{a}}{\sqrt{b^2}} = \sqrt{\dfrac{a}{b^2}}$

Example

$$3\sqrt{5} = \sqrt{9} \cdot \sqrt{5} = \sqrt{45} \qquad \dfrac{\sqrt{5}}{3} = \dfrac{\sqrt{5}}{\sqrt{9}} = \sqrt{\dfrac{5}{9}}$$

4) Rationalizing Denominators

To express $\dfrac{1}{\sqrt{2}}$ as a decimal, we have to divide 1 by $\sqrt{2}$.

Since $\dfrac{1}{\sqrt{2}} = \dfrac{1}{1.414\cdots}$, it is quite awkward to calculate. So, you use a simpler procedure to eliminate the radical in the denominator.

That is, $\dfrac{1}{\sqrt{2}} = \dfrac{1}{\sqrt{2}} \cdot 1 = \dfrac{1}{\sqrt{2}} \cdot \left(\dfrac{\sqrt{2}}{\sqrt{2}}\right) = \dfrac{1 \cdot \sqrt{2}}{\sqrt{2} \cdot \sqrt{2}} = \dfrac{\sqrt{2}}{(\sqrt{2})^2} = \dfrac{\sqrt{2}}{2} \approx \dfrac{1.414}{2} = 0.707$

For any $a > 0,\ b > 0,\ c > 0,$

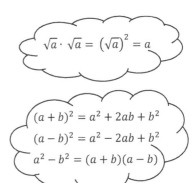

$$① \quad \dfrac{1}{\sqrt{a}} = \dfrac{1}{\sqrt{a}} \cdot 1 = \dfrac{1 \cdot \sqrt{a}}{\sqrt{a} \cdot \sqrt{a}} = \dfrac{\sqrt{a}}{(\sqrt{a})^2} = \dfrac{\sqrt{a}}{a}$$

$$② \quad \dfrac{\sqrt{b} + \sqrt{c}}{\sqrt{a}} = \dfrac{(\sqrt{b} + \sqrt{c}) \cdot \sqrt{a}}{\sqrt{a} \cdot \sqrt{a}} = \dfrac{\sqrt{ab} + \sqrt{ac}}{a}$$

$$③ \quad \dfrac{1}{\sqrt{a} + \sqrt{b}} = \dfrac{1 \cdot (\sqrt{a} - \sqrt{b})}{(\sqrt{a} + \sqrt{b}) \cdot (\sqrt{a} - \sqrt{b})} = \dfrac{\sqrt{a} - \sqrt{b}}{a - b}, a \neq b$$

$\sqrt{a} \cdot \sqrt{a} = \left(\sqrt{a}\right)^2 = a$

$(a + b)^2 = a^2 + 2ab + b^2$
$(a - b)^2 = a^2 - 2ab + b^2$
$a^2 - b^2 = (a + b)(a - b)$

Example

$$\dfrac{1}{\sqrt{2}} + \dfrac{1}{\sqrt{3}} = \dfrac{1}{\sqrt{2}} \cdot \left(\dfrac{\sqrt{2}}{\sqrt{2}}\right) + \dfrac{1}{\sqrt{3}} \cdot \left(\dfrac{\sqrt{3}}{\sqrt{3}}\right) = \dfrac{\sqrt{2}}{2} + \dfrac{\sqrt{3}}{3} = \dfrac{3\sqrt{2} + 2\sqrt{3}}{6}$$

$$\dfrac{1}{\sqrt{2} - \sqrt{3}} = \dfrac{1 \cdot (\sqrt{2} + \sqrt{3})}{(\sqrt{2} - \sqrt{3}) \cdot (\sqrt{2} + \sqrt{3})} = \dfrac{\sqrt{2} + \sqrt{3}}{(\sqrt{2})^2 - (\sqrt{3})^2} = \dfrac{\sqrt{2} + \sqrt{3}}{2 - 3} = -\sqrt{2} - \sqrt{3}$$

5) Adding and Subtracting Square Roots

To add and subtract square roots, the radicands of the square roots must be the same.

A. If the radicands of the square roots are the same, then just add or subtract the numbers which are multiplied by the square roots of the same radicands.

That is, for any real numbers $a(a > 0),\ m,$ and $n,$

$$m\sqrt{a} + n\sqrt{a} = (m + n)\sqrt{a} \qquad m\sqrt{a} - n\sqrt{a} = (m - n)\sqrt{a}$$

Note :

Treat the square root of the same radicands as a term which is an expression of a single number or a product of numbers and variables. $a\sqrt{x} + b\sqrt{x} = (a + b)\sqrt{x}$

Example

$$5\sqrt{3} + 4\sqrt{3} = (5 + 4)\sqrt{3} = 9\sqrt{3}$$

$$5\sqrt{3} - 4\sqrt{3} = (5 - 4)\sqrt{3} = \sqrt{3}$$

$\sqrt{a} + \sqrt{b} \neq \sqrt{a + b}$
$\sqrt{a} - \sqrt{b} \neq \sqrt{a - b}$

B. If the radicands of the square roots are different, simplify each of the radicands.

If each of the radicands has a perfect square factor, they can be simplified.

Example

$$2\sqrt{12} + 3\sqrt{27} = 2\sqrt{4\cdot 3} + 3\sqrt{9\cdot 3} = 2\left(\sqrt{4}\cdot\sqrt{3}\right) + 3\left(\sqrt{9}\cdot\sqrt{3}\right)$$

$$= 2\left(2\sqrt{3}\right) + 3\left(3\sqrt{3}\right) = 4\sqrt{3} + 9\sqrt{3} = (4+9)\sqrt{3} = 13\sqrt{3}$$

> $\sqrt{a\cdot b} = \sqrt{a}\cdot\sqrt{b}$
> $\sqrt{12} = \sqrt{4\cdot 3} = \sqrt{4}\cdot\sqrt{3} = 2\cdot\sqrt{3} = 2\sqrt{3}$
> $a\sqrt{b}$ means $a\times\sqrt{b}$

5. Number Operations

(1) Magnitude of Real Numbers

A. For any real numbers a and b,

① $a > b$ or $a = b$ or $a < b$

② $a > 0,\ b > 0 \Rightarrow a + b > 0,\ ab > 0$

③ $a > 0 \Rightarrow -a < 0$

$a < 0 \Rightarrow -a > 0$

B. For any real numbers $a, b,$ and c,

① $a > b,\ b > c \Rightarrow a > c$

② $a > b \Rightarrow a + c > b + c$

$a - c > b - c$

③ $a > b,\ c > 0 \Rightarrow ac > bc$

$\dfrac{a}{c} > \dfrac{b}{c}\ (c \neq 0)$

④ $a > b,\ c < 0 \Rightarrow ac < bc$

$\dfrac{a}{c} < \dfrac{b}{c}\ (c \neq 0)$

> $a > b > 0 \Rightarrow \dfrac{1}{a} < \dfrac{1}{b}$
> $(\because \dfrac{1}{a} - \dfrac{1}{b} = \dfrac{b-a}{ab} < 0$
> $(\because a > b > 0 \Rightarrow b - a < 0$ and $ab > 0)$
> Thus $\dfrac{1}{a} < \dfrac{1}{b}.)$

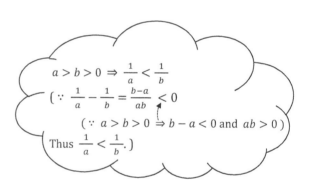

(2) Properties of Real Numbers

For any real numbers $a, b,$ and c,

1) Closure Law

① The sum $a + b$ is a unique real number c. (addition)

② The product $a \times b$ is a unique real number c. (multiplication)

2) Commutative Law

① $a + b = b + a$ (addition)

② $a \times b = b \times a$ (multiplication)

3) Associative Law

① $(a + b) + c = a + (b + c)$ (addition)

② $(a \times b) \times c = a \times (b \times c)$ (multiplication)

4) Identity

① $a + 0 = 0 + a = a$ (addition)

② $a \times 1 = 1 \times a = a$ (multiplication)

Note : 0 is called the identity element in the addition of real numbers.

1 is called the identity element in the multiplication of real numbers.

5) Inverse

① $a + (-a) = (-a) + a = 0$ (addition)

② $a \times \left(\frac{1}{a}\right) = \left(\frac{1}{a}\right) \times a = 1, \ a \neq 0$ (multiplication)

Note : The real number – a is called the additive inverse (or opposite) of the real number a.

The real number $\frac{1}{a}$ (a ≠ 0) is called the multiplicative inverse (or reciprocal) of the real number

a (a ≠ 0).

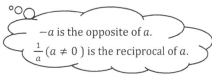

−a is the opposite of a.

$\frac{1}{a}$ (a ≠ 0) is the reciprocal of a.

Note : ① $a \times b = 1 \Rightarrow$ a and b are reciprocals of each other.

② The reciprocal of a fraction is obtained by reversing the numerator and denominator.

For any a(a ≠ 0) and b(b ≠ 0), the reciprocal of $\frac{a}{b}$ is $\frac{b}{a}$.

For example, the reciprocal of a is $\frac{1}{a}$, because $a = \frac{a}{1}$.

③ If a(a ≠ 0) is an integer, $\frac{1}{a}$ is not the reciprocal of a . This is because $\frac{1}{a}$ (fraction) is not

included in integers. Thus, there is no reciprocal of a in integers. However, $\frac{1}{a}$ is the

reciprocal of a in rational numbers (fractions) and real numbers.

6) Distributive Law

$a \times (b + c) = (a \times b) + (a \times c)$

$(a + b) \times c = (a \times c) + (b \times c)$

6. The Absolute Value of a Real Number

(1) Definition of Absolute Value

For any real number a, the absolute value of a is denoted by

$$|a| = \begin{cases} a, & \text{if } a \geq 0 \\ -a, & \text{if } a < 0 \end{cases}$$

Note : The absolute value of a real number is always greater than or equal to 0 ($|a| \geq 0$).

For example, $|0| = 0,\ |-2| = -(-2) = 2$

(2) Properties of Absolute Values

For any real numbers a and b,

① $|a| \geq 0$

② $|a| = |-a|,\ |a|^2 = a^2$

③ $|ab| = |a||b|,\ \left|\dfrac{a}{b}\right| = \dfrac{|a|}{|b|}\ (b \neq 0)$

④ $|a| + |b| = 0 \iff a = 0$ and $b = 0$

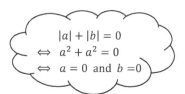

$|a| + |b| = 0$
$\iff a^2 + a^2 = 0$
$\iff a = 0$ and $b = 0$

Note : $|a| = |-a|$ for any real number a.

    (Case 1. $a \geq 0$

        $\Rightarrow -a \leq 0$

        $\therefore |-a| = -(-a) = a$ $\therefore |a| = |-a|$

    Case 2. $a < 0$

        $\Rightarrow -a > 0$

        $\therefore |a| = -a,\ |-a| = -a$ $\therefore |a| = |-a|$

    Therefore, $|a| = |-a|$)

Note : $|a| + |b| = 0 \Rightarrow a = 0$ and $b = 0$

    ($\because$ *Prove* $[a \neq 0$ *or* $b \neq 0 \Rightarrow |a| + |b| \neq 0$]

    Case 1. $a \neq 0 \Rightarrow |a| > 0$

        Since $|b| \geq 0$ *by the definition of absolute values,* $|a| + |b| > 0$

        $\therefore |a| + |b| \neq 0$

    Case 2. $b \neq 0 \Rightarrow |b| > 0$

        Since $|a| \geq 0$ *by the definition of absolute values,* $|a| + |b| > 0$

        $\therefore |a| + |b| \neq 0$

    Thus if $a \neq 0$ *or* $b \neq 0$, *then* $|a| + |b| \neq 0$

    Therefore, $|a| + |b| = 0 \Rightarrow a = 0$ *and* $b = 0$ *i.e.,* $a = b = 0$)

7. Correspondence between Real Numbers and Points on a Line

(1) Coordinates

The coordinate is the number associated with a point. We use coordinates to identify points. The length between two points is always a positive number. Thus, we use the absolute value to express the length.

(2) The Distance between Two Points in the Plane

For any real numbers a and b, the distance d between a and b is defined by a real number :
$d = |a - b| = |b - a|$.

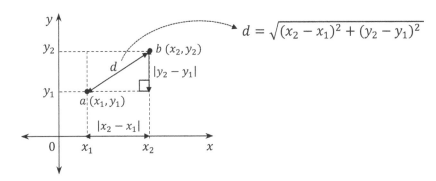

By the Pythagorean Theorem, the square of the distance d between $a(x_1, y_1)$ and $b(x_2, y_2)$ is

$$d^2 = |x_2 - x_1|^2 + |y_2 - y_1|^2.$$

Since the distance d is always a positive number, choose the positive square root. That is,

$$d = \sqrt{|x_2 - x_1|^2 + |y_2 - y_1|^2} = \sqrt{(x_2 - x_1)^2 + (y_2 - y_1)^2}$$

(3) The Midpoint Formula

The coordinates of the midpoint of a line segment are the average values of the corresponding coordinates of the endpoints. The *midpoint* m of the line segment joining the $a(x_1, y_1)$ and $b(x_2, y_2)$ in the coordinate plane is

$$m = \left(\frac{x_1 + x_2}{2} , \frac{y_1 + y_2}{2} \right)$$

1-2 Complex Numbers

1. Imaginary Unit and Complex Number

(1) The Imaginary Unit i

1) Definition

The square roots of positive numbers, negative numbers, and zero are never negative numbers ; i.e., (Real number)$^2 \geq 0$. Therefore, the quadratic equation $x^2 = -1$ has no real solution because there is no real number x which can be squared to produce -1 (Negative). To work with square roots of negative numbers, the expanded system of numbers using the imaginary unit i was created.

The square of imaginary unit i is -1. That is, *imaginary unit i* is defined as

$$i = \sqrt{-1} \quad \text{where } i^2 = -1$$

Note : Square roots of negative numbers other than -1 can be expressed in terms of i by factoring out $\sqrt{-1}$ and replacing it with i.

2) Properties of the Imaginary Unit i

For any positive real number a,

$$\text{①} \quad \sqrt{-a} = \sqrt{-1 \cdot a} = \sqrt{-1}\sqrt{a} = i\sqrt{a} = \sqrt{a}\, i$$
$$\text{②} \quad (i\sqrt{a})^2 = i^2(\sqrt{a})^2 = -1 \cdot a = -a$$

3) Properties of i^n

For any integer n ($n > 0$),

$$\text{①} \quad i^{4n} = 1$$
$$\text{②} \quad i^{4n+1} = i$$
$$\text{③} \quad i^{4n+2} = -1$$
$$\text{④} \quad i^{4n+3} = -i$$

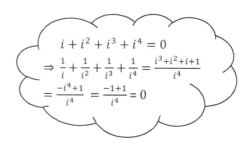

$$i + i^2 + i^3 + i^4 = 0$$
$$\Rightarrow \frac{1}{i} + \frac{1}{i^2} + \frac{1}{i^3} + \frac{1}{i^4} = \frac{i^3 + i^2 + i + 1}{i^4}$$
$$= \frac{-i^4 + 1}{i^4} = \frac{-1 + 1}{i^4} = 0$$

Note : The pattern of values $i, -1, -i,$ and 1 is repeated for power greater than 4.

$i^1 = i$ $\qquad\qquad i^2 = -1 \qquad\qquad i^3 = i^2 \cdot i = (-1) \cdot i = -i \qquad\qquad i^4 = i^2 \cdot i^2 = (-1) \cdot (-1) = 1$

$i^5 = i^4 \cdot i = 1 \cdot i = i \qquad\qquad i^6 = i^4 \cdot i^2 = 1 \cdot (-1) = -1 \qquad\qquad i^7 = i^4 \cdot i^3 = 1 \cdot (-i) = -i$

$i^8 = i^4 \cdot i^4 = (1) \cdot (1) = 1$

(2) Complex Numbers

1) Definition of a Complex Number

For any real numbers *a and b*, the *number* $a + bi$ is a *complex number* written in *standard form*. The number *a* is the *real part* and the number *bi* is the *imaginary part* of the complex number.

If $b = 0$, then $a + bi$ is a real number.

If $b \neq 0$, then $a + bi$ is an imaginary number.

If $a = 0$ and $b \neq 0$, then $a + bi = bi$ is a *pure imaginary number.*

Complex Numbers ($a + bi$)

Real Numbers
($b = 0$)

Imaginary Numbers
($b \neq 0$)

Pure Imaginary Numbers
($a = 0$)

> If two expressions have the same value for all values of their variables, the expressions are called *equivalent.*
> For example,
> the expressions $2x + 3x - 4$ and $5x - 4$ are equivalent.
> The statement
> $2x + 3x - 4 = 5x - 4$
> is called an *identity.*

2) Equality of Two Complex Numbers

For any real numbers *a, b, c,* and *d,*

① $a + bi = 0 \xleftrightarrow[\text{if and only if}]{} a = 0,\ b = 0$

② $a + bi = c + di \xleftrightarrow[\text{if and only if}]{} a = c,\ b = d$

  (Two complex numbers $a + bi$ and $c + di$ are equal to each other :

  i.e., $a + bi = c + di$)

  if and only if ($a = c$ and $b = d$)

> For any real number *a,*
> $a = a + 0i$

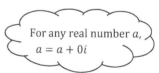

Note : a and b are rational numbers. If $a + bi = 0$, then $a = b = 0$.

  ($\because$ Suppose $b \neq 0$. Then $i = -\dfrac{a}{b}$

  Since a and b are rational numbers, $-\dfrac{a}{b}$ is a rational number.

  Since i is an imaginary number, $i = -\dfrac{a}{b}$ is impossible. Thus b must be equal to 0.

  Therefore, $a = b = 0$)

2. Adding and Subtracting Complex Numbers

For any real numbers a, b, c, and d,

> (1) $(a + bi) + (c + di) = (a + c) + (b + d)i$
>
> (2) $(a + bi) - (c + di) = (a - c) + (b - d)i$

Note : The additive identity in the complex system is zero and the additive inverse of the complex number

$a + bi$ is $-(a + bi) = -a - bi$.

Therefore, $(a + bi) + (-a - bi) = 0 + 0i = 0$

Example Write each expression as a complex number in standard form.

$$(3 + 2i) + (5 - 7i) = (3 + 5) + (2 - 7)i = 8 - 5i$$

$$-4 + (-2 + 6i) - (3 - 5i) = -4 + (-2 - 3) + (6 + 5)i = -9 + 11i$$

3. Multiplying and Dividing Complex Numbers

(1) Multiplying Complex Numbers

> ① $bi \cdot di = (bd) \cdot (i^2) = (bd)(-1) = -bd$
>
> ② $(a + bi) \cdot (c + di) = (ac - bd) + (ad + bc)i$

Note : $(a + bi) \cdot (c + di) = a(c + di) + bi(c + di)$ (Distributive)

$\qquad\qquad\qquad = ac + adi + bci + bd\, i^2$ (Distributive)

$\qquad\qquad\qquad = ac + (ad + bc)i + bd(-1)$ (Distributive, $i^2 = -1$)

$\qquad\qquad\qquad = (ac - bd) + (ad + bc)i$ (Commutative)

Example Write each expression as a complex number in standard form.

$$3i \cdot -4i = (3) \cdot (-4) \cdot (i^2) = -12 \cdot (-1) = 12$$

$$(2 + 6i) \cdot (3 - 5i) = (2 \cdot 3 + 6 \cdot 5) + (2 \cdot (-5) + 6 \cdot 3)i = 36 + 8i$$

(2) Multiplying Complex Conjugates

For a complex number $z = a + bi$ (a and b are real numbers), the *complex conjugate* $\bar{z}$ of z is

defined to be the complex number $\bar{z} = \overline{a + bi} = a - bi$.

Multiplying complex conjugates, we have

> $z\,\bar{z} = (a + bi)(a - bi) = a^2 + b^2$

Example Express each product in simplest form.

$$(2 + 3i)(2 - 3i) = 2^2 + 3^2 = 13$$

$2 + 3i$ and $2 - 3i$ are complex conjugates

$$(-3 + i)(-3 - i) = (-3)^2 + 1^2 = 10$$

$-3 + i$ and $-3 - i$ are complex conjugates

(3) Dividing Complex Numbers

If the denominator of a fraction is a complex number $+bi$, then multiply the numerator and the denominator by the complex conjugate of the denominator.

Note that the product of complex conjugates is always a real number.

$$\frac{c+di}{a+bi} = \frac{c+di}{a+bi} \cdot 1 = \frac{c+di}{a+bi} \cdot \frac{a-bi}{a-bi} = \frac{(c+di)(a-bi)}{(a+bi)(a-bi)} = \frac{(ac+bd)+(ad-bc)i}{a^2+b^2}$$

$$= \frac{(ac+bd)}{a^2+b^2} + \frac{(ad-bc)}{a^2+b^2}i, \quad a + bi \neq 0$$

Example Find the multiplicative inverse of $2 - 3i$ in standard form.

Since he multiplicative inverse of $2 - 3i$ is its reciprocal $\frac{1}{2-3i}$,

$$\frac{1}{2-3i} = \frac{2+3i}{(2-3i)(2+3i)} = \frac{2+3i}{4+9} = \frac{2}{13} + \frac{3}{13}i$$

∴ The multiplicative inverse of $2 - 3i$ in standard form is $\frac{2}{13} + \frac{3}{13}i$.

Example Express the value of x in standard form : $(3 + 4i)x + 2 - i = (5 + 2i)x + 6 + 3i$

$$(3 + 4i)x + 2 - i = (5 + 2i)x + 6 + 3i$$

$$\Rightarrow (3 + 4i)x - (5 + 2i)x = 6 + 3i - (2 - i) \quad \longleftarrow \quad \text{Separate all the } x's \text{ on one side and everything else on the other.}$$

$$\Rightarrow (-2 + 2i)x = 4 + 4i \quad \longleftarrow \quad \text{Combine similar terms.}$$

$$\Rightarrow x = \frac{4(1+i)}{-2(1-i)} = \frac{-2(1+i)}{1-i} \quad \longleftarrow \quad \text{Divide by the coefficient of .}$$

$$= \frac{-2(1+i)^2}{(1-i)(1+i)} \quad \longleftarrow \quad \text{Multiply numerator and denominator by the conjugate of the denominator.}$$

$$= \frac{-2(1+2i-1)}{1+1} = -2i \quad \longleftarrow \quad \text{Simplify}$$

(4) Properties of Complex Conjugates

For complex numbers z_1 and z_2,

① $\overline{(\overline{z_1})} = z_1$

② $\overline{z_1 \pm z_2} = \overline{z_1} \pm \overline{z_2}$

③ $\overline{z_1 \cdot z_2} = \overline{z_1} \cdot \overline{z_2}$

④ $\overline{\left(\dfrac{z_1}{z_2}\right)} = \dfrac{\overline{z_1}}{\overline{z_2}}$, $z_2 \neq 0$

⑤ $z_1 \cdot \overline{z_1}$ is a real number.

⑥ $z_1 + \overline{z_1}$ is a real number.

Note : For any real numbers a, b, c, and d, let $z_1 = a + bi$ *and* $z_2 = c + di$. *Then,*

① $\overline{z_1} = \overline{a + bi} = a - bi$ $\therefore \overline{(\overline{z_1})} = \overline{a - bi} = a + bi = z_1$

② $\overline{z_1 \pm z_2} = \overline{(a + bi) \pm (c + di)} = \overline{(a \pm c) + (b \pm d)i} = (a \pm c) - (b \pm d)i$

$\overline{z_1} \pm \overline{z_2} = \overline{a + bi} \pm \overline{c + di} = (a - bi) \pm (c - di) = (a \pm c) - (b \pm d)i$

$\therefore \overline{z_1 \pm z_2} = \overline{z_1} \pm \overline{z_2}$

③ $\overline{z_1 \cdot z_2} = \overline{(a + bi)(c + di)} = \overline{(ac - bd) + (ad + bc)i} = (ac - bd) - (ad + bc)i$

$\overline{z_1} \cdot \overline{z_2} = \overline{(a + bi)} \cdot \overline{(c + di)} = (a - bi)(c - di) = (ac - bd) - (ad + bc)i$

$\therefore \overline{z_1 \cdot z_2} = \overline{z_1} \cdot \overline{z_2}$

④ *Since* $\dfrac{z_1}{z_2} = \dfrac{a + bi}{c + di} = \dfrac{(a + bi)(c - di)}{c^2 + d^2} = \dfrac{ac + bd}{c^2 + d^2} + \dfrac{bc - ad}{c^2 + d^2} i$, $\overline{\left(\dfrac{z_1}{z_2}\right)} = \dfrac{ac + bd}{c^2 + d^2} - \dfrac{bc - ad}{c^2 + d^2} i$

Since $\dfrac{\overline{z_1}}{\overline{z_2}} = \dfrac{a - bi}{c - di} = \dfrac{(a - bi)(c + di)}{c^2 + d^2} = \dfrac{ac + bd}{c^2 + d^2} + \dfrac{ad - bc}{c^2 + d^2} i = \dfrac{ac + bd}{c^2 + d^2} - \dfrac{bc - ad}{c^2 + d^2} i$, $\overline{\left(\dfrac{z_1}{z_2}\right)} = \dfrac{\overline{z_1}}{\overline{z_2}}$

⑤ $z_1 \cdot \overline{z_1} = (a + bi)(a - bi) = a^2 + b^2$: *Real number*

⑥ $z_1 + \overline{z_1} = (a + bi) + (a - bi) = 2a$: *Real number*

For a complex number $z = a + bi$,

1) $z = \overline{z} \xLeftrightarrow{\text{if and only if}} z$ is a real number.

2) If $\overline{z} = -z$, $z \neq 0$, then z is a pure imaginary number.

3) If $z^2 < 0$, then z is a pure imaginary number.

Note :

① $z = \bar{z} \iff a + bi = \overline{a + bi} = a - bi$

$\qquad \iff a = a \ and \ b = -b \ ; \ i.e., \ b = 0$

$\qquad \iff$ z is a real number.

② $\bar{z} = -z, z \neq 0 \implies a - bi = -a - bi \implies a = 0$

$\qquad\qquad\qquad \implies z = 0 + bi = bi \ (Pure \ imaginary \ number)$

③ $z^2 = (a + bi)^2 = a^2 + 2abi - b^2 = (a^2 - b^2) + 2abi < 0$

$\qquad \implies a = 0, \ b \neq 0 \implies z = bi \ (Pure \ imaginary \ number)$

4. Square Root of a Negative Number

> (1) $a > 0$
>
> $\qquad$ Since $\sqrt{-a} = \sqrt{a}\,i$, the square root of the negative number $-a$ is $\pm\sqrt{a}\,i$.

Example $\qquad (2i)^2 = -4$ and $(-2i)^2 = -4$

$\qquad\qquad$ Therefore, the square roots of -4 are $2i$ and $-2i$.

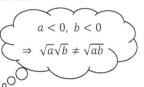

$a < 0, \ b < 0$
$\implies \sqrt{a}\sqrt{b} \neq \sqrt{ab}$

> (2) $a \leq 0, \ b \leq 0$ (a and b are both negative numbers) $\implies \sqrt{a}\sqrt{b} = -\sqrt{ab}$

Example $\qquad \sqrt{-2}\sqrt{-2} = (\sqrt{2}\,i\,)(\sqrt{2}i) = (\sqrt{2})^2 (i)^2 = 2 \cdot (-1) = -2$

$\qquad\qquad$ whereas $\sqrt{(-2)(-2)} = \sqrt{4} = 2$

$\qquad\qquad$ Therefore, $\sqrt{-2}\sqrt{-2} \neq \sqrt{(-2)(-2)}$

> (3) $a \geq 0, \ b < 0 \implies \dfrac{\sqrt{a}}{\sqrt{b}} = -\sqrt{\dfrac{a}{b}}$

Before multiplying
square roots of negative numbers,
be sure to convert to standard form $a + bi$.

Example $\qquad \dfrac{\sqrt{2}}{\sqrt{-3}} = \dfrac{\sqrt{2}}{\sqrt{3}\,i} = \dfrac{\sqrt{2} \cdot \sqrt{3}\,i}{\sqrt{3}\,i \cdot \sqrt{3}\,i} = \dfrac{\sqrt{6}\,i}{(\sqrt{3})^2 (i)^2} = -\dfrac{\sqrt{6}}{3}\,i$

$\qquad\qquad$ whereas $\sqrt{\dfrac{2}{-3}} = \sqrt{\dfrac{2}{3}}\,i = \dfrac{\sqrt{2}}{\sqrt{3}}\,i = \dfrac{\sqrt{2} \cdot \sqrt{3}}{\sqrt{3} \cdot \sqrt{3}}\,i = \dfrac{\sqrt{6}}{3}\,i$

$\qquad\qquad$ Therefore, $\dfrac{\sqrt{2}}{\sqrt{-3}} \neq \sqrt{\dfrac{2}{-3}}$

5. Absolute Values of Complex Numbers

(1) Definition

For a complex number $z = a + bi$, the *absolute value* of z is defined by a non-negative real number $|z| = |a + bi| = \sqrt{a^2 + b^2}$.

A complex number $z = a + bi$ can be graphed in the complex plane by using the ordered pair (a, b).

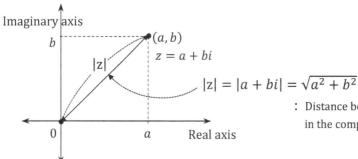

$$|z| = |a + bi| = \sqrt{a^2 + b^2}$$

: Distance between a number and the origin in the complex plane

Note : (*One-to-One Correspondence*)

   Every complex number corresponds to a point in the complex plane.

Example Find the absolute value of each complex number.

$$z = 3 + 4i \;\Rightarrow\; |z| = |3 + 4i| = \sqrt{3^2 + 4^2} = \sqrt{25} = 5$$

$$z = -2i \;\Rightarrow\; |z| = |0 - 2i| = \sqrt{0^2 + (-2)^2} = \sqrt{4} = 2$$

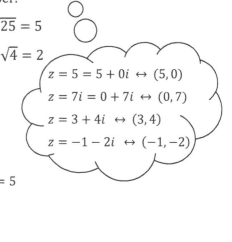

$$z = 5 = 5 + 0i \;\leftrightarrow\; (5, 0)$$
$$z = 7i = 0 + 7i \;\leftrightarrow\; (0, 7)$$
$$z = 3 + 4i \;\leftrightarrow\; (3, 4)$$
$$z = -1 - 2i \;\leftrightarrow\; (-1, -2)$$

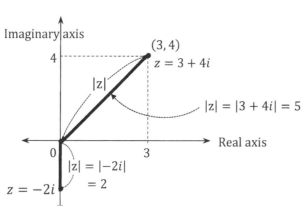

(2) Properties of Absolute Values of Complex Numbers

For a complex number $z = a + bi$

1) $|z| = |\bar{z}|$

2) $z\,\bar{z} = |z|^2 = |\bar{z}|^2$

Note :

① *Since* $|z| = |a + bi| = \sqrt{a^2 + b^2}$ *and* $|\bar{z}| = |a - bi| = \sqrt{a^2 + (-b)^2} = \sqrt{a^2 + b^2}$,

$|z| = |\bar{z}|$

② *Since* $z\,\bar{z} = (a + bi)(a - bi) = a^2 - (bi)^2 = a^2 + b^2$,

$|z|^2 = |a + bi|^2 = \left(\sqrt{a^2 + b^2}\,\right)^2 = a^2 + b^2$, *and*

$|\bar{z}|^2 = |a - bi|^2 = \left(\sqrt{a^2 + b^2}\,\right)^2 = a^2 + b^2$,

$z\,\bar{z} = |z|^2 = |\bar{z}|^2$

6. Classification of Numbers

$$
\text{Complex Numbers } (a + bi)
\begin{cases}
\text{Real Numbers } (b = 0)
\begin{cases}
\text{Rational Numbers}
\begin{cases}
\text{Integers}
\begin{cases}
\text{Positive} \\
0 \\
\text{Negative}
\end{cases} \\
\text{Fractions}
\end{cases} \\
\text{Irrational Numbers}
\end{cases} \\
\text{Imaginary Numbers } (b \neq 0)
\begin{cases}
\text{Imaginary Numbers } (a \neq 0,\ b \neq 0) \\
\text{Pure Imaginary Numbers } (a = 0,\ b \neq 0)
\end{cases}
\end{cases}
$$

Exercises

#1 Simplify the following given expressions :

(1) When $-1 < a < 1$, $|a+1| + |1-a|$

(2) When $-1 < a < 1$, $|a-1| - |1+a|$

(3) When a is a real number, $\left|a + |a|\right| - \left|a - |a|\right|$

(4) When $-2 < a < 2$, $\sqrt{(a-2)^2} + |a+2|$

(5) When $a < b < c$, $\sqrt{(a-b)^2} - \sqrt{(b-c)^2} - \sqrt{(c-a)^2}$

(6) When $a + b < 0$, $ab > 0$, $\sqrt{(-a)^2} - \sqrt{(-b)^2} - \sqrt{(-2a)^2} + \sqrt{(-2b)^2}$

(7) When $a < 0$ and $b < 0$, $\sqrt{(-a)^2} + |b| - \sqrt{(-b)^2} - 2|a+b|$

#2 Find the value of $a + b$ for two real numbers a and b which satisfy the following :

(1) $\sqrt{(a+5)^2} + |2a - 3b + 4| = 0$

(2) $|2a - b + 3| + |a - 2b - 3| = 0$

#3 Find the square roots of each number.

(1) 2   (2) 0.01   (3) $\dfrac{16}{9}$ (4) 0

#4 Evaluate the expression.

(1) $\left(\sqrt{3}\right)^2$

(2) $\left(-\sqrt{5}\right)^2$

(3) $\sqrt{(-8)^2}$

(4) $-\sqrt{36}$

(5) $-\sqrt{\left(-\dfrac{2}{3}\right)^2}$

(6) $\sqrt[3]{(-2)^2}$

(7) $\sqrt[3]{(-5)^3}$

(8) $-\sqrt[3]{-27}$

(9) $\sqrt{(-2)^2} + \sqrt[3]{(-2)^3} - 5\sqrt{(-2)^2}$

(10) $\sqrt[3]{(-5)^3} - \sqrt{(-5)^2} + \sqrt{5^3}$

(11) $3\sqrt[4]{4^2} - \sqrt[4]{(-2)^4}$

(12) $\dfrac{\sqrt[3]{20}}{\sqrt[3]{10}}$

(13) $\dfrac{\sqrt[3]{20}}{\sqrt{10}}$

#5 Simplify each of the following :

(1) $\sqrt{12} \times \sqrt{\dfrac{3}{4}}$

(2) $2\sqrt{10} \div 8\sqrt{5}$

(3) $\dfrac{3}{\sqrt{5}} \times \dfrac{\sqrt{15}}{9} \div \dfrac{\sqrt{18}}{12}$

(4) $\sqrt[3]{20} \cdot \sqrt{15}$

(5) $\dfrac{\sqrt{15}}{\sqrt{5}} \div \dfrac{5}{3\sqrt{2}}$

(6) $\sqrt{\dfrac{3}{4}} \div \sqrt{\dfrac{15}{10}} \div \dfrac{1}{\sqrt{6}}$

(7) $\sqrt[3]{6} \cdot \sqrt[3]{9}$

(8) $\dfrac{|2-4|}{|-3\sqrt{2}|-2}$

(9) $\left| \dfrac{-|\sqrt{2}-\sqrt{5}|}{\sqrt{5}-\sqrt{2}} \right|$

#6 Rationalize the denominator. Then, simplify the result.

(1) $\dfrac{3}{\sqrt{2}}$

(2) $\dfrac{3}{2\sqrt{5}}$

(3) $\dfrac{\sqrt{18}}{\sqrt{12}}$

(4) $\dfrac{3}{2\sqrt{3}} - 4\sqrt{3}$

(5) $\dfrac{3\sqrt{5}-\sqrt{3}}{\sqrt{3}}$

(6) $\dfrac{5}{\sqrt[3]{2}}$

(7) $\sqrt{\dfrac{2}{3}} - \sqrt{5} + \sqrt{\dfrac{3}{2}}$

(8) $\dfrac{4}{2+\sqrt{3}}$

(9) $\dfrac{2\sqrt{3}}{\sqrt{3}-\sqrt{2}} + \dfrac{3\sqrt{2}}{\sqrt{3}+\sqrt{2}}$

#7 Find the distance d between the points and find the midpoint m of the line segment joining the points.

(1) $a(1,2),\ b(3,4)$

(2) $a(-2,3),\ b(3,-5)$

(3) $a(3,4),\ b(-2,-3)$

(4) $a(-\sqrt{2},3),\ b(3,\sqrt{2})$

(5) $a(0,-1),\ b(-4,\tfrac{1}{2})$

#8 For rational numbers a and b, each expression is a rational number. Find the value of ab.

(1) $\dfrac{3+b\sqrt{2}}{\sqrt{2}-a}$

(2) $\dfrac{\sqrt{3}+b}{a\sqrt{3}+2}$

#9 Find the real numbers a and b so that the statement is identity.

(1) $(3a-2b)+(a+3b)i = -5+2i$

(2) $a(1+2i)-b(2-5i) = 10+8i$

(3) $\dfrac{a}{1+i} - \dfrac{b}{1-i} = 2+i$

(4) $\dfrac{a}{1+i} + \dfrac{b}{1-i} = \dfrac{3}{2-i}$

(5) $|a - b| + (a - 3)i = 3b - 2 - bi$

#10 Simplify each of the following in standard form.

(1) $2 + \sqrt{-9}$

(2) $\sqrt{-80}$

(3) $-3i^2 + 4i$

(4) $(12 - 3i) + (-4 + 8i)$

(5) $\sqrt{-5} \cdot \sqrt{-4}$

(6) $\dfrac{1}{2+\sqrt{-2}} + \dfrac{1}{2-\sqrt{-2}}$

(7) $(\sqrt{-75})^2$

(8) $(3 - 2i)(5 - 3i)$

(9) $\dfrac{2+3i}{1-2i}$

(10) $\dfrac{\sqrt{2}}{\sqrt{-3}}$

(11) $\dfrac{\sqrt{-3}}{\sqrt{-5}}$

(12) $\dfrac{\sqrt{2}-\sqrt{-3}}{\sqrt{2}+\sqrt{-3}}$

(13) $(1 + i)^2(1 - i)^2 - (2 + 3i)(2 - 3i)$

(14) $\left(\dfrac{1+i}{1-i}\right)^9 - \left(\dfrac{1-i}{1+i}\right)^{10}$

#11 For a complex number $z = -1 + \sqrt{3}i$, find the value of each number.

(1) z^2

(2) z^3

(3) $\left(\dfrac{z}{2}\right)^{100}$

#12 Find the value of a $(a > 0)$ so that the complex number $z = (a + \sqrt{3}i)^2 i$ is a real number.

#13 For a complex number $z = (2 - 3i)a - 2i(4 + 2i)$, find the value of a so that z^2 is a real number.

#14 For a non-zero complex number $z = (1 - i)x^2 - (5 + i)x + 6 + 6i$, find the value of real number x so that $z + \bar{z} = 0$.

#15 For any non-zero real numbers $a, b, c,$ and d, $\sqrt{a}\sqrt{b} = -\sqrt{ab}$, $\dfrac{\sqrt{c}}{\sqrt{d}} = -\sqrt{\dfrac{c}{d}}$.

Simplify the expression $\sqrt{a^2} - |a + b| + \sqrt{b^2} - |d| - \sqrt{(d - c)^2}$.

#16 Find all possible integers for x which satisfies $\dfrac{\sqrt{x+2}}{\sqrt{x-6}} = -\sqrt{\dfrac{x+2}{x-6}}$.

#17 Find the absolute value of each complex number.

(1) $-2i$ (2) $-3 + 4i$ (3) $\sqrt{5} + \sqrt{3}i$

#18 Solve the following :

(1) When $z = 2 - i$, find the value of $\left| z - \dfrac{1}{z} \right|$.

(2) For any complex numbers z_1 and z_2, $|z_1| = 1$ and $z_1 \neq z_2$.

Find the value of $\left| \dfrac{z_1 - z_2}{1 - \overline{z_1}\, z_2} \right|$.

#19 Find the values of a and b for the complex number $z = a + bi$ such that $\overline{z - zi} = 3 - 5i$.

#20 For any real numbers a and b, $\sqrt{(a - b - 2)^2} + |a - b - 2ab| = 0$.

Find the value of $\dfrac{b}{a} + \dfrac{a}{b}$.

Chapter 2. Polynomials

2-1 Simplifying Expressions by Combining Like Terms

1. Definition

(1) Monomial Expressions

A monomial expression is an expression of only one term which is the product of numbers and variables.

For example, $\frac{2}{3}x^2y$, $\sqrt{5}\,x$

(2) Polynomial Expressions

A polynomial expression is an expression of two or more terms combined by addition and/or subtraction.

For example, $\frac{2}{3}x^2y + \sqrt{5}\,x - \frac{1}{4}x^3$

Note : ① *A polynomial with one term is called a monomial. Ex.* $-3x^2y$

　　　② *A polynomial with two terms is called a binomial. Ex.* $-3x^2y + 5y$

　　　③ *A polynomial with three terms is called a trinomial. Ex.* $-3x^2y + 5y - \frac{2}{5}$

Note : $\frac{2x^2}{y} = 2 \cdot x \cdot x \div y$ *(combined by* $\div$ *)* $\quad \therefore \frac{2x^2}{y}$ *is not a monomial. ;* $\frac{2x^2}{y}$ *is not a polynomial.*

　　　$\frac{1}{2}x^2y + \frac{3}{x}$ *is not a polynomial.*

> Since $\frac{2}{3}x^2$ and $-5x^2$ have the same variable part, these terms are like terms.

2. Polynomials in Standard Form

Like terms are terms which differ only in their numerical coefficient. In other words, like terms are terms having the same variable to the same powers.

For example, $2x^2y$ and $-5x^2y$ are like terms. $3x^2y$ and $3xy^2$ are not like terms.

So a polynomial is expressed as a finite sum of unlike terms.

Consider a polynomial $P = 3x^2 + 2xy + 5y^2 - x + 3y - 4$ in x.

In standard form, a polynomial in one variable x is written with descending powers of x from left to right. So we combine the like terms using the commutative and associative laws.

Then, $P = 3x^2 + (2y - 1)x + 5y^2 + 3y - 4$

Define a polynomial P in x to be an expression of the form

a_0 is the constant term.

$$P(x) = a_n x^n + a_{n-1} x^{n-1} + \cdots\cdots\cdots + a_1 x + a_0$$

where n is a non-negative integer and the coefficients $a_0, a_1, \cdots\cdots\cdots, a_n$ are real or complex numbers.

The degree of the polynomial in one variable is the value of the highest power of the variable. If the polynomial is of degree n, then the leading coefficient $a_n \neq 0$.

Note : Linear form in x : $ax + b$ $(a \neq 0)$

Quadratic form in x : $ax^2 + bx + c$ $(a \neq 0)$

Cubic form in x : $ax^3 + bx^2 + cx + d$ $(a \neq 0)$

Example

$P = 2x^2 - 3x^5 + 6 - 4x \ \Rightarrow\ P = -3x^5 + 2x^2 - 4x + 6$ (a polynomial of degree 5)

$P = -3 + x^3 - 2x \ \Rightarrow\ P = x^3 - 2x - 3$ (a polynomial of degree 3)

$P = 5 \ \Rightarrow\ P = 5 = 5x^0$ (a polynomial of degree 0)

Note : Two polynomials are equal. $\xleftrightarrow[\text{if and only if}]{}$ The coefficients of the corresponding powers are equal.

That is, ① $ax^2 + bx + c = 0 \ \Leftrightarrow\ a = 0, \ b = 0, \ c = 0$

② $ax^2 + bx + c = px^2 + qx + r \ \Leftrightarrow\ a = p, \ b = q, \ c = r$

Example

$ax^3 + bx^2 + cx + d = -2x^3 + 4x - 5 \ \xleftrightarrow[\text{if and only if}]{}\ a = -2, \ b = 0, \ c = 4, \ d = -5$

Note : <u>Proof of Identities and Equalities (Coefficient Comparison Method)</u>

To show that an equality is an identity, expand both left hand side (LHS) and right hand side (RHS), and arrange with x as variable. That is,

[If $(ax + b)(x + 1) = (cx + d)(x + 2)$, then $ax^2 + (a + b)x + b = cx^2 + (2c + d)x + 2d$]

Now, match the coefficients of x^2, x, and the constants of the LHS and RHS. That is,

[$a = c, \ a + b = 2c + d, \ b = 2d$]

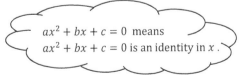

$ax^2 + bx + c = 0$ means
$ax^2 + bx + c = 0$ is an identity in x .

2-2 Operations with Polynomials

1. Adding and Subtracting Polynomials

To add or subtract polynomials, add or subtract the coefficients of like terms.

(1) Order of Operations

Step 1. Remove the parentheses ( ) or other enclosure marks (brace{ }, bracket[ ], absolute value sign | |, etc.) in order from the innermost enclosure marks to the outermost ones.

Step 2. Regroup the like terms.

Step 3. Simplify by combining the expressions.

Example

1. $2x - [3y - \{x - (2x - 3y) + 2y\} - 5] = 2x - [3y - \{x - 2x + 3y + 2y\} - 5]$

$$= 2x - [3y - x + 2x - 3y - 2y - 5]$$

$$= 2x - 3y + x - 2x + 3y + 2y + 5$$

To group like terms together, using the commutative and associative properties

$$= (2x + x - 2x) + (-3y + 3y + 2y) + 5$$

$$= x + 2y + 5$$ To combine the like terms using the distibutive property

2. $2(x^2 - 3x + 5) - 3(2x^2 - x + 3) = 2x^2 - 6x + 10 - 6x^2 + 3x - 9$

To group like terms together using the commutative and associative properties

$$= (2x^2 - 6x^2) + (-6x + 3x) + (10 - 9)$$

$$= -4x^2 - 3x + 1$$ To combine the like terms using the distibutive property

(2) Properties of Polynomials

For any polynomials A, B, and C,

1) Commutative Property

$A + B = B + A$

2) Associative Property

$(A + B) + C = A + (B + C)$

3) Identity

$A + 0 = 0 + A = A$ (0 is the identity in the addition of polynomial.)

4) Inverse

$A + (-A) = (-A) + A = 0$ ($-A$ is the additive inverse of the polynomial A .)

2. Multiplying Polynomials

To multiply two polynomials, use the distributive property and the rules of exponents.

(1) Rules of Exponents

For any real number a ($a \neq 0$) and positive integers m and n,

① Addition of Exponents

$$a^m \cdot a^n = \underbrace{a \cdot a \cdots a \cdot a}_{m \text{ times}} \cdot \underbrace{a \cdot a \cdots a \cdot a}_{n \text{ times}} = a^{m+n}$$

② Multiplication of Exponents

$$(a^m)^n = \underbrace{a^m \cdot a^m \cdots a^m \cdot a^m}_{n \text{ times}} = a^{\overbrace{m+m+\cdots+m}^{n \text{ times}}} = a^{mn}$$

③ Division of Exponents

$$a^m \div a^n = \frac{a^m}{a^n} = \frac{\overbrace{a \cdots a}^{m}}{\underbrace{a \cdots a}_{n}} = \begin{cases} a^{m-n}, & m > n \\ 1, & m = n \\ \dfrac{1}{a^{n-m}}, & m < n \end{cases}$$

④ Distributive Properties of Exponents

$$(ab)^m = \underbrace{(ab) \cdot (ab) \cdots (ab) \cdot (ab)}_{m \text{ times}}$$

$$= \underbrace{(a \cdot b) \cdot (a \cdot b) \cdots (a \cdot b) \cdot (a \cdot b)}_{m \text{ times}} = \underbrace{a \cdot a \cdots a \cdot a}_{m \text{ times}} \cdot \underbrace{b \cdot b \cdots b \cdot b}_{m \text{ times}}$$

$$= a^m \cdot b^m$$

$$\left(\frac{a}{b}\right)^m = \underbrace{\left(\frac{a}{b}\right) \cdot \left(\frac{a}{b}\right) \cdots \left(\frac{a}{b}\right) \cdot \left(\frac{a}{b}\right)}_{m \text{ times}} = \frac{\overbrace{a \cdots a}^{m}}{\underbrace{b \cdots b}_{m}} = \frac{a^m}{b^m}$$

Note : Expanding Exponents

(1) $a^0 = 1$

(2) $a^{-m} = \frac{1}{a^m}$

(3) $a^{m+1} - a^m = a^m(a-1)$

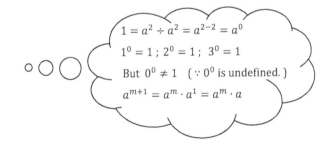

$1 = a^2 \div a^2 = a^{2-2} = a^0$

$1^0 = 1 \; ; \; 2^0 = 1 \; ; \; 3^0 = 1$

But $0^0 \neq 1$ ($\because 0^0$ is undefined.)

$a^{m+1} = a^m \cdot a^1 = a^m \cdot a$

(2) FOIL Method

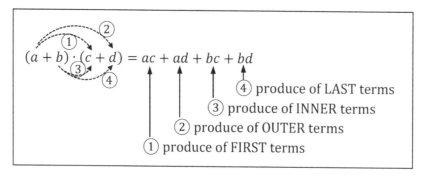

$$(a + b) \cdot (c + d) = ac + ad + bc + bd$$

④ produce of LAST terms

③ produce of INNER terms

② produce of OUTER terms

① produce of FIRST terms

Note : For any polynomials A, B, and C, and a monomial m,

(1) $m \cdot (A + B) = mA + mB$ $\qquad (A + B)m = mA + mB$

(2) $(A + B) \div m = (A + B) \cdot \frac{1}{m} = \left(A \cdot \frac{1}{m}\right) + \left(B \cdot \frac{1}{m}\right) = \frac{A}{m} + \frac{B}{m}$

Note : (1) $(a + b) \cdot (c + d) = a(c + d) + b(c + d) = ac + ad + bc + bd$

(2) $(x + a)(x + b) = x^2 + (a + b)x + ab$

(3) $(ax + b)(cx + d) = acx^2 + (ad + bc)x + bd$

(4) $(a + b + 1)(a + b - 1) = (A + 1)(A - 1)$ *letting* $a + b = A$

$$= A^2 - A + A - 1 = A^2 - 1 = (a + b)^2 - 1$$

$$= (a + b)(a + b) - 1 = a^2 + ab + ba + b^2 - 1$$

$$= a^2 + 2ab + b^2 - 1$$

(5) $(a + 1)(a + 2)(a + 3)(a + 4)$

$$= \underline{(a + 1)(a + 4)}\,\underline{(a + 2)(a + 3)} \ \ regroup \ to \ get \ a^2 + 5a$$

$$= (\underline{a^2 + 5a} + 4)(\underline{a^2 + 5a} + 6)$$

$$= (A + 4)(A + 6) \ \ letting \ a^2 + 5a = A$$

$$= A^2 + 10A + 24$$

$$= (a^2 + 5a)^2 + 10(a^2 + 5a) + 24 \ \ replace \ A \ as \ a^2 + 5a$$

$$= a^4 + 10a^3 + 35a^2 + 50a + 24$$

(3) Formulas of Special Products

Special Products :

1) $(a + b)^2 = a^2 + 2ab + b^2$
   $(a - b)^2 = a^2 - 2ab + b^2$

2) $(a + b)(a - b) = a^2 - b^2$

3) $(x + a)(x + b) = x^2 + (a + b)x + ab$

4) $(ax + b)(cx + d) = acx^2 + (ad + bc)x + bd$

5) $(x + a)(x + b)(x + c) = x^3 + (a + b + c)x^2 + (ab + bc + ca)x + abc$
   $(x - a)(x - b)(x - c) = x^3 - (a + b + c)x^2 + (ab + bc + ca)x - abc$

6) $(a + b + c)^2 = a^2 + b^2 + c^2 + 2ab + 2bc + 2ca$

7) $(a + b)^3 = a^3 + 3a^2b + 3ab^2 + b^3 = a^3 + 3ab(a + b) + b^3$
   $(a - b)^3 = a^3 - 3a^2b + 3ab^2 - b^3 = a^3 - 3ab(a - b) - b^3$

8) $(a + b)(a^2 - ab + b^2) = a^3 + b^3$
   $(a - b)(a^2 + ab + b^2) = a^3 - b^3$

9) $(a + b + c)(a^2 + b^2 + c^2 - ab - bc - ca) = a^3 + b^3 + c^3 - 3abc$

10) $(a^2 + ab + b^2)(a^2 - ab + b^2) = a^4 + a^2b^2 + b^4$

(4) Properties of Polynomials

For any polynomials A, B, and C,

1) Commutative Property

   $A \cdot B = B \cdot A$

2) Associative Property

   $(A \cdot B) \cdot C = A \cdot (B \cdot C)$

3) Distributive Property

   $A \cdot (B + C) = A \cdot B + A \cdot C \quad (A + B) \cdot C = A \cdot C + B \cdot C$

2-3 Factoring Polynomials

1. Definition

(1) The Greatest Common Factor (GCF)

If a polynomial $P(x)$ is expressed as

$P(x) = Q_1(x) \cdot Q_2(x) \cdots Q_n(x)$ where $Q_1(x),\ Q_2(x), \cdots, Q_n(x)$ are polynomials, then

$Q_1(x),\ Q_2(x), \cdots, Q_n(x)$ are called factors of $P(x)$.

The Greatest Common Factor (GCF) is the product of their common factors raised to their lowest powers.

Example

$90x^2y^3 + 12x^4y^5$ It cannot be factored further.

Since $90 = 2 \cdot 3^2 \cdot 5$ (prime factorization) and $12 = 2^2 \cdot 3$ (prime factorization),

the common factors are $2,\ 3, x,$ and y.

$\therefore\ \text{GCF} = 2^1 \cdot 3^1 \cdot x^2 \cdot y^3 = 6x^2y^3$

(2) Factorization

Factorization is an expression of a polynomial as a product of its greatest common factor and two or more prime polynomials.

Example

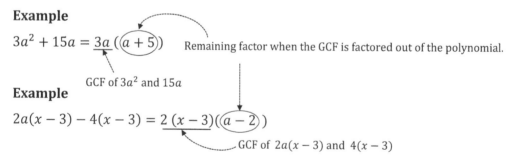

$3a^2 + 15a = 3a\ (a + 5)$ Remaining factor when the GCF is factored out of the polynomial.

GCF of $3a^2$ and $15a$

Example

$2a(x - 3) - 4(x - 3) = 2\ (x - 3)(a - 2)$

GCF of $2a(x - 3)$ and $4(x - 3)$

Note :

$(x + 2)(x + 3) = x^2 + 5x + 6$: *Multiplying two polynomials using the distributive property*

(Expanding)

$x^2 + 5x + 6 = (x + 2)(x + 3)$: *Factoring polynomials as a product of two prime polynomials*

Note :   The relationship of GCF and LCM

Let A and B be two polynomials whose leading coefficients are 1.

If GCF of A and B is G and LCM of A and B is L, then

1) $A = aG, B = bG$

2) $L = abG = aB = bA$

3) $AB = abG^2 = LG$

4) $A + B = (a + b)G, \quad A - B = (a - b)G$

for any a and b which don't have any common factor except 1.

2. Factorization Formulas

(1) The Form of a Perfect Square

$$a^2 + 2ab + b^2 = (a + b)^2$$

$$a^2 - 2ab + b^2 = (a - b)^2$$

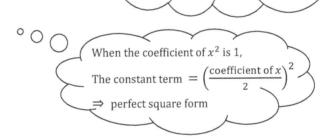

Example

$$x^2 + \boxed{6}x + \boxed{9} = (x + 3)^2$$
$$ {}_{2 \cdot 3} \qquad {}_{3^2}$$

$$x^2 + \boxed{5}x + \boxed{6} \neq (\cdots)^2$$

$$2 \cdot \frac{5}{2} \qquad 6 \neq (\frac{5}{2})^2$$

When the coefficient of x^2 is 1,

The constant term $= \left(\dfrac{\text{coefficient of } x}{2}\right)^2$

$\Rightarrow$ perfect square form

Note :
1) $a - b = -(b - a)$
2) $(a - b)^2 = (-(b - a))^2 = (b - a)^2$
3) $(a - b)^3 = (-(b - a))^3 = -(b - a)^3$
4) $(a - b)^4 = (-(b - a))^4 = (b - a)^4$

(2) The Form of $(a + b)(a - b)$

$$a^2 - b^2 = (a + b)(a - b)$$

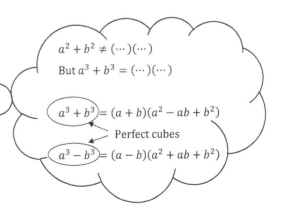

$$\frac{\bullet^2 - \blacksquare^2}{} = (\bullet + \blacksquare)(\bullet - \blacksquare)$$

Subtraction for   addition   subtraction
Perfect squares

$a^2 + b^2 \neq (\cdots)(\cdots)$

But $a^3 + b^3 = (\cdots)(\cdots)$

$a^3 + b^3 = (a + b)(a^2 - ab + b^2)$

Perfect cubes

$a^3 - b^3 = (a - b)(a^2 + ab + b^2)$

Note : $-a^2 + b^2 = b^2 - a^2 = (b - a)(b + a)$

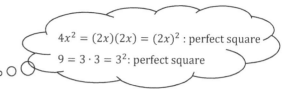

$4x^2 = (2x)(2x) = (2x)^2$: perfect square

$9 = 3 \cdot 3 = 3^2$: perfect square

Example

$4x^2 - 9 = (2x)^2 - (3)^2 = (2x + 3)(2x - 3)$

(3) Quadratic Trinomials with Binomial Factors

Factors of c

$$ax^2 + bx + c = (\blacksquare x + \bullet)(\blacktriangle x + \blacklozenge)$$

Factors of a

$x^2 + (a + b)x + ab = (x + a)(x + b)$

① If the constant term $> 0 \Rightarrow \begin{cases} (a > 0 \text{ and } b > 0) \\ \text{or} \\ (a < 0 \text{ and } b < 0) \end{cases}$

② If the constant term $< 0 \Rightarrow \begin{cases} (a > 0 \text{ and } b < 0) \\ \text{or} \\ (a < 0 \text{ and } b > 0) \end{cases}$

1) If the Leading Coefficient is 1,

$$x^2 + (\boldsymbol{a + b})x + \boldsymbol{ab} = (x + a)(x + b)$$

$= 1$ sum of a and b

product of a and b

① Step 1: Find all the possible numbers which factor the constant ab.

② Step 2: Check to see if the sum of the two numbers found in Step 1 is the same as the coefficient of x.

③ Step 3: Factor $(x + a)(x + b)$.

Example 1

$x^2 + ③x + ②= (x + 1)(x + 2)$

$1 + 2$ 1×2

Example 2

$x^2 - ③x - ④= (x - 4)(x + 1)$

$-4 + 1$ -4×1

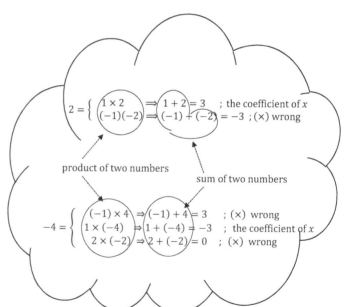

$2 = \begin{cases} 1 \times 2 & \Rightarrow 1 + 2 = 3 & \text{; the coefficient of } x \\ (-1)(-2) & \Rightarrow (-1) + (-2) = -3 & \text{; } (\times) \text{ wrong} \end{cases}$

product of two numbers

sum of two numbers

$-4 = \begin{cases} (-1) \times 4 & \Rightarrow (-1) + 4 = 3 & \text{; } (\times) \text{ wrong} \\ 1 \times (-4) & \Rightarrow 1 + (-4) = -3 & \text{; the coefficient of } x \\ 2 \times (-2) & \Rightarrow 2 + (-2) = 0 & \text{; } (\times) \text{ wrong} \end{cases}$

2) If the Leading Coefficient is not 1,

$$acx^2 + (ad + bc)x + bd = (ax + b)(cx + d)$$

$\neq 1$

Note :

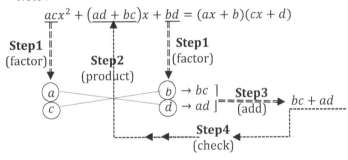

① Step 1. Find all the possible numbers which factor the coefficient of x^2 and the constant

② Step 2. Product the numbers diagonally

③ Step 3. Add the numbers obtained from step 2

④ Step 4. Check to see if the sum obtained from Step 3 is the same as the coefficient of x

⑤ Step 5. Factor as $(ax + b)(cx + d)$

Example 1

$$2x^2 + 7x + 6 = (x + 2)(2x + 3)$$

$\begin{matrix} 1 \\ 2 \end{matrix} \times \begin{matrix} 2 \to 4 \\ 3 \to 3 \end{matrix} \Big] \xrightarrow{+} 7$

$2 = \mathbf{1} \times \mathbf{2}$ or 2×1
    or $(-1) \times (-2)$
    or $(-2) \times (-1)$
$6 = 1 \times 6$ or 6×1
    or $\mathbf{2} \times \mathbf{3}$ or 3×2
    or $(-1) \times (-6)$ or $(-6) \times (-1)$
    or $(-2) \times (-3)$ or $(-3) \times (-2)$

Example 2

$$3x^2 + 2x - 5 = (x - 1)(3x + 5)$$

$\begin{matrix} 1 \\ 3 \end{matrix} \times \begin{matrix} -1 \to -3 \\ 5 \to 5 \end{matrix} \Big] \xrightarrow{+} 2$

Consider all possibilities:

$3x^2 + ②x - 5 = (x - 1)(3x + 5)$

$\begin{matrix} 1 \\ 3 \end{matrix} \times \begin{matrix} 1 \to 3 \\ -5 \to -5 \end{matrix} \Big] \xrightarrow{+} -2$

$\begin{matrix} 1 \\ 3 \end{matrix} \times \begin{matrix} -1 \to -3 \\ 5 \to 5 \end{matrix} \Big] \xrightarrow{+} ②$

$\begin{matrix} 1 \\ 3 \end{matrix} \times \begin{matrix} 5 \to 15 \\ -1 \to -1 \end{matrix} \Big] \xrightarrow{+} 14$

$\begin{matrix} 1 \\ 3 \end{matrix} \times \begin{matrix} -5 \to -15 \\ 1 \to 1 \end{matrix} \Big] \xrightarrow{+} -14$

Example 3

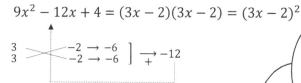

$$9x^2 - 12x + 4 = (3x-2)(3x-2) = (3x-2)^2$$

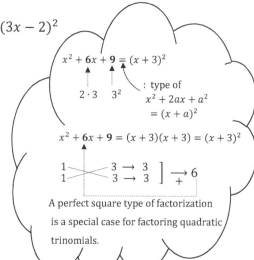

: type of
$x^2 + 2ax + a^2$
$= (x+a)^2$

$x^2 + 6x + 9 = (x+3)(x+3) = (x+3)^2$

A perfect square type of factorization is a special case for factoring quadratic trinomials.

3. More Factorization

(1) If each term has common factors,

⇒ Find the GCF first and

express the polynomial as a product of the GCF and prime polynomials.

Note :

$$ax^2 + 4ax + 4a$$

$$\xrightarrow[\text{find the GCF}]{} a(x^2 + 4x + 4)$$

$$\xrightarrow[\text{factor}]{} a(x+2)^2$$

Example

$$2x^2 + 6x + 4$$

$$= (x+2)(2x+2)$$

$$= 2(x+2)(x+1)$$

Begin by removing a common factor, and then factoring further.

That is, $2x^2 + 6x + 4 = 2(x^2 + 3x + 2)$

$$x^2 + 3x + 2$$

Common factors should be removed before other methods of factoring are used .

$$= (x+1)(x+2)$$

$$\therefore \ 2x^2 + 6x + 4 = 2(x^2 + 3x + 2) = 2(x+1)(x+2)$$

(2) If a polynomial contains common terms,

⇒ Substitute the common terms with other variables. Then factor the resulting polynomial.

Rewrite it with common terms.

Example

$$(a - 1)^2 + 2(a - 1) - 3 = A^2 + 2A - 3 \quad \text{Substitute } a - 1 \text{ as } A$$
$$= (A + 3)(A - 1) \quad \text{Factor}$$
$$= (a - 1 + 3)(a - 1 - 1) \quad \text{Replace } A \text{ with } a - 1$$
$$= (a + 2)(a - 2)$$

(3) If polynomials containing more than 3 terms have no common factors,

⇒ Factor by groups considering the combined terms have a common term.

Example

$$x^3 + 3x^2 - 4x - 12 = (x^3 + 3x^2) + (-4x - 12) \quad \text{Grouping}$$
$$= x^2(x + 3) - 4(x + 3)$$
$$= (x + 3)(x^2 - 4) \quad \text{Find GCF}$$
$$= (x + 3)(x + 2)(x - 2) \quad \text{Factorization}$$

The GCF of the first group $(x^3 + 3x^2)$ is x^2.
Factor it out of the expression to get $x^2(x + 3)$.
The GCF of the second group $(-4x - 12)$ is -4.
Factor it out of the expression to get $-4(x + 3)$.

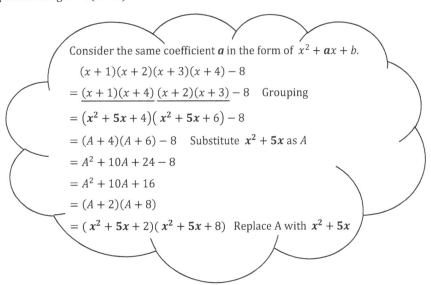

Consider the same coefficient a in the form of $x^2 + ax + b$.

$$(x + 1)(x + 2)(x + 3)(x + 4) - 8$$
$$= \underline{(x + 1)(x + 4)}\ \underline{(x + 2)(x + 3)} - 8 \quad \text{Grouping}$$
$$= (x^2 + 5x + 4)(x^2 + 5x + 6) - 8$$
$$= (A + 4)(A + 6) - 8 \quad \text{Substitute } x^2 + 5x \text{ as } A$$
$$= A^2 + 10A + 24 - 8$$
$$= A^2 + 10A + 16$$
$$= (A + 2)(A + 8)$$
$$= (x^2 + 5x + 2)(x^2 + 5x + 8) \quad \text{Replace A with } x^2 + 5x$$

(4) If polynomials contain more than 2 variables,

⇒ Rewrite the polynomials in the descending order of powers for the variable with the lowest power. Consider the other terms (which don't contain the variable) as constants. Then factor the pretended constant term.

Example

$x^2 + 2x + xy - y - 3 = (xy - y) + \underline{(x^2 + 2x - 3)}$ Descending order of powers of y

pretended constant term

$= y(x - 1) + (x^2 + 2x - 3)$ Factor a group

$= y(x - 1) + (x + 3)(x - 1)$ Factor a group

$= (x - 1)(y + x + 3)$ Common factor

$= (x - 1)(x + y + 3)$

> x has a power of 2.
> y has a power of 1.
> Because y has the lowest power, choose it for the descending order of powers.

4. Factoring Special Polynomial Forms

Formulas

1) $ax + bx + cx = (a + b + c)x$

2) $a^2 + b^2 = (a + b)^2 - 2ab = (a - b)^2 + 2ab$

3) $a^2 - b^2 = (a + b)(a - b)$

4) $x^2 + (a + b)x + ab = (x + a)(x + b)$

5) $acx^2 + (ad + bc)x + bd = (ax + b)(cx + d)$

6) $a^3 + b^3 = (a + b)^3 - 3ab(a + b)$ Sum of two cubes

   $a^3 - b^3 = (a - b)^3 + 3ab(a - b)$ Difference of cubes

7) $a^3 + 3a^2b + 3ab^2 + b^3 = (a + b)^3$

   $a^3 - 3a^2b + 3ab^2 - b^3 = (a - b)^3$

8) $a^2 + b^2 + c^2 = (a + b + c)^2 - 2(ab + bc + ca)$

   $a^2 + b^2 + c^2 + 2ab + 2bc + 2ca = (a + b + c)^2$

9) $a^4 + a^2b^2 + b^4 = (a^2 + ab + b^2)(a^2 - ab + b^2)$

   $a^4 + a^2 + 1 = (a^2 + a + 1)(a^2 - a + 1)$

10) $a^3 + b^3 + c^3 = (a + b + c)(a^2 + b^2 + c^2 - ab - bc - ca) + 3abc$

   $a^3 + b^3 + c^3 - 3abc = (a + b + c)(a^2 + b^2 + c^2 - ab - bc - ca)$

Note: $a^2 + b^2 + c^2 + ab + bc + ca$

$= \frac{1}{2}(2a^2 + 2b^2 + 2c^2 + 2ab + 2bc + 2ca)$

$= \frac{1}{2}\{(a^2 + 2ab + b^2) + (b^2 + 2bc + c^2) + (c^2 + 2ca + a^2)\}$

$= \frac{1}{2}\{(a + b)^2 + (b + c)^2 + (c + a)^2\}$

$\therefore\ a^2 + b^2 + c^2 \pm ab \pm bc \pm ca = \frac{1}{2}\{(a \pm b)^2 + (b \pm c)^2 + (c \pm a)^2\}$

2-4 Rational Expressions and Irrational Expressions

1. Simplifying Rational Expressions

(1) Rational numbers and Rational Expressions

A rational number is a fraction whose numerator and denominator are both integers and denoted by

$$\frac{a}{b} \quad \text{for any integers } a \text{ and } b \ (b \neq 0).$$

A rational expression is a fraction whose numerator and denominator are both polynomials and denoted by

$$\frac{P(x)}{Q(x)} \quad \text{for any polynomials } P(x) \text{ and } Q(x) \ (Q(x) \neq 0).$$

Note : ① $y = \frac{1}{x}$ *is not defined at* $x = 0$.

② $y = \frac{x+1}{x+2}$ *is not defined at* $x = -2$.

③ $y = \frac{x^2}{(x-1)(x+3)}$ *is not defined at either* $x = 1$ *or at* $x = -3$.

④ $y = \frac{x}{x}$ *is not defined at* $x = 0$ *, but for other variables of* x, $\frac{x}{x} = 1$.

Thus, the two expressions $\frac{x}{x}$ *and* 1 *are consequently not identical.*

Similarly, $y = \frac{x(x+1)}{x+1}$ *and* $y = x$ *are not the same.*

$(\because$ *Both* $y = \frac{x(x+1)}{x+1}$ *and* $y = x$ *have the same variables for* $x \neq -1$.

But $y = \frac{x(x+1)}{x+1}$ *is not defined at* $x = -1$, *whereas* $y = x$ *has the value* -1.)

(2) Reducing Fractions to Lowest Terms

To reduce rational expressions to lowest terms, simplify the rational expressions and identify restrictions on the variables.

The domain of an expression is the set of real numbers for which the expression is defined.

Since division by zero is undefined, we must restrict the domain of the reduced expression by excluding the value, which would produce an undefined division by zero.

1) When the quotients (fractions) of exponential expressions have the same base :

To simplify the rational expression, reduce the fraction to its lowest term by the properties of the exponents. Then, reduce coefficients by prime factorizations.

Example

$$\frac{2x^3}{24x} = \frac{2 \cdot x^3}{2^3 \cdot 3 \cdot x} = \frac{x^2}{2^2 \cdot 3} = \frac{x^2}{12} \qquad \therefore \ \frac{2x^3}{24x} = \frac{x^2}{12}, \ \ x \neq 0$$

$$\frac{a^m}{a^n} = a^{m-n}, \quad a^{-m} = \frac{1}{a^m}$$

$$\frac{6x^2y^4}{15x^5y} = \frac{2 \cdot 3 \cdot x^2 \cdot y^4}{3 \cdot 5 \cdot x^5 y} = \frac{2}{5} x^{2-5} y^{4-1} = \frac{2}{5} x^{-3} y^3 = \frac{2y^3}{5x^3} \qquad \therefore \ \frac{6x^2y^4}{15x^5y} = \frac{2y^3}{5x^3}, \ \ y \neq 0$$

$$\frac{x^2}{x^2 y} = \frac{1}{y} \ \Rightarrow \ \text{restriction } x \neq 0 \text{ is required.}$$

However, $y \neq 0$ is not necessary

($\because$ if $y = 0$, then both of the expressions are undefined.)

2) When the polynomials can be factorized :

To simplify the rational expression, factor the numerator and the denominator. If the expression contains common factors, eliminate the common factors to reduce the fraction.

Example 1

$$\frac{2x^2 - 8}{x + 2} = \frac{2(x^2 - 4)}{x + 2} = \frac{2(x+2)(x-2)}{x+2} = 2(x - 2)$$

Therefore, $\frac{2x^2 - 8}{x + 2} = 2(x - 2), \ \ x \neq -2$

$$x^2 - (a + b)x + ab = (x - a)(x - b)$$

Example 2

$$\frac{2x + 4}{6x^2 - 18x - 60} = \frac{2(x+2)}{6(x^2 - 3x - 10)} = \frac{2(x+2)}{2 \cdot 3(x-5)(x+2)} = \frac{1}{3(x-5)}$$

Therefore, $\frac{2x + 4}{6x^2 - 18x - 60} = \frac{1}{3(x-5)}, \ \ x \neq -2$

Example 3

$$\frac{2x^2 - 2}{6x^2 + 24x - 30} = \frac{2(x^2 - 1)}{6(x^2 + 4x - 5)} = \frac{2(x+1)(x-1)}{2 \cdot 3(x+5)(x-1)} = \frac{x+1}{3(x+5)}$$

Therefore, $\frac{2x^2 - 2}{6x^2 + 24x - 30} = \frac{x+1}{3(x+5)}, \ \ x \neq 1$

2. Operations with Rational Expressions

(1) Adding and Subtracting Rational Expressions

For any polynomials $P(x)$, $Q(x)$, $R(x)(R(x) \neq 0)$, and $S(x)(S(x) \neq 0)$,

1) If the denominators are the same :

Keep the denominator and add or subtract the numerators

$$\frac{P(x)}{R(x)} + \frac{Q(x)}{R(x)} = \frac{P(x) + Q(x)}{R(x)}, \quad \frac{P(x)}{R(x)} - \frac{Q(x)}{R(x)} = \frac{P(x) - Q(x)}{R(x)}$$

Example

$$\frac{3x}{x-2} + \frac{5}{x-2} = \frac{3x+5}{x-2}$$

$$\frac{x+1}{2x-1} - \frac{3x+5}{2x-1} = \frac{(x+1)-(3x+5)}{2x-1} = \frac{-2x-4}{2x-1} = \frac{-2(x+2)}{2x-1}$$

2) If the denominators are different :

Identify the least common denominator (LCD) of the fractions and simplify the expression by adding or subtracting the numerators, then dividing by the least common denominator.

$$\frac{P(x)}{R(x)} + \frac{Q(x)}{S(x)} = \frac{P(x)\cdot S(x)}{R(x)\cdot S(x)} + \frac{R(x)\cdot Q(x)}{R(x)\cdot S(x)} = \frac{P(x)\cdot S(x)+R(x)\cdot Q(x)}{R(x)\cdot S(x)}$$

$$\frac{P(x)}{R(x)} - \frac{Q(x)}{S(x)} = \frac{P(x)\cdot S(x)}{R(x)\cdot S(x)} - \frac{R(x)\cdot Q(x)}{R(x)\cdot S(x)} = \frac{P(x)\cdot S(x)-R(x)\cdot Q(x)}{R(x)\cdot S(x)}$$

> The least common denominator (LCD) is the least common multiple (LCM) of denominators .

Example 1

$$\frac{x-1}{2x(x+1)} + \frac{2}{3x(x-1)} = \frac{3\cdot(x-1)(x-1)}{6x(x+1)(x-1)} + \frac{2\cdot 2(x+1)}{6x(x-1)(x+1)}$$

$$= \frac{3(x^2-2x+1)+4(x+1)}{6x(x+1)(x-1)} = \frac{3x^2-2x+7}{6x(x+1)(x-1)}$$

Example 2

$$\frac{3}{x+2} - \frac{2}{x-4} = \frac{3(x-4)}{(x+2)(x-4)} - \frac{2(x+2)}{(x-4)(x+2)} = \frac{3(x-4)-2(x+2)}{(x+2)(x-4)} = \frac{x-16}{(x+2)(x-4)}$$

(2) Multiplying and Dividing Rational Expressions

For any polynomials $P(x)$, $Q(x)$, $R(x)(R(x) \neq 0)$, and $S(x)(S(x) \neq 0)$,

1) To multiply rational expressions,

Multiply the numerators and then multiply the denominators. If possible, simplify the expressions before multiplying.

$$\frac{P(x)}{R(x)} \cdot \frac{Q(x)}{S(x)} = \frac{P(x)\cdot Q(x)}{R(x)\cdot S(x)}$$

Example 1

$$\frac{x+1}{3x} \cdot \frac{4x^2}{2(x+1)} = \frac{4(x+1)x^2}{3x\cdot 2(x+1)} = \frac{4\cdot(x+1)\cdot x\cdot x}{3\cdot x\cdot 2\cdot(x+1)} = \frac{2x}{3}$$

Example 2

$$\frac{2x^2-7x+3}{x^2+x-2} \cdot \frac{x^3+2x^2-3x}{4x^2-2x} = \frac{(2x-1)(x-3)}{(x+2)(x-1)} \cdot \frac{x(x+3)(x-1)}{2x(2x-1)} \quad \text{Factor and reduce}$$

$$= \frac{(x-3)(x+3)}{2(x+2)} \quad x \neq 0, \ x \neq 1, \ x \neq \frac{1}{2}$$

2) To divide rational expressions (Dividing Polynomials),

Convert the division into multiplication using the reciprocal of the 2nd fraction (the fraction we are dividing by). If possible, simplify the expressions before multiplying.

$$\boxed{\frac{P(x)}{R(x)} \div \frac{Q(x)}{S(x)} = \frac{P(x)}{R(x)} \cdot \frac{S(x)}{Q(x)} = \frac{P(x)S(x)}{R(x)Q(x)} \quad , Q(x) \neq 0}$$

Example

$$\frac{2x}{(x+1)^2} \div \frac{1}{x^2-2x-3} = \frac{2x}{(x+1)^2} \cdot \frac{x^2-2x-3}{1} = \frac{2x(x^2-2x-3)}{(x+1)^2}$$

$$= \frac{2x(x-3)(x+1)}{(x+1)(x+1)} = \frac{2x(x-3)}{(x+1)}$$

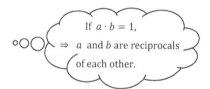

If $a \cdot b = 1$,
$\Rightarrow$ a and b are reciprocals of each other.

3) To divide long rational expressions (Long division of polynomials),

<u>The Division Algorithm</u>

Let $P(x)$ and $D(x)$ be polynomials of degrees n and r, respectively, where $n \geq r$.
Then there exist polynomials $Q(x)$, called the *quotient polynomial*, and $R(x)$, called the *remainder polynomial*, such that :

    ① $P(x) = D(x) \cdot Q(x) + R(x)$ for all x

    ② The degree of the remainder $R(x)$ must be less than the degree of the divisor $D(x)$, that is, $\deg R(x) < \deg D(x)$ or $R(x) = 0$

Note : If the degree of the divisor $D(x)$ is 1, then the degree of the remainder $R(x)$ is 0,

    i.e., $R(x) = a$ (a is a constant).

    If the degree of the divisor $D(x)$ is 2, then the degree of the remainder $R(x)$ is 1,

    i.e., $R(x) = ax + b$ ($a \neq 0$).

    If the degree of the divisor $D(x)$ is 3, then the degree of the remainder $R(x)$ is 2,

    i.e., $R(x) = ax^2 + bx + c$ ($a \neq 0$).

Example

Divide $x^3 + 2x^2 + 4x + 5$ by $x + 2$ using polynomial long division.

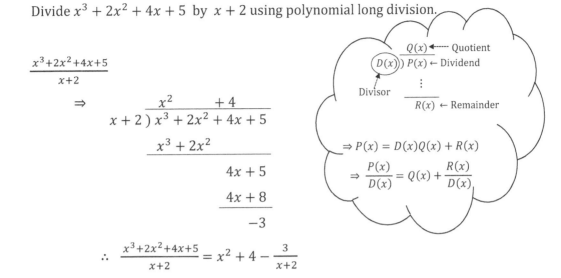

$$\frac{x^3 + 2x^2 + 4x + 5}{x + 2}$$

$$\Rightarrow \quad \begin{array}{r} x^2 \qquad\quad + 4 \\ x + 2 \overline{)\; x^3 + 2x^2 + 4x + 5} \\ \underline{x^3 + 2x^2} \\ 4x + 5 \\ \underline{4x + 8} \\ -3 \end{array}$$

$\Rightarrow P(x) = D(x)Q(x) + R(x)$

$\Rightarrow \dfrac{P(x)}{D(x)} = Q(x) + \dfrac{R(x)}{D(x)}$

$\therefore \quad \dfrac{x^3 + 2x^2 + 4x + 5}{x + 2} = x^2 + 4 - \dfrac{3}{x + 2}$

Note : If terms are missing from the dividend or the divisor, supply the missing terms with zero

coefficients.

Example Divide $4x^3 + 6x^2 - 3$ by $2x - 1$ using polynomial long division.

$$\frac{4x^3 + 6x^2 - 3}{2x - 1} \quad \Rightarrow \quad \frac{4x^3 + 6x^2 + 0 \cdot x - 3}{2x - 1}$$

$$\Rightarrow \quad \begin{array}{r} 2x^2 + 4x + 2 \\ 2x - 1 \overline{)\; 4x^3 + 6x^2 + 0 \cdot x - 3} \\ \underline{4x^3 - 2x^2} \qquad\qquad \longleftarrow 2x^2(2x - 1) \\ 8x^2 + 0 \cdot x - 3 \\ \underline{8x^2 - 4x} \qquad\qquad \longleftarrow 4x(2x - 1) \\ 4x - 3 \\ \underline{4x - 2} \qquad \longleftarrow 2(2x - 1) \\ -1 \qquad \longleftarrow \text{Remainder} \end{array}$$

$$\therefore \quad \frac{4x^3 + 6x^2 - 3}{2x - 1} = 2x^2 + 4x + 2 - \frac{1}{2x - 1}$$

4) To simplify complex fractions,

Fractional expressions with fractions in the numerator or denominator, or both, are called *complex fractions* (or *compound fractions*).

Convert the complex fraction into a rational quotient. Rewrite the rational quotient as a product and simplify the expressions.

Example

$$\frac{\frac{x(x+1)}{x^2}}{\frac{x^2+6x+5}{x-2}} = \frac{x(x+1)}{x^2} \div \frac{x^2+6x+5}{x-2} = \frac{x(x+1)}{x^2} \cdot \frac{x-2}{x^2+6x+5} = \frac{x \cdot (x+1) \cdot (x-2)}{x \cdot x \cdot (x+5) \cdot (x+1)}$$

$$= \frac{x-2}{x(x+5)} \quad , x \neq 0, \ x \neq -1$$

$$\frac{\frac{a}{b}}{\frac{c}{d}} = \frac{ad}{bc}$$

5) Synthetic Division

When the divisor is a linear binomial (of the form $x + \text{constant}$), the long division of polynomials is solved by synthetic division.

Example Divide $x^3 + 2x^2 + 4x + 5$ by $x + 2$ using synthetic division.

Write the coefficients (1, 2, 4, 5) of the dividend, $x^3 + 2x^2 + 4x + 5$, in a line :

$$\begin{array}{|cccc} 1 & 2 & 4 & 5 \\ \hline \end{array}$$

To the left side of the coefficients, write the opposite (-2) of the constant term of the divisor, $x + 2$:

$$-2\begin{array}{|cccc} 1 & 2 & 4 & 5 \\ \hline \end{array}$$

Bring first coefficient (1) below the line :

$$-2\begin{array}{|cccc} 1 & 2 & 4 & 5 \\ \hline 1 \end{array}$$

Multiply the left number (-2) by the number (1) below the line and place the product $(-2 \cdot 1 = -2)$ under the second coefficient (2):

$$-2\begin{array}{|cccc} 1 & 2 & 4 & 5 \\ & -2 & & \\ \hline 1 \end{array}$$

Add the second coefficient (2) with the product (-2). Write the sum below the line to complete the column :

$$-2\begin{array}{|cccc} 1 & 2 & 4 & 5 \\ & -2 & & \\ \hline 1 & 0 \end{array}$$

Multiply the left number (-2) by the number (0) and place the product ($-2 \cdot 0 = 0$) under the third coefficient (4). Add the third coefficient (4) and the product (0). Write the sum below the line to complete the column :

$$
\begin{array}{r|rrrr}
-2 & 1 & 2 & 4 & 5 \\
& & -2 & 0 & \\
\hline
& 1 & 0 & 4 &
\end{array}
$$

Repeat the steps, multiply, place, and add .

$$
\begin{array}{r|rrrr}
-2 & 1 & 2 & 4 & 5 \\
& & -2 & 0 & -8 \\
\hline
& \underbrace{1 \quad 0 \quad 4} & & & \;\Big|\; -3
\end{array}
$$

Coefficients of quotient

All the numbers below the line except the last number (-3) represent the coefficients of the quotient . The last number (-3) represents the remainder. Since we divided the dividend by a linear binomial which has a power of 1, the power of the quotient is 1 less than the power of the dividend. Since the dividend has a power of 3, the quotient must have a power of 2. Therefore, the quotient is $1 \cdot x^2 + 0 \cdot x + 4$ and remainder is -3.

That is, $\dfrac{x^3+2x^2+4x+5}{x+2} = x^2 + 4 - \dfrac{3}{x+2}$

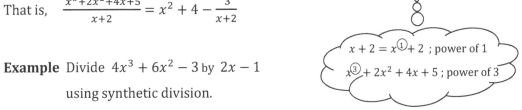

$x + 2 = x^{①} + 2$; power of 1
$x^{③} + 2x^2 + 4x + 5$; power of 3

Example Divide $4x^3 + 6x^2 - 3$ by $2x - 1$ using synthetic division.

Since the divisor is $2x - 1$, express it as the form of $x +$constant (linear binomial).

Dividing the expression by 2, we get $\dfrac{4x^3+6x^2-3}{2x-1} = \dfrac{2x^3+3x^2-\frac{3}{2}}{x-\frac{1}{2}}$.

Now we can use synthetic division.

Since x-term is missing, supply the x-term with a zero coefficient.

$$
\begin{array}{r|rrrr}
\frac{1}{2} & 2 & 3 & 0 & -\frac{3}{2} \\
& & 1 & 2 & 1 \\
\hline
& 2 & 4 & 2 & -\frac{1}{2}
\end{array}
$$

$$
\therefore \quad \frac{2x^3+3x^2-\frac{3}{2}}{x-\frac{1}{2}} = 2x^2 + 4x + 2 - \frac{\frac{1}{2}}{x-\frac{1}{2}} = 2x^2 + 4x + 2 - \frac{1}{2x-1}
$$

Therefore, $\dfrac{4x^3+6x^2-3}{2x-1} = 2x^2 + 4x + 2 - \dfrac{1}{2x-1}$

3. The Remainder and Factor Theorems

(1) Remainder Theorem

Let a polynomial $P(x) = x^3 - 2x^2 + x - 3$ and divide $P(x)$ by $x - 2$.

To find the remainder, we can consider 3 different methods:

1) Using Polynomial Long Division

Divide $x^3 - 2x^2 + x - 3$ by $x - 2$

$$\Rightarrow \quad \begin{array}{r} x^2 \qquad\quad + 1 \\ \hline x - 2 \,)\, \overline{x^3 - 2x^2 + x - 3} \end{array}$$

$$\underline{x^3 - 2x^2} \qquad \longleftarrow \quad x^2(x-2)$$

$$x - 3$$

$$\underline{x - 2} \qquad \longleftarrow \quad 1(x-2)$$

$$-1 \qquad \longleftarrow \quad \text{Remainder}$$

∴ The remainder is -1.

2) Using Synthetic Division

$$\begin{array}{c|cccc} 2 & 1 & -2 & 1 & -3 \\ & & 2 & 0 & 2 \\ \hline & 1 & 0 & 1 & -1 \end{array}$$

∴ The remainder is -1.

3) Using Division Algorithm

Let $Q(x)$ and R be the quotient and remainder of $P(x)$, respectively.

Then $x^3 - 2x^2 + x - 3 = (x-2)Q(x) + R$.

Subtracting $x = 2$ into both sides of the equal sign, we get

$2^3 - 2(2^2) + 2 - 3 = (2-2)Q(2) + R \quad \therefore R = -1$

∴ The remainder is -1.

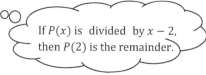

If $P(x)$ is divided by $x - 2$, then $P(2)$ is the remainder.

This third observation leads to the following Remainder Theorem.

Remainder Theorem

① If a polynomial $P(x)$ is divided by $x - k$ where k is a number, then the remainder is $P(k)$.

② In general, if a polynomial $P(x)$ is divided by $ax + b$, $a \neq 0$, then the remainder is $P\left(-\frac{b}{a}\right)$.

Note : (Proof of Remainder Theorem)

If $P(x)$ is divided by $x - k$ where k is a number, then $P(x)$ is the dividend and $x - k$ is the divisor.

Letting $Q(x)$ be the quotient and R be the remainder, we have $P(x) = (x - k)Q(x) + R$.

Thus, $P(k) = (k - k)Q(k) + R = 0 \cdot Q(x) + R = R$.

Therefore, the remainder R is $P(k)$.

Note : General Form of Remainder

    ① *If the divisor is a second-degree expression (a quadratic), then*

    the remainder must be of degree 1 or less, in the form of $ax + b$.

    ② *If the divisor is a third-degree expression (a cubic), then*

    the remainder must be of degree 2 or less, in the form of $ax^2 + bx + c$.

Note : The remainder theorem can only be used when dividing by a linear expression.

    If dividing by a quadratic expression, use long division.

Example 1   When $x^3 + 2x^2 + 4x + 5$ is divided by $x + 2$, find the remainder.

Let $P(x) = x^3 + 2x^2 + 4x + 5$ and $k = -2$. By the remainder theorem,

$P(-2) = (-2)^3 + 2(-2)^2 + 4(-2) + 5 = -8 + 8 - 8 + 5 = -3$.

Thus, the remainder is -3.

Example 2

For a polynomial $P(x)$, the remainder of $\frac{P(x)}{x+1}$ is 2 and the remainder of $\frac{P(x)}{x+2}$ is 3.

Find the remainder of $\frac{P(x)}{(x+1)(x+2)}$.

Let $Q(x)$ and $R(x)$ be the quotient and remainder of $\frac{P(x)}{(x+1)(x+2)}$, respectively.

Then $P(x) = (x + 1)(x + 2)Q(x) + R(x)$.

Since the divisor is degree of 2, $R(x) = ax + b$, $a \neq 0$ (degree of 1).

Thus $P(x) = (x + 1)(x + 2)Q(x) + ax + b$

Substituting $x = -1$ and $x = -2$ into $P(x)$, we have

$P(-1) = -a + b$, $P(-2) = -2a + b$.

By the remainder theorem, $-a + b = 2$ and $-2a + b = 3$

$\therefore a = -1$ and $b = 1$ $\therefore R(x) = ax + b = -x + 1$

Thus, the remainder of $\dfrac{P(x)}{(x+1)(x+2)}$ is $-x + 1$.

(2) Factor theorem

> **Factor Theorem**
>
> A polynomial $P(x)$ has a factor $x - k$ if and only if $P(k) = 0$.
>
> $$P(x) = (x - k)Q(x) \underset{\text{if and only if}}{\Longleftrightarrow} P(k) = 0$$

Note : (Proof of factor Theorem)

($\Rightarrow$) : If $x - k$ is a factor of $P(x)$, then the division of $P(x)$ by $x - k$ is exact, so that the remainder is 0.

By the remaining theorem, $P(k) = 0$

($\Leftarrow$) : If $P(k) = 0$, then $P(x) = (x - k)Q(x) + 0 = (x - k)Q(x)$. That is, $x - k$ is a factor of $P(x)$.

Example 1

Show that $x + 3$ is a factor of the polynomial $P(x) = x^3 + 27$.

Since $x + 3 = x - (-3)$, $k = -3$

So, $P(-3) = (-3)^3 + 27 = -27 + 27 = 0$.

By the factor theorem, $x + 3$ is a factor of $x^3 + 27$.

Example 2

For a polynomial $P(x) = x^3 - 2x^2 - 16x + 32$, $P(2) = 0$.

Find the value of k $(k \neq 2)$ such that $P(k) = 0$.

Since $P(2) = 0$, $x - 2$ is a factor of $P(x)$.

Using synthetic division to find the other factors, we have

$$
\begin{array}{r|rrrr}
2 & 1 & -2 & -16 & 32 \\
  &   & 2 & 0   & -32 \\
\hline
  & 1 & 0 & -16 & 0
\end{array}
$$

$\therefore P(x) = (x - 2)(x^2 - 16) = (x - 2)(x + 4)(x - 4)$

By the factor theorem, $P(2) = P(-4) = P(4) = 0$ $\therefore k = -4$ or $k = 4$ $(\because k \neq 2)$

4. The Rational Zero (Root) Theorem

For a polynomial $P(x)$, the number k satisfying $P(k) = 0$ is called a *rational zero* of $P(x)$.

If the polynomial $P(x) = a_n x^n + a_{n-1} x^{n-1} + \cdots\cdots\cdots + a_1 x + a_0$, $a_n \neq 0$

where n is a non-negative integer, has integer coefficients $a_0, a_1, \cdots\cdots\cdots, a_n$,

then $P\left(\dfrac{\text{factor of constant term } a_0}{\text{factor of leading coefficient } a_n}\right) = 0$.

That is, if $P(x) = 0$ has a rational zero (root) $\dfrac{s}{t}$ and $\dfrac{s}{t}$ is in lowest terms,

then s is a factor of a_0 and t is a factor of a_n.

Note : If the leading coefficient is 1, then the possible rational zeros are simply the factors of the constant

term.

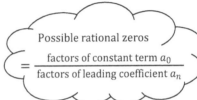

Possible rational zeros

$= \dfrac{\text{factors of constant term } a_0}{\text{factors of leading coefficient } a_n}$

Example

Find the rational zeros of $P(x) = x^3 - 3x^2 - 6x + 8$.

Consider the possible rational zeros of the polynomial $P(x)$.

Since the leading coefficient is 1 and the constant term is 8,

the possible rational zeros are $x = \pm\dfrac{1}{1},\ \pm\dfrac{2}{1}, \pm\dfrac{4}{1}, \pm\dfrac{8}{1}$.

To determine which of these possible zeros are actual zeros of $P(x)$, use synthetic

division.

When $x = 1$:

$$
\begin{array}{r|rrrr}
1 & 1 & -3 & -6 & 8 \\
  &   & 1 & -2 & -8 \\
\hline
  & 1 & -2 & -8 & 0 \\
\end{array}
$$

When $x = -1$:

$$
\begin{array}{r|rrr|r}
-1 & 1 & -3 & -6 & 8 \\
   &   & -1 & 4 & 2 \\
\hline
   & 1 & -4 & -2 & 10 \\
\end{array}
$$

Since $x = 1$ is a zero of $P(x)$, $p(x) = (x - 1)(x^2 - 2x - 8)$.

Now, factor the trinomial.

By the factor theorem, $p(x) = (x - 1)(x - 4)(x + 2)$.

Therefore, the zeros of $p(x)$ are 1, 4, and -2.

5. Irrational Expressions

(1) Parts of Irrational Numbers

For any irrational number $A = n + \alpha$ (n is an integer, $0 \leq \alpha < 1$), n is called an *integer part* and α is called a *decimal part* of A.

The decimal part α of the irrational number A is expressed as $\alpha = A - n$.

For example, $\sqrt{2} = 1.414\cdots = 1 + 0.414\cdots = 1 + (\sqrt{2} - 1)$

Note : An irrational number is separated into an integer part and a non-repeating infinite decimal part.
   Thus, the decimal part can be obtained by removing the integer part from the irrational number.

Example 1

Separate a number $5 - \sqrt{15}$ into an integer part and a decimal part.

Since $\sqrt{9} = 3$ and $\sqrt{16} = 4$, $3 < \sqrt{15} < 4$ $\therefore \sqrt{15} = 3.\times\times\times$

Since $-4 < -\sqrt{15} < -3$, $5 - 4 < 5 - \sqrt{15} < 5 - 3$ $\therefore 1 < 5 - \sqrt{15} < 2$

Thus $5 - \sqrt{15} = 1.\times\times\times = 1 + 0.\times\times\times = 1 + \text{decimal part}$

Therefore, the integer part of $5 - \sqrt{15}$ is 1 and the decimal part is $(5 - \sqrt{15}) - 1 = 4 - \sqrt{15}$.

Note : For $a > 0$; $\sqrt{a} > 0$,

$\sqrt{a} = (\text{Positive integer part}) + (\text{Decimal part})$

$\therefore \text{Decimal part} = \sqrt{a} - (\text{Positive integer part})$

For a negative irrational number :

Negative irrational number
$= (\text{negative integer part}) - (\text{decimal part})$
$\therefore \text{decimal part}$
$= (\text{negative integer part})$
$\quad -(\text{negative irrational number})$

Example 2

Separate a number $5 - 2\sqrt{15}$ into an integer part and a decimal part.

Since $3 < \sqrt{15} < 4$, $6 < 2\sqrt{15} < 8$ $\therefore 2\sqrt{15} = 6.\times\times\times$ or $7.\times\times\times$

In this case, you have to consider another way to solve the problem.

Note that $2\sqrt{15} = \sqrt{4 \cdot 15} = \sqrt{60}$.

Since $7^2 = 49$ and $8^2 = 64$, $7 < 2\sqrt{15} < 8$ $\therefore 2\sqrt{15} = 7.\times\times\times$

Thus, $7 < 2\sqrt{15} < 8$ $\Rightarrow -8 < -2\sqrt{15} < -7$ $\Rightarrow 5 - 8 < 5 - 2\sqrt{15} < 5 - 7$

$\Rightarrow -3 < 5 - 2\sqrt{15} < -2$

$\therefore 5 - 2\sqrt{15} = -2.\times\times\times = -2 - 0.\times\times\times = -2 - (\text{decimal part})$

Therefore, the integer part of $5 - 2\sqrt{15}$ is -2 and

the decimal part is $-2 - (5 - 2\sqrt{15}) = -7 + 2\sqrt{15}$.

(2) Operations of Irrational Expressions

For any real numbers $a, b,$ and $c,$

1) $\sqrt{a^2} = |a| = \begin{cases} a, & a \geq 0 \\ -a, & a < 0 \end{cases}$

$$\sqrt{(a-b)^2} = \begin{cases} a-b, & a \geq b \\ -(a-b), & a < b \end{cases}$$

2) $a \geq 0, \ b \geq 0 \ \Rightarrow \ \sqrt{a}\sqrt{b} = \sqrt{ab}$

$$\frac{\sqrt{a}}{\sqrt{b}} = \sqrt{\frac{a}{b}}, \ b \neq 0$$

3) $a \leq 0, \ b \leq 0 \ \Rightarrow \ \sqrt{a}\sqrt{b} = -\sqrt{ab}$

4) $a \geq 0, \ b < 0 \ \Rightarrow \ \dfrac{\sqrt{a}}{\sqrt{b}} = -\sqrt{\dfrac{a}{b}}, \ b \neq 0$

5) Rationalizing denominators :

$$a > 0, \ b > 0 \ (a \neq b) \ \Rightarrow \ \frac{a}{\sqrt{b}} = \frac{a\sqrt{b}}{\sqrt{b}\sqrt{b}} = \frac{a\sqrt{b}}{b}$$

$$\frac{c}{\sqrt{a} \pm \sqrt{b}} = \frac{c\,(\sqrt{a} \mp \sqrt{b})}{(\sqrt{a} \pm \sqrt{b})(\sqrt{a} \mp \sqrt{b})}$$

(3) Equalities of Two Irrational Numbers

For any rational numbers $a, b,$ and $c,$ and irrational number $\sqrt{m}$,

1) $a + b\sqrt{m} = 0 \xleftrightarrow[\text{if and only if}]{} a = b = 0$

2) $a + b\sqrt{m} = c + d\sqrt{m} \xleftrightarrow[\text{if and only if}]{} a = c, \ b = d$

(4) Nested Radical

Consider an irrational number $\sqrt{2} + \sqrt{3}$.

$$\left(\sqrt{2} + \sqrt{3}\right)^2 = 2 + 2\sqrt{2}\sqrt{3} + 3 = 5 + 2\sqrt{6}$$

Since $\sqrt{2} + \sqrt{3} > 0, \ \sqrt{5 + 2\sqrt{6}} = \sqrt{2} + \sqrt{3}$

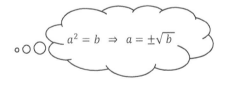
$a^2 = b \ \Rightarrow \ a = \pm\sqrt{b}$

In general, if $a > 0$ and $b > 0$, then $\left(\sqrt{a} + \sqrt{b}\right)^2 = a + b + 2\sqrt{ab}$

Since $\sqrt{a} + \sqrt{b} > 0, \ \sqrt{a + b + 2\sqrt{ab}} = \sqrt{a} + \sqrt{b}$

Similarly, if $a > b > 0$, then $\left(\sqrt{a} - \sqrt{b}\right)^2 = a + b - 2\sqrt{ab}$

Since $\sqrt{a} - \sqrt{b} > 0, \ \sqrt{a + b - 2\sqrt{ab}} = \sqrt{a} - \sqrt{b}$

A nested radical is a radical expression that contains another radical expression.

Some nested radicals can be written in a form that is not nested.

Square roots of the form $m + 2\sqrt{n}$ are converted into expressions without nested square roots.

De-nesting Nested Radicals (Double Radicals)

For rational numbers $a(\geq 0)$, $b(\geq 0)$,

$$① \quad \sqrt{a + b + 2\sqrt{ab}} = \sqrt{a} + \sqrt{b}$$

$$② \quad \sqrt{a + b - 2\sqrt{ab}} = \sqrt{a} - \sqrt{b} \quad (a \geq b)$$

Example

$$\sqrt{3 + 2\sqrt{2}} = 1 + \sqrt{2}$$

$$(\because 3 = 1 + 2 \ \text{and} \ 2 = 1 \cdot 2)$$

$$\sqrt{6 + 2\sqrt{5}} = 1 + \sqrt{5}$$

$$(\because 6 = 1 + 5 \ \text{and} \ 5 = 1 \cdot 5)$$

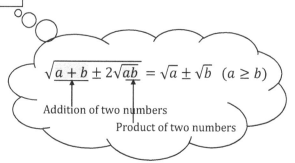

$$\sqrt{a + b \pm 2\sqrt{ab}} = \sqrt{a} \pm \sqrt{b} \quad (a \geq b)$$

Addition of two numbers

Product of two numbers

$$\sqrt{4 - \sqrt{7}} = \sqrt{\frac{8 - 2\sqrt{7}}{2}} = \frac{\sqrt{8 - 2\sqrt{7}}}{\sqrt{2}} = \frac{\sqrt{7} - 1}{\sqrt{2}} = \frac{\sqrt{14} - \sqrt{2}}{2}$$

$$\left(\begin{array}{c} \because 8 = 1 + 7 \\ 7 = 1 \cdot 7 \\ 7 > 1 \end{array} \right)$$

Exercises

#1 Perform the indicated operation.

(1) $(2x + y) + (3x - 4y)$

(2) $3(x - 2y) - (5x + y)$

(3) $(5x^3 - 2x + 8) + (2x^2 - x + 3)$

(4) $(x - 2)(3x^2 - x + 1)$

(5) $(x + 3)(x - 1)(2x - 5)$

(6) $(x + 2y + 1)(x - 2y - 1)$

(7) $(x^2 + x + 3)(x^2 + x - 1)$

(8) $(x + 2)(x - 2)(x - 1)(x - 5)$

#2 $a = 2^{x+1}$, $b = 3^{x-1}$. Express 6^x using a and b.

#3 Simplify $6^{\frac{x-y}{x+y}}$ where $10^x = 2$, $10^y = 3$.

#4 For a positive integer n, compute $(-1)^{2n+1} \cdot (-1)^{3n-1} \cdot (-1)^{2n-1} \div (-1)^{3n}$

#5 $(2x^2y^3)^a \div 4xy \cdot \frac{1}{2}x^2y = bx^3y^3$. Find the value of $a + b$, where a and b are constants.

#6 Determine values :

(1) Two polynomials $f(x) = (k^2 + k)x - (k^2 - k)y + (k - 1)z$ and $g(x) = -3k + 5$ are always equal, not depending on the values of k. Determine the values of real numbers x, y, and z.

(2) Two polynomials $f(x) = x^2 - 3x + 6$ and
$g(x) = a(x - 1)(x - 2) + b(x - 2)(x - 3) + c(x - 3)(x - 1)$ are equal. Determine the values of constants a, b, and c.

#7 Factor each polynomial.

(1) $x^4 - 6x^2 + 5$

(2) $x^4 + 7x^2 + 16$

(3) $x^4 + 4x^2 + 16$

(4) $x^4 + x^2y^2 + y^4$

(5) $(x^2 - 2x)^2 - 2(x^2 - 2x) - 8$

(6) $x(x - 1)(x^2 - x + 1) - 30$

(7) $(x - 1)(x - 2)(x + 3)(x + 4) - 84$

(8) $x^2 + xy - 4x - 2y + 4$

(9) $x^3 - x^2y - xz^2 + yz^2$

(10) $x^2(y - z) + y^2(z - x) + z^2(x - y)$

(11) $x^2y^2 - 3xyz - xy + 3z$

(12) $9x^2 + y^2 + 4 - 6xy - 12x + 4y$

(13) $-8x^3 + 12x^2 - 6x + 1$

#8 Find the value of each expression :

(1) When $a - b = 3$ and $ab = -2$,

    ① $a^2 + b^2$ ② $\dfrac{b}{a} + \dfrac{a}{b}$ ③ $(a + b)^2$ ④ $a^3 - b^3$

(2) When $a + b = 2$ and $a^3 + b^3 = 32$, find the value of $a^2 + b^2$.

(3) For any real number a $(a > 3)$, $a^2 + \dfrac{9}{a^2} = 15$. Find the value of $a^3 - \dfrac{27}{a^3}$.

(4) When $A = x^3 + x - 2$ and $B = x - 2$, find the coefficient of x^3 in $A^3 - B^3$.

(5) When $a + b + c = 0$ and $a^2 + b^2 + c^2 = 1$, find the value of $a^2 b^2 + b^2 c^2 + c^2 a^2$.

(6) For non-zero numbers $a, b,$ and $c,$

    $a + b + c = 0$, $\dfrac{1}{a} + \dfrac{1}{b} + \dfrac{1}{c} = 3$, and $a^2 + b^2 + c^2 = 12$. Find the value of $a^3 + b^3 + c^3$.

(7) For positive real numbers a and b, $a - b = 2$, $a^2 + ab + b^2 = 10$.

    Find the value of $a^3 + b^3$.

(8) When $a^2 - 3a + 1 = 0$, find the value of $3a^2 + 4a^3 + a^4 + \dfrac{3}{a^2} + \dfrac{4}{a^3} + \dfrac{1}{a^4}$.

(9) For real numbers $a, b,$ and c, $a + b + c = \sqrt{3}$ and $a^2 + b^2 + c^2 = 1$. Find the value of abc.

(10) For real numbers $a, b,$ and c, $a + b + c = 1$, $a^2 + b^2 + c^2 = 9$, and $a^3 + b^3 + c^3 = 28$.

    Find the value of $(a + b)(b + c)(c + a)$.

(11) For real numbers $a, b,$ and c, $a + b + c = -2$, $abc = 6$, and $\dfrac{1}{a} + \dfrac{1}{b} + \dfrac{1}{c} = -\dfrac{3}{2}$.

    Find the value of $(1 - a)(1 - b)(1 - c)$.

#9 Evaluate each expression.

(1) When $x - \dfrac{1}{x} = 3$, find the values of $x^2 + \dfrac{1}{x^2}$ and $x^3 - \dfrac{1}{x^3}$.

(2) When $\dfrac{x}{2} = \dfrac{y}{3}$, $xy \neq 0$, find the values of $\dfrac{xy}{x^2 + y^2}$ and $\dfrac{2x^2 + 5xy}{x^2 + y^2}$.

(3) When $x^2 - 3x - 1 = 0$, find the value of $x + \dfrac{1}{x}$.

(4) When $x^2 - 5x + 1 = 0$, find the value of $x^3 - 3x + 1 - \dfrac{3}{x} + \dfrac{1}{x^3}$.

(5) When $x^4 - 3x^2 + 1 = 0$ $(0 < x < 1)$, find the value of $x - \dfrac{1}{x}$.

(6) When $x = \dfrac{3 + \sqrt{5}}{2}$, find the value of $x^4 - x^3 - 6x^2 + 9x - 4$.

(7) When $x = \dfrac{1 + \sqrt{3}i}{2}$, find the value of $x^4 + 2x^2 - 3x - 1$.

(8) When $x : y : z = 2 : 3 : 4$, find the value of $\dfrac{x + y + z}{2x + 3y - 4z}$.

#10 When $x + y + z = 1$, factor the polynomial $x^2 - 3xy + 2y^2 - 4x - y - 2z - 1$.

#11 Find the value of k that will make each polynomial perfect square form.

(1) $x^2 + 5x + k$

(2) $9x^2 - 12x + k$

(3) $2x^2 - 6x + k$

(4) $25x^2 + 4x + k$

(5) $\frac{1}{25}x^2 + kx + 4$

(6) $4x^2 - kx + 25$

(7) $2x^2 + kx + 8$

(8) $9x^2 + (2k - 4)x + 4$

(9) $4x^2 + (k + 5)xy + 9y^2$

(10) $k - \frac{1}{4}xy + \frac{1}{4}y^2$

(11) $9x^2 + (k - 1)xy + 25y^2$

(12) $kx^2 + 3x + 9$

(13) $(x + 3)(x - 4) - k$

(14) $(2x + 1)(2x - 4) + k$

(15) $(x - 1)(x - 2)(x + 4)(x + 5) + k$

#12. Find the value of a for the following polynomials. Each polynomial has a given factor.

(1) $x^2 + 2x + a$ has the factor $(x + 3)$.

(2) $3x^2 + ax - 8$ has the factor $(x - 2)$.

(3) $4x^2 + ax - 6$ has the factor $(3 - 2x)$.

(4) $2x^2 + (3a - 1)x - 15$ has the factor $(2x + 3)$.

(5) $2ax^2 - 5x + 2$ has the factor $(3x - 2)$.

#13 Find the value of $a + b$ for any constants a and b.

(1) $3ax^2 - 6x + ab$ has the factor $(3x - 1)^2$.

(2) $ax^2 + 8x + 4b$ has two factors $(3x + 2)$ and $(x - 2)$.

(3) $(4x - 3)^2 - (3x - 2)^2$ has two factors $(ax + 5)$ and $(b - x)$.

(4) $2x^2 + ax - 4$ and $bx^2 - x - 2$ have the same factor $(2x + 1)$.

#14 Factor the polynomial given that $P(k) = 0$.

(1) $P(x) = 2x^3 + 3x^2 + x + 6$; $k = -2$

(2) $P(x) = x^3 - 3x - 2$; $k = -1$

(3) $P(x) = 6x^4 - x^3 - 7x^2 + x + 1$; $k = 1$

(4) $P(x) = 2x^3 - x^2 - 7x + 6$; $k = -2$

#15 Find all real zeros of each polynomial.

    (1) $P(x) = x^4 + 4x^3 - 2x^2 - 12x + 9$

    (2) $P(x) = 2x^3 + 3x^2 - 6x + 2$

    (3) $P(x) = 4x^4 - 4x^3 + x^2 - 2x + 1$

#16 Find each polynomial $P(x)$.

    (1) Let the quotient of $P(x)$ be $2x - 1$.

        If $P(x)$ is divided by $x^2 - 2x + 3$, then the remainder is $-3x + 1$.

    (2) Let $P(x) = x^2 - a$ $(a \neq 0)$. Then $P(x^2) = P(x)Q(x)$ for a polynomial $Q(x)$.

    (3) For constants a and b,

        $P(x) = x^3 + 4x^2 + ax + b$ is divided by $(x + 1)^2$ with no remainder.

#17 Using synthetic division, divide $2x^3 + 4x^2 - 3x - 5$ by

    (1) $2x - 1$

    (2) $2x + 3$

    Find the quotient and remainder for each division.

#18 Obtain the remainder.

    (1) When a polynomial $P(x) = 3x^4 - 2x^3 - x + 1$ is divided by a polynomial $x^2 - 1$,

        find the remainder.

    (2) When $x + 1$ is a factor of a polynomial $P(x) = x^4 - 2x + a$,

        divide $P(x)$ by $x - 1$ and find the remainder.

    (3) For all real number x, a polynomial $P(x)$ satisfies $P(1 + x) = P(1 - x)$ and $P(0) = -4$.

        Find the remainder of $\dfrac{P(x)}{x(x-2)}$.

    (4) The remainder of $\dfrac{P(x)}{(x+1)^2}$ is $-3x + 1$ and the remainder of $\dfrac{P(x)}{x-2}$ is 4.

        Find the remainder of $\dfrac{P(x)}{(x+1)^2(x-2)}$.

    (5) When $P(x)$ is divided by $(x - 1)(x - 2)$, the remainder is $-5x + 2$.

        Find the remainder when the polynomial $P(2x)$ is divided by $x - 1$.

    (6) When $P(x)$ is divided by $x - 1$, the remainder is 1, and when divided by $(x - 2)(x - 3)$,

        the remainder is 5. Find the remainder when $P(x)$ is divided by $(x - 1)(x - 2)(x - 3)$.

#19 When a polynomial $P(x) = x^5 - 2ax + b$ is divided by $(x-1)^2$, the remainder is -3.

Find the values of real numbers a and b.

#20 For a triangle with lengths $a, b,$ and $c,$ classify each triangle.

(1) $a^3 - ac^2 + a^2b - bc^2 = 0$

(2) $a^3 + b^3 + c^3 = 3abc$

#21 Evaluate each expression using factorization.

(1) $99^2 - 1$

(2) $99^2 - 89^2$

(3) $49^2 - 51^2$

(4) $3^8 - 1$

(5) $6^2 - 5^2 + 4^2 - 3^2 + 2^2 - 1$

(6) $\left(1 - \frac{1}{2^2}\right)\left(1 - \frac{1}{3^2}\right)\left(1 - \frac{1}{4^2}\right) \cdots \cdots \left(1 - \frac{1}{99^2}\right)\left(1 - \frac{1}{100^2}\right)$

(7) $3(2^2 + 1)(2^4 + 1)(2^8 + 1) + 1$

(8) $\dfrac{99 \times 101 + 99 \times 2}{101^2 - 4}$

(9) $36 \times 34 - 35 \times 34$

(10) $87 \times 56 + 87 \times 44$

(11) $65^2 - 2 \times 65 \times 35 + 35^2$

(12) $25^2 + 30 \times 25 + 15^2$

(13) $\dfrac{1000^3 + 1}{1000^2 - 999}$

(14) $(3+1)(3^2+1)(3^4+1)(3^8+1)$

(15) $\left(\frac{97}{100}\right)^3 + \left(\frac{3}{100}\right)^3 - 1$

#22 When a polynomial $P(x) = x^3 + ax^2 + bx - 2$ is divided by $x+2$ and $x-1$, respectively, there are no remainders. For the constants a and $b,$ find the value of $a^2 - b^2$.

#23 Simplify each expression.

(1) $\dfrac{\dfrac{1}{x+y}+\dfrac{1}{x-y}}{\dfrac{1}{x+y}-\dfrac{1}{x-y}}$

(2) $1-\dfrac{1}{1-\dfrac{1}{1-x}}$

(3) $\dfrac{1-\dfrac{x-y}{x+y}}{1+\dfrac{x-y}{x+y}}$

(4) $1+\dfrac{1}{1+\dfrac{1}{1+\dfrac{1}{1+x}}}$

#24 For two polynomials A and B whose leading coefficients are 1, G and L are GCF and LCM of A and B, respectively. Find each polynomial.

(1) When $G = x - 1$, $L = (x-1)^2(x-3)$,

(2) When $G = x - 1$, $L = x^3 - 7x + 6$,

(3) When $G = x + 1$, $AB = x^4 - 4x^3 - 3x^2 + 10x + 8$,

#25 For three polynomials $x^2 - 1$, $x^2 - 6x + 5$, and $x^2 + 3x + 2a$, the GCF is a polynomial of degree 1. Find the value of a.

#26 For two polynomials $f(x) = x^3 - ax^2 + 2x + 4$ and $g(x) = x^2 + bx + 6$, $R(x)$ is the remainder of $\dfrac{f(x)}{g(x)}$. When the GCF of $g(x)$ and $R(x)$ is $x + 2$, find the value of $R(-1)$.

#27 For two polynomials $f(x)$ and $g(x)$ whose leading coefficients are 1, the GCF and LCM of $f(x)$ and $g(x)$ are $x - 2$ and $x^3 - 3x^2 - 4x + 12$, respectively. Find the value of $|f(0) - g(0)|$.

#28 Simplify each expression.

(1) $\sqrt{7 + 2\sqrt{10}}$

(2) $\sqrt{2 - \sqrt{3}}$

(3) $\sqrt{4 + \sqrt{7}} - \sqrt{4 - \sqrt{7}}$

(4) $\sqrt{9 + \sqrt{72}} - \sqrt{9 - \sqrt{72}}$

(5) $\dfrac{\sqrt{x+2}-\sqrt{x-2}}{\sqrt{x+2}+\sqrt{x-2}}$

(6) $\dfrac{1-\sqrt{2}-\sqrt{3}}{1+\sqrt{2}+\sqrt{3}}$

#29 Evaluate each expression.

(1) When $x = 2 + \sqrt{3}$, $\dfrac{1}{1+\sqrt{x+2}} + \dfrac{1}{1-\sqrt{x+2}}$

(2) When $x = \sqrt{3 + \sqrt{8}}$, $x^3 + \dfrac{1}{x^3}$

(3) When $x = \sqrt{5 + 2\sqrt{6}}$ and $y = \sqrt{5 - 2\sqrt{6}}$, $\dfrac{\sqrt{x}-\sqrt{y}}{\sqrt{x}+\sqrt{y}}$

(4) When $x = \sqrt{2}$, $\dfrac{1}{\sqrt{x+1+2\sqrt{x}}} - \dfrac{1}{\sqrt{x+1-2\sqrt{x}}}$

#30 Simplify the expression :

$$\dfrac{1}{x^2+x} + \dfrac{1}{x^2+3x+2} + \dfrac{1}{x^2+5x+6} + \cdots\cdots + \dfrac{1}{x^2+19x+90}$$

#31 Find the value of k such that $\dfrac{x+y}{3} = \dfrac{2y-z}{4} = \dfrac{z}{5} = \dfrac{3x+7y-4z}{k}$

#32 When a is the integer and b is the decimal part of an irrational number $\sqrt{11 + \sqrt{72}}$,

(1) Find the value of $a + \dfrac{1}{b-1}$.

(2) Find the value of $\sqrt{\dfrac{b+1-\sqrt{b^2+2b}}{b+1+\sqrt{b^2+2b}}}$.

(3) For a polynomial $P(x) = x^4 + 2x^3 + x^2 + 4x + 1$, find the value of $P(b)$.

#33 When a is the decimal part of $\sqrt{5}$, find the value of $\sqrt{\left(a - \dfrac{1}{2}\right)^2} + \sqrt{\left(a - \dfrac{1}{5}\right)^2}$.

#34 For a real number a $(0 < a < 1)$, $x = a^2 + \dfrac{1}{a^2}$.

Express $\sqrt{x + 2} - \sqrt{x - 2}$ as a polynomial in a.

#35 When $\dfrac{1}{4} < x < 1$, simplify the expression $\sqrt{4x + 1 - 4\sqrt{x}} + \left|\sqrt{x} - 1\right|$.

Chapter 3. Equations and Inequalities

3-1 Solving Linear Equations

1. Linear Equations and their Solutions

(1) Linear Equations

A *linear equation* is an equation with one variable which has the highest power of 1.

For example, the equation $ax = b$, where a and b are constants, is a linear equation with a variable x.

(2) Solutions

The *solution* (*root*) of an equation is a number which makes the equation a true expression.

For example, the linear equation $2x = 10$ is true when $x = 5$.

Thus, $x = 5$ is the solution of the equation $2x = 10$.

For the linear equation $ax = b$, where a and b are constants, the solution is

---

1) $a \neq 0 \Rightarrow x = \dfrac{b}{a}$ (Only one solution)

2) $a = 0 \Rightarrow$ ① $b \neq 0 \Rightarrow$ No solution ($\because$ Division is not defined.)

 ② $b = 0 \Rightarrow$ All real numbers

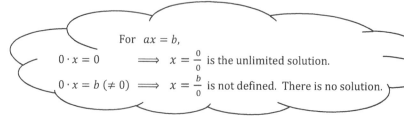

For $ax = b$,

$0 \cdot x = 0 \implies x = \dfrac{0}{0}$ is the unlimited solution.

$0 \cdot x = b \, (\neq 0) \implies x = \dfrac{b}{0}$ is not defined. There is no solution.

Note : The linear equation $ax + b = 0$, $a \neq 0$, has only one solution $x = -\dfrac{b}{a}$.

Example Solve the equation for x.

1. $a(x - 1) = x + 3 \Rightarrow ax - a = x + 3 \quad \therefore (a - 1)x = a + 3$

 If $a \neq 1$, then $x = \dfrac{a+3}{a-1}$

 If $a = 1$, then $0 \cdot x = 4 \quad \therefore x$ is not defined.

2. $(a^2 + 2)x - 1 = a(3x - 1) \Rightarrow a^2 x + 2x - 1 = 3ax - a$

$\Rightarrow (a^2 - 3a + 2)x = 1 - a$

$\Rightarrow (a - 2)(a - 1)x = 1 - a$

If $a \neq 1$, $a \neq 2$, then $x = -\dfrac{1}{a-2}$

If $a = 2$, then $0 \cdot x = -1$ $\quad \therefore x$ is not defined.

If $a = 1$, then $0 \cdot x = 0$ $\quad \therefore x$ is unlimited solution.

(The equation is always true for all x.)

2. Properties of Equality

If $A = B$, then

(1) $A + m = B + m$ $\qquad$ (Add the same number to both sides)

(2) $A - m = B - m$ $\qquad$ (Subtract the same number from both sides)

(3) $A \times m = B \times m$ $\qquad$ (Multiply both sides by the same non-zero number)

(4) $\dfrac{A}{m} = \dfrac{B}{m}$, $m \neq 0$ $\qquad$ (Divide both sides by the same non-zero number)

3. Solving Linear Equations

Steps for Solving Equations with One Variable

Step 1. If the coefficient of the equation is a fraction or decimal, change it to an integer by multiplying a proper number by both sides of the equation.

Step 2. If there are parentheses, remove them by using the distributive property.

Step 3. Convert it into a simpler equation; transfer all variables to one side of the equation and transfer all numbers to the other side of the equation in the form of $ax = b$, $a \neq 0$.

Step 4. Solve for the variable x by dividing each side of the equation ($ax = b$) by a.

This gives $x = \dfrac{b}{a}$, $a \neq 0$. Therefore, $\dfrac{b}{a}$ is the one and only solution of $ax = b$.

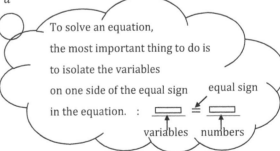

To solve an equation, the most important thing to do is to isolate the variables on one side of the equal sign in the equation.

Example 1

$2x + 3 = 7 \xRightarrow{\text{Subtract 3 from both sides}} 2x + 3 - 3 = 7 - 3$

$\xRightarrow{\text{Form of Variable=Number}} 2x = 4$

$\xRightarrow{\text{Multiply each side by } \frac{1}{2} \text{ or divide each side by 2}} 2x \times \dfrac{1}{2} = 4 \times \dfrac{1}{2}$ or $\dfrac{2x}{2} = \dfrac{4}{2}$

$\xRightarrow{\text{Reduce to lowest terms}} x = 2$

Example 2

$$\frac{5}{2}x + 4 = x - 2 \Longrightarrow 5x + 8 = 2x - 4$$
$$\underset{\text{Multiply each side by 2}}{}$$

$$\underset{\text{Transfer(Step 3)}}{\Longrightarrow} 5x - 2x = -4 - 8$$

$$\underset{\text{Simplify}}{\Longrightarrow} 3x = -12$$

$$\underset{\substack{\text{Divide each side by 3 or multiply each side by } \frac{1}{2}}}{\Longrightarrow} \frac{3x}{3} = \frac{-12}{3} \text{ or } 3x \times \frac{1}{3} = -12 \times \frac{1}{3}$$

$$\underset{\text{Simplify}}{\Longrightarrow} x = -4$$

4. Solving Absolute Value Equations

The absolute value of a number a, denoted by $|a|$, is the distance of a from zero on a number line.

Since a distance is always a positive number, absolute value is never negative.

For example,

$|2| = 2$: two units to the right of zero on a number line.

$|-2| = 2$: two units to the left of zero on a number line.

$|-2| = -(-2) = 2$
$-|-2| = -2$

If the number has no minus sign in front of it, the absolute value is not changed.

But if the number has minus sign in front of it, we remove the minus sign to find the absolute Value. For example,

if $a = -1$, then $-a = -(-1) = 1$ and if $a = -2$, then $-a = -(-2) = 2$.

Therefore, the absolute value of a real number a, denoted by $|a|$, is defines as follows:

$$|a| = \begin{cases} a & \text{if } a \geq 0 \\ -a & \text{if } a < 0 \end{cases}$$

Example 1

$|x - 1| = 2x + 3$

$$\Longrightarrow \begin{cases} x - 1 \geq 0 \ (x \geq 1) \Rightarrow |x - 1| = x - 1 = 2x + 3 & \therefore x = -4 \text{ ; Not possible} \\ x - 1 < 0 \ (x < 1) \Rightarrow |x - 1| = -(x - 1) = 2x + 3 & \therefore x = -\frac{2}{3} \end{cases}$$

$\therefore x = -\frac{2}{3}$

Example 2

$|x + 1| + |x + 2| = 5$

Since $(x + 1 = 0 \Rightarrow x = -1)$ and $(x + 2 = 0 \Rightarrow x = -2)$,

Consider $x < -2$, $-2 \leq x < -1$, and $x \geq -1$

$$\Longrightarrow \begin{cases} ① \quad x < -2 \,;\; -(x + 1) - (x + 2) = 5 \,;\; x = -4 \\ ② \; -2 \leq x < -1 \,;\; -(x + 1) + (x + 2) = 5 \,;\; 0 \cdot x = 4 \,;\; \text{No solution} \\ ③ \quad x \geq -1 \,;\; x + 1 + x + 2 = 5 \,;\; x = 1 \end{cases}$$

$\therefore \; x = -4, \; x = 1$

Example 3

$$|x + 1| = |x + 3|$$

$$\Longrightarrow \; x + 1 = \pm (x + 3)$$

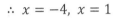

$\circ \circ \circ$ $\quad |a| = |b| \Rightarrow a = \pm b$

$$\Longrightarrow \begin{cases} x + 1 = x + 3 \Longrightarrow 0 \cdot x = 2 \,;\; \text{No solution} \\ x + 1 = -(x + 3) \Longrightarrow 2x = -4 \,;\; x = -2 \end{cases}$$

A quadratic equation is the equation containing a variable raised to the second power.

3-2 Quadratic Equations and Their Solutions

1. Quadratic Equations

(1) Definition

1) A *quadratic equation* which has transferred all the terms on the right side of the equal sign to the left side of the equal sign is formed by

$$ax^2 + bx + c = 0, \text{ for any constants } a \neq 0, \; b, \text{ and } c.$$

Note : Quadratic equations : $ax^2 + bx + c = 0, \; a \neq 0$
$$ax^2 + bx = 0, \; a \neq 0 \; (When \; c = 0)$$
$$ax^2 + c = 0, \; a \neq 0 \; (When \; b = 0)$$
$$ax^2 = 0, \; a \neq 0 \; (When \; b = 0 \; and \; c = 0)$$

2) The *solution (root)* of an unknown variable x, must make the quadratic equation

$(ax^2 + bx + c = 0, a \neq 0)$ always true.

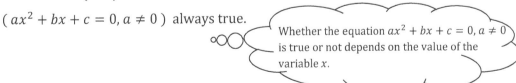

Whether the equation $ax^2 + bx + c = 0, a \neq 0$ is true or not depends on the value of the variable x.

Example

$x^2 - 2x + 1 = 0$

If $x = 1$, then $1 - 2 + 1 = 0$ (This makes the quadratic equation true.)

$\therefore x = 1$ is a solution.

If $x = -1$, then $1 + 2 + 1 \neq 0$ (This makes the quadratic equation untrue.)

$\therefore x = -1$ is not a solution.

2. Solving Quadratic Equations

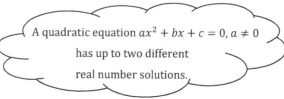

A quadratic equation $ax^2 + bx + c = 0, a \neq 0$ has up to two different real number solutions.

(1) Solutions of Quadratic Equations

1) Using Factorization

If a quadratic equation in one variable is factored, then the equation can be solved using the zero product property.

The Zero product property

For any expressions or numbers A and B,

$AB = 0 \xLeftrightarrow[\text{if and only if}]{} A = 0$ or $B = 0$

$AB \neq 0 \Leftrightarrow A \neq 0$ and $B \neq 0$

Note : $A = 0$ or $B = 0 \Leftrightarrow \begin{cases} A = 0 \ and \ B \neq 0 \\ A \neq 0 \ and \ B = 0 \\ A = 0 \ and \ B = 0 \end{cases}$

Steps for Solving Quadratic Equations

Step 1. Simplify the equation as a form of $ax^2 + bx + c = 0, a \neq 0$

Step 2. Factor the left side of the equation into the form of $a(x - \alpha)(x - \beta) = 0$

Step 3. Solve the equation using the zero product property :

$(x - \alpha) = 0$ or $(x - \beta) = 0$ $\therefore x = \alpha$ or $x = \beta$

Example

$2x^2 - 4x + 6 = 0 \xRightarrow{\text{Step 1}} 2(x^2 - 2x + 3) = 0$

$\xRightarrow{\text{Step 2}} 2(x - 3)(x + 1) = 0$

$\xRightarrow{\text{Step 3}} x = 3$ or $x = -1$

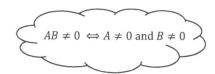

$ax^2 + bx + c = 0 \Leftrightarrow (mx + p)(nx + q) = 0,$

(where $mn = a, \ mq + np = b, \ pq = c$)

$\Rightarrow mx + p = 0$ or $nx + q = 0$

$\Rightarrow mx = -p$ or $nx = -q$

$\Rightarrow x = -\dfrac{p}{m}$ or $x = -\dfrac{q}{n}$

Double roots

If a quadratic equation has the same factor twice, then the equation has a double root.

$ax^2 + bx + c = 0 \Rightarrow a(x - \underline{\alpha})(x - \underline{\alpha}) = 0$: Two identical factors.

$\Rightarrow a(x - \alpha)^2 = 0$

$\Rightarrow x - \alpha = 0 \ (\because a \neq 0)$

$\Rightarrow x = \alpha \ \text{(A double root)}$

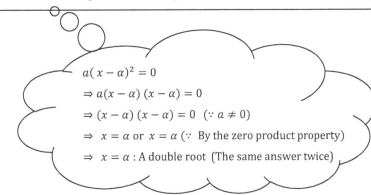

$a(x - \alpha)^2 = 0$

$\Rightarrow a(x - \alpha)(x - \alpha) = 0$

$\Rightarrow (x - \alpha)(x - \alpha) = 0 \ (\because a \neq 0)$

$\Rightarrow x = \alpha \text{ or } x = \alpha \ (\because \text{ By the zero product property})$

$\Rightarrow x = \alpha : \text{A double root} \ \text{(The same answer twice)}$

Example

$x^2 + 4x + 4 = 0 \Rightarrow (x + 2)^2 = 0 \Rightarrow x = -2 \ \text{(A double root)}$

A double root: Form of $a(\ \ )^2 = \mathbf{0}$

Example

If the equation $x^2 - 3x + k = 0$ has a double root, $k = \left(\dfrac{-3}{2}\right)^2 = \dfrac{9}{4}$.

So $x^2 - 3x + k = x^2 - 3x + \dfrac{9}{4} = \left(x - \dfrac{3}{2}\right)^2 = 0$

$\therefore \ x = \dfrac{3}{2} \ \text{(A double root)}$

$x^2 + ax + (\dfrac{a}{2})^2 = (x + \dfrac{a}{2})^2 = \mathbf{0}$

Constant $= (\dfrac{x - \text{coefficient}}{2})^2$, when x^2-coefficient $= 1$

① $a(x - \alpha)^2 = 0, \ a \neq 0 \Rightarrow x = \alpha \ \text{(A double root)}$

② To find a double root for the quadratic equation $a(x - p)^2 = q$,

$q = 0 \ (q \text{ must be } 0)$. So we get $x = p \ \text{(A double root)}$.

Solutions of Quadratic Equations

Every quadratic equation with real number coefficients has two solutions (roots) in complex number system.

Given that a, b, and c are rational numbers,

> (1) if a quadratic equation in the form $ax^2 + bx + c = 0$ has irrational solutions
>
> and one of them is $x = \alpha + \sqrt{\beta}$, then the other solution is $x = \alpha - \sqrt{\beta}$.
>
> (2) if a quadratic equation has imaginary solutions (complex number solutions),
>
> the solutions are complex conjugates
>
> ; i.e., if $\alpha + \beta i$ is a solution, then $\alpha - \beta i$ is the other solution.

Example　When $1 + \sqrt{2}$ is a root of a quadratic equation $x^2 + bx + c = 0$,

find the value of the other root. (b and c are rational numbers.)

Substituting $x = 1 + \sqrt{2}$ in the quadratic equation, we have

$(1 + \sqrt{2})^2 + b(1 + \sqrt{2}) + c = 0$

$\therefore\ 3 + 2\sqrt{2} + b + b\sqrt{2} + c = 0$

$\therefore\ (3 + b + c) + (2 + b)\sqrt{2} = 0$

$\therefore\ 3 + b + c = 0,\ 2 + b = 0$

$\therefore\ b = -2,\ c = -1$

Substituting $b = -2$ and $c = -1$ in the original quadratic equation, we have

$x^2 - 2x - 1 = 0 \qquad \therefore\ x = \frac{1 \pm \sqrt{1+1}}{1} = 1 \pm \sqrt{2}$

Thererfore, the other root is $1 - \sqrt{2}$.

2) Using Perfect Squares (Completing Squares)

Consider the quadratic equation $x^2 = p$, $p \geq 0$.

$x^2 = p,\ p \geq 0 \ \Rightarrow\ x^2 - p = 0,\ p \geq 0$

$\Rightarrow\ \left(x - \sqrt{p}\right)\left(x + \sqrt{p}\right) = 0 \qquad$ Factor

$\Rightarrow\ x - \sqrt{p} = 0\ \text{ or }\ x + \sqrt{p} = 0 \quad$ Zero property

$\Rightarrow\ x = \sqrt{p}\ \text{ or }\ x = -\sqrt{p}$

$\Rightarrow\ x = \pm\sqrt{p}$

Therefore, the solutions to $x^2 = p$, $p \geq 0$, are $x = \sqrt{p}$ and $x = -\sqrt{p}$.

Example   Solve the quadratic equation $2x^2 + 4 = -32$

$2x^2 + 4 = -32 \quad \Rightarrow \quad 2x^2 = -36 \qquad$ Subtract 4 from each side

$\Rightarrow \quad x^2 = -18 \qquad$ Divide each side by 2

$\Rightarrow \quad x = \pm\sqrt{-18} \qquad$ Square roots of each side

$\Rightarrow \quad x = \pm\sqrt{18}\,i \qquad$ Express in the form of i

$\Rightarrow \quad x = \pm 3\sqrt{2}\,i \qquad$ Simplify the radical

Therefore, the solutions are $3\sqrt{2}\,i$ and $-3\sqrt{2}\,i$.

Solve the equation in the form of a perfect square : $(x + p)^2 = q$

<u>Square roots</u>

If the quadratic equation $ax^2 + bx + c = 0$ is formed by $(x + p)^2 = q$,

$\Rightarrow$ The solution is

$\begin{cases} x = -p \pm \sqrt{q} & , \ q > 0 \ \text{(Two distinct real roots)} \\ x = -p \ \text{(A double root)} & , \ q = 0 \ \text{(One real root)} \\ \text{No real solution} & , \ q < 0 \ \text{(Two distinct complex roots)} \end{cases}$

Note : ① *If the quadratic equation $x^2 = q$ has solutions, then $q \geq 0$ (q must be positive or zero).*

② *If the quadratic equation $x^2 = q$ has 2 different solutions,*

   then $q > 0$ (q must be positive).

③ *If the quadratic equation $x^2 = q$ has a double root, then $q = 0$ (q must be zero).*

> If the real solution for the quadratic equation $x^2 = q$ does not exist (No solution), then $q < 0$ (q must be negative).

Note:

① $x^2 = p, \ p \geq 0 \quad \Rightarrow x = \pm\sqrt{p}$

② $ax^2 = p, \ a \neq 0, \ \frac{p}{a} \geq 0 \quad \Rightarrow x^2 = \frac{p}{a} \quad \Rightarrow \ x = \pm\sqrt{\frac{p}{a}}$

③ $(x + p)^2 = q, \ q \geq 0 \quad \Rightarrow x + p = \pm\sqrt{q} \quad \Rightarrow \ x = -p \pm \sqrt{q}$

④ $(ax + p)^2 = q, \ a \neq 0, \ q \geq 0 \quad \Rightarrow ax + p = \pm\sqrt{q} \quad \Rightarrow \ x = \frac{-p \pm \sqrt{q}}{a}$

Note : ① Number $\begin{cases} \text{Real number} \begin{cases} \text{positive} \\ \quad 0 \\ \text{negative} \end{cases} \\ \text{Imaginary number} \end{cases}$

$$(\text{Real number})^2 \geq 0 \quad \text{but} \quad (\underbrace{\text{Imaginary number}}_{\text{Not a Real number}})^2 < 0$$

② $i = A$ unit of imaginary numbers

$$i = \sqrt{-1} \; ; \; i^2 = -1 < 0$$

③ If there is no condition about the number,

then we consider only real number solutions not complex number solutions.

Steps for Solving Quadratic Equations Using Perfect Squares

Step 1. Make the coefficient of x^2 equal to 1 .

$$ax^2 + bx + c = 0 \quad \Rightarrow \quad a\left(x^2 + \frac{b}{a}x + \frac{c}{a}\right) = 0$$

Step 2. Divide by a on both sides or use the zero product property.

$$\Rightarrow \; x^2 + \frac{b}{a}x + \frac{c}{a} = 0$$

Step 3. Transfer the constant to the right side of the equal sign by subtracting $\frac{c}{a}$ from both

sides.

$$\Rightarrow \; x^2 + \frac{b}{a}x = -\frac{c}{a}$$

Step 4. (Completing the square)

Square of half the coefficient of x. Add the result to both sides of the equal sign.

$$x^2 + \frac{b}{a}x + \left(\frac{b}{2a}\right)^2 = -\frac{c}{a} + \left(\frac{b}{2a}\right)^2$$

$$\Rightarrow \left(x + \frac{b}{2a}\right)^2 = -\frac{c}{a} + \left(\frac{b}{2a}\right)^2 \quad \text{Form of perfect square, } (x+p)^2 = q$$

OR

$$x^2 + \frac{b}{a}x = \left(x + \frac{1}{2}\cdot\frac{b}{a}\right)^2 - \left(\frac{b}{2a}\right)^2 = -\frac{c}{a}$$

$$\Rightarrow \left(x + \frac{1}{2}\cdot\frac{b}{a}\right)^2 = -\frac{c}{a} + \left(\frac{b}{2a}\right)^2 \quad \text{Form of perfect square, } (x+p)^2 = q$$

Step 5. Solve the equation using square root.

$$\frac{1}{2}\cdot\frac{b}{a} = \frac{1}{2}\cdot \text{ (the coefficient of } x)$$

$$x^2 + mx = \left(x + \frac{1}{2}m\right)^2 - \left(\frac{1}{2}m\right)^2$$

$$\because \text{ Since } (x \pm m)^2 = x^2 \pm 2mx + m^2,$$

$$\left(x + \frac{1}{2}m\right)^2 = x^2 + 2\cdot\frac{1}{2}mx + \left(\frac{1}{2}m\right)^2 = x^2 + mx + \left(\frac{1}{2}m\right)^2$$

$$\therefore \; x^2 + mx = \left(x + \frac{1}{2}m\right)^2 - \left(\frac{1}{2}m\right)^2$$

Example

$2x^2 + 5x + 3 = 0$

$\Longrightarrow \qquad 2\left(x^2 + \dfrac{5}{2}x + \dfrac{3}{2}\right) = 0$

$\underset{\text{Divide by 2}}{\Longrightarrow} \qquad x^2 + \dfrac{5}{2}x + \dfrac{3}{2} = 0$

$\underset{\text{Transfer constant}}{\Longrightarrow} \qquad x^2 + \dfrac{5}{2}x = -\dfrac{3}{2}$ \qquad Move the constant to the right side of the equation

$\Longrightarrow \qquad \left(x + \dfrac{1}{2}\cdot\dfrac{5}{2}\right)^2 - \left(\dfrac{1}{2}\cdot\dfrac{5}{2}\right)^2 = -\dfrac{3}{2}$ \qquad To find the perfect square

$\Longrightarrow \qquad \left(x + \dfrac{5}{4}\right)^2 = -\dfrac{3}{2} + \left(\dfrac{5}{4}\right)^2$

$\Longrightarrow \qquad \left(x + \dfrac{5}{4}\right)^2 = \dfrac{1}{16}$ \qquad Perfect square form

$\underset{\text{Use the square root}}{\Longrightarrow} \qquad x + \dfrac{5}{4} = \pm\sqrt{\dfrac{1}{16}}$ \qquad To solve the equation for x

$\Longrightarrow \qquad x = -\dfrac{5}{4} \pm \dfrac{1}{4}$

$\Longrightarrow \qquad x = -\dfrac{5}{4} + \dfrac{1}{4} = -1 \ \text{ or } \ x = -\dfrac{5}{4} - \dfrac{1}{4} = -\dfrac{3}{2}$ \qquad The solution to the equation

3) Using Quadratic Formulas

If an equation is not factorized, solve it using quadratic formulas.

Before using a quadratic formula to a quadratic equation, you must have the equation in standard form, $ax^2 + bx + c = 0, \ a \neq 0$.

Quadratic Formula I

$ax^2 + bx + c = 0, \ a \neq 0$

$\Rightarrow \quad x = \dfrac{-b \pm \sqrt{b^2 - 4ac}}{2a} \ , \ \ b^2 - 4ac \geq 0$

Quadratic Formula II (When $b = 2b'$ is an even number)

$ax^2 + 2b'x + c = 0, \ a \neq 0$

$\Rightarrow \quad x = \dfrac{-b' \pm \sqrt{b'^2 - ac}}{a} \ , \ \ b'^2 - ac \geq 0$

Note : When the coefficient of x is an even number, the Quadratic Formula II can be easily used.

Note : $ax^2 + bx + c = 0, \ a \neq 0$ *Standard form of general equation*

$\Rightarrow \ a\left(x^2 + \dfrac{b}{a}x + \dfrac{c}{a}\right) = 0$

$\Rightarrow \ x^2 + \dfrac{b}{a}x + \dfrac{c}{a} = 0$ *Divide each side by a*

$\Rightarrow \ x^2 + \dfrac{b}{a}x = -\dfrac{c}{a}$ *Subtract $-\dfrac{c}{a}$ from each side*

$\Rightarrow \ \left(x + \dfrac{1}{2}\cdot\dfrac{b}{a}\right)^2 - \left(\dfrac{b}{2a}\right)^2 = -\dfrac{c}{a}$ ($\left(\dfrac{b}{2a}\right)^2$ *: The square of half the coefficient of x*)

$\Rightarrow \ \left(x + \dfrac{b}{2a}\right)^2 = \dfrac{b^2-4ac}{4a^2}$ *Add $\left(\dfrac{b}{2a}\right)^2$ to each side*

$\Rightarrow \ x + \dfrac{b}{2a} = \pm\sqrt{\dfrac{b^2-4ac}{4a^2}}$ *Take square roots of each side*

$\Rightarrow \ x = -\dfrac{b}{2a} \pm \sqrt{\dfrac{b^2-4ac}{4a^2}}$ *Subtract $-\dfrac{b}{2a}$ from each side*

$\Rightarrow \ x = \dfrac{-b \pm \sqrt{b^2-4ac}}{2a}$ *Simplify*

Example 1

$x^2 + 5x + 3 = 0$ (This equation can't be factorized.)

$\xRightarrow{a=1,b=5,c=3} \ x = \dfrac{-5\pm\sqrt{(5)^2-4\cdot1\cdot3}}{2\cdot1} = \dfrac{-5\pm\sqrt{13}}{2}$ Using Quadratic Formula I

$\therefore \ x = \dfrac{-5+\sqrt{13}}{2}$ or $x = \dfrac{-5-\sqrt{13}}{2}$

Example 2

$4x^2 + 4x - 3 = 0$

① Using Quadratic Formula I :

$\xRightarrow{a=4,b=4,c=-3} \ x = \dfrac{-4\pm\sqrt{(4)^2-4\cdot4\cdot(-3)}}{2\cdot4} = \dfrac{-4\pm\sqrt{64}}{8} = \dfrac{-4\pm8}{8}$

$\therefore \ x = \dfrac{1}{2}$ or $x = -\dfrac{3}{2}$

② Using Quadratic Formula II :

$\xRightarrow{a=4,b\prime=2,c=-3} \ x = \dfrac{-2\pm\sqrt{(2)^2-4\cdot(-3)}}{4} = \dfrac{-2\pm\sqrt{16}}{4} = \dfrac{-2\pm4}{4}$

$\therefore \ x = \dfrac{1}{2}$ or $x = -\dfrac{3}{2}$

③ Using Factorization :

$4x^2 + 4x - 3 = (2x - 1)(2x + 3) = 0$

$$\therefore \; x = \frac{1}{2} \; \text{ or } \; x = -\frac{3}{2}$$

④ Using a Perfect Square :

$$4x^2 + 4x - 3 = 0 \Rightarrow 4\left(x^2 + x - \frac{3}{4}\right) = 0 \Rightarrow x^2 + x - \frac{3}{4} = 0$$

$$\Rightarrow \left(x + \frac{1}{2}\right)^2 - \frac{1}{4} = \frac{3}{4} \;\; \Rightarrow \left(x + \frac{1}{2}\right)^2 = 1$$

$$\Rightarrow x + \frac{1}{2} = \pm 1 \;\; \Rightarrow x = -\frac{1}{2} \pm 1$$

$$\therefore \; x = \frac{1}{2} \; \text{ or } \; x = -\frac{3}{2}$$

(2) Number and Type of Solutions (The Discriminant $D = b^2 - 4ac$)

For a quadratic equation of the form $ax^2 + bx + c = 0,$ the solution is

$$x = \frac{-b \pm \sqrt{b^2 - 4ac}}{2a} \;, \;\; b^2 - 4ac \geq 0$$

In the quadratic formula, the expression $b^2 - 4ac$ (denoted by D), underneath the radical sign, is called the *Discriminant* of the quadratic equation $ax^2 + bx + c = 0$ and the discriminant determines the number and the different types of solutions for the equation.

① If $D > 0$, then the equation has two distinct roots

$$x = \frac{-b + \sqrt{b^2 - 4ac}}{2a} \;\; \text{ and } \;\; x = \frac{-b - \sqrt{b^2 - 4ac}}{2a} \quad (2 \text{ different real number solutions })$$

② If $D = 0$, then the equation has a double root

$$x = \frac{-b}{2a} \quad (\text{Only 1 real number solution; the same real number solution twice })$$

③ If $D < 0$, then the equation has two distinct complex number solutions

(No real number solution)

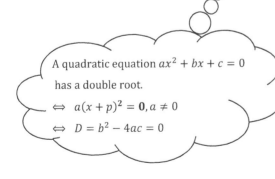

A quadratic equation $ax^2 + bx + c = 0$ has a double root.

$\Leftrightarrow \; a(x + p)^2 = 0, a \neq 0$

$\Leftrightarrow \; D = b^2 - 4ac = 0$

The solutions of a quadratic equation depend on the sign of the number D under the symbol $\sqrt{}$.

In order to have possible real number solutions, $D \geq 0$.

$a + b\sqrt{k} = 0$

$\Rightarrow a = 0$ and $b\sqrt{k} = 0$

$\Rightarrow a = 0$ and $b = 0$ $(\because \sqrt{k} \neq 0)$

Note : *Given a quadratic equation of the form* $ax^2 + bx + c = 0$, *the solution is* $x = \dfrac{-b \pm \sqrt{b^2 - 4ac}}{2a}$ *, and the discriminant is calculated as* $D = b^2 - 4ac$.

For quadratic equations of the form $ax^2 + 2b'x + c = 0$ *(The coefficient of x is an even number.),*

the solution is $x = \dfrac{-b' \pm \sqrt{b'^2 - ac}}{a}$ *and the discriminant is calculated as* $\dfrac{D}{4} = b'^2 - ac$.

Example 1 (Quadratic equation with two real number solutions)

Solve the equation $2x^2 + x = 4$.

$2x^2 + x = 4 \;\Rightarrow\; 2x^2 + x - 4 = 0$ (Standard form)

Since $D = b^2 - 4ac = 33 > 0$, the equation has two different real number solutions.

$\therefore \;\; x = \dfrac{-1 \pm \sqrt{1^2 - 4(2)(-4)}}{2(2)}$ (Quadratic Formula I)

$= \dfrac{-1 \pm \sqrt{33}}{4}$ (Simplify) $\qquad \therefore \;\; x = \dfrac{-1 + \sqrt{33}}{4}$ or $x = \dfrac{-1 - \sqrt{33}}{4}$

Therefore, the solutions are $x = \dfrac{-1 + \sqrt{33}}{4}$ and $x = \dfrac{-1 - \sqrt{33}}{4}$

Example 2 (Quadratic equation with one real number solution)

Solve the equation $x^2 + 3x = 7x - 4$.

$x^2 + 3x = 7x - 4 \;\Rightarrow\; x^2 - 4x + 4 = 0$ (Standard form)

Since $\dfrac{D}{4} = b'^2 - ac = 4 - 4 = 0$, the equation has only one real number solution.

$\therefore \;\; x = \dfrac{2 \pm \sqrt{(-2)^2 - (1)(4)}}{1}$ (Quadratic Formula II)

$= 2 \pm \sqrt{0} = 2$ (Simplify) $\qquad \therefore \; x = 2$ (A double root)

Therefore, the solution is $x = 2$ (A double root)

Example 3 (Quadratic equation with two complex number solutions)

Solve the equation $-x^2 + 3x = 3$.

$-x^2 + 3x = 3 \;\Rightarrow\; x^2 - 3x + 3 = 0$ (Standard form)

Since $D = b^2 - 4ac = -3 < 0$, the equation has no real number solution.

$\therefore \;\; x = \dfrac{3 \pm \sqrt{(-3)^2 - 4(1)(3)}}{2(1)}$ (Quadratic Formula I)

$= \dfrac{3 \pm \sqrt{-3}}{2} = \dfrac{3 \pm \sqrt{3}i}{2}$ (Simplify) $\qquad \therefore \;\; x = \dfrac{3 + \sqrt{3}i}{2}$ or $x = \dfrac{3 - \sqrt{3}i}{2}$

Therefore, the solutions are $x = \dfrac{3}{2} + \dfrac{\sqrt{3}}{2}i$ and $x = \dfrac{3}{2} - \dfrac{\sqrt{3}}{2}i$

(3) The Roots-Coefficients Relationship

The solutions (Roots) and the coefficients of a quadratic equation are related to the obtaining the sum and the product of the solutions. We can also obtain a quadratic equation from the solutions.

Dividing both sides of the quadratic equation $ax^2 + bx + c = 0$ by a,

we have $x^2 + \frac{b}{a}x + \frac{c}{a} = 0$. This equation can be expressed as $x^2 - \left(-\frac{b}{a}x\right) + \frac{c}{a} = 0$.

Note that : $x^2 - (Sum\ of\ \alpha\ and\ \beta)x + (Product\ of\ \alpha\ and\ \beta) = (x - \alpha)(x - \beta)$

Since $-\frac{b}{a}$ represents the sum of the solutions and $\frac{c}{a}$ represents the product of the solutions for the quadratic equation, it leads to the following results.

1) Sum and Product of Solutions (Roots)

Given a quadratic equation $ax^2 + bx + c = 0$, suppose the two solutions are $x = \alpha$ and $x = \beta$. Then we obtain :

① $\alpha + \beta = -\dfrac{b}{a}$ (Sum of solutions)

② $\alpha \cdot \beta = \dfrac{c}{a}$ (Product of solutions)

If $\alpha\beta = \dfrac{c}{a} < 0 \Rightarrow ac < 0$

$\Rightarrow D = b^2 - 4ac > 0$

$(\because b^2 > 0 \text{ and } -(ac) > 0)$

$\therefore$ 2 different real solutions exist.

$x^2 - (\alpha + \beta)x + (\alpha\beta) = 0$

Sum of solutions   Product of solutions

Note : $ax^2 + bx + c = 0$, $a \neq 0$

$\Rightarrow x = \dfrac{-b + \sqrt{b^2 - 4ac}}{2a}$ *or* $x = \dfrac{-b - \sqrt{b^2 - 4ac}}{2a}$

$(\alpha \pm \beta)^2 = \alpha^2 \pm 2\alpha\beta + \beta^2$
$\alpha^2 + \beta^2 = (\alpha + \beta)^2 - 2\alpha\beta$
$(\alpha + \beta)^2 = (\alpha - \beta)^2 + 4\alpha\beta$
$(\alpha - \beta)^2 = (\alpha + \beta)^2 - 4\alpha\beta$
$\alpha^2 - \beta^2 = (\alpha + \beta)(\alpha - \beta)$

Let $\alpha = \dfrac{-b + \sqrt{b^2 - 4ac}}{2a}$ *and* $\beta = \dfrac{-b - \sqrt{b^2 - 4ac}}{2a}$

Then, $\alpha + \beta = \dfrac{-b + \sqrt{b^2 - 4ac}}{2a} + \dfrac{-b - \sqrt{b^2 - 4ac}}{2a} = \dfrac{-2b}{2a} = -\dfrac{b}{a}$

$\alpha \cdot \beta = \dfrac{-b + \sqrt{b^2 - 4ac}}{2a} \cdot \dfrac{-b - \sqrt{b^2 - 4ac}}{2a} = \dfrac{(-b)^2 - \left(\sqrt{b^2 - 4ac}\right)^2}{4a^2} = \dfrac{b^2 - (b^2 - 4ac)}{4a^2} = \dfrac{4ac}{4a^2} = \dfrac{c}{a}$

2) Forming Quadratic Equations whose solutions are given

If the solutions are given, then a quadratic equation can be obtained from them.

① $\begin{pmatrix} x^2\text{-coefficient} = a \ (\neq 0) \\ \text{Solutions}: x = \alpha \ \text{and} \ x = \beta \end{pmatrix}$ $\Rightarrow$ The quadratic equation is
$a(x - \alpha)(x - \beta) = 0$

② $\begin{pmatrix} x^2\text{-coefficient} = a \ (\neq 0) \\ \text{Solution}: x = \alpha \ (\text{A double root}) \end{pmatrix}$ $\Rightarrow$ The quadratic equation is
$a(x - \alpha)^2 = 0$

③ $\begin{pmatrix} x^2\text{-coefficient} = a \ (\neq 0) \\ \text{The sum of the solutions} = p, \\ \text{The product of the solutions} = q \end{pmatrix}$ $\Rightarrow$ The quadratic equation is
$a(x^2 - px + q) = 0$

$a(x - \alpha)(x - \beta) = 0 \Leftrightarrow a(x^2 - (\alpha + \beta)x + \alpha\beta) = 0$

where $(\alpha + \beta)$ is the sum of the solutions and

$\alpha\beta$ is the product of the solutions

Example For two sets $A = \{3, 4\}$ and $B = \left\{\frac{a+b}{2}, \sqrt{ab}\right\}$, $A = B$.

When the quadratic equation $x^2 + mx + n = 0$ has solutions $a + 1$ and $b + 1$,

find the value of $m + n$.

> *Note : For positive real numbers a and b,* $\frac{a+b}{2} \geq \sqrt{ab}$
>
> ($\because$ *Since* $\frac{a+b}{2} - \sqrt{ab} = \frac{a+b-2\sqrt{ab}}{2} = \frac{(\sqrt{a}-\sqrt{b})^2}{2} \geq 0$, $\frac{a+b}{2} \geq \sqrt{ab}$)

Since $\frac{a+b}{2} \geq \sqrt{ab}$, $\frac{a+b}{2} = 4$ and $\sqrt{ab} = 3$

$\therefore \ a + b = 8, \ ab = 9$

Since $a + 1$ and $b + 1$ are the solutions of a quadratic equation with leading coefficient 1,

we obtain the equation $x^2 - (a + 1 + b + 1)x + (a + 1)(b + 1) = 0$.

$\therefore \ x^2 - (a + b + 2)x + (ab + a + b + 1) = 0$

$\therefore \ x^2 - 10x + 18 = 0$

$\therefore \ m = -10$ and $n = 18$.

Therefore, $m + n = -10 + 18 = 8$.

Be sure you see that this is true only if the polynomial equation has "real" coefficients.

(4) Conjugate Pairs

For a quadratic equation $ax^2 + bx + c = 0$,

Case 1. When a, b, and c are rational numbers,

if $\alpha + \beta\sqrt{m}$ is a solution, then the conjugate $\alpha - \beta\sqrt{m}$ is also a solution

where α, β $(\beta \neq 0)$ are rational numbers and $\sqrt{m}$ is an irrational number.

Case 2. When a, b, and c are real numbers,

if $\alpha + \beta i$ is a solution, then the conjugate $\alpha - \beta i$ is also a solution

where α, β $(\beta \neq 0)$ are real numbers and $i = \sqrt{-1}$.

Note : If a quadratic equation has a solution $\alpha + \beta\sqrt{m}$, then the other solution is $\alpha - \beta\sqrt{m}$.

$\because \quad ax^2 + bx + c = 0, \ a \neq 0$

$\Rightarrow \quad x = \dfrac{-b + \sqrt{b^2 - 4ac}}{2a}$ (*Form of* $\alpha + \beta\sqrt{m}$)

$or \quad x = \dfrac{-b - \sqrt{b^2 - 4ac}}{2a}$ (*Form of* $\alpha - \beta\sqrt{m}$)

Example 1

With given solutions, we can find a quadratic equation.

If the x^2-coefficient is 3 and the solutions are $x = 1$ and $x = 2$, then the quadratic equation is

$3(x - 1)(x - 2) = 0 \quad \therefore \ 3(x^2 - 3x + 2) = 0$ or

$3(x^2 - (1 + 2)x + (1 \cdot 2)) = 0 \quad \therefore \ 3(x^2 - 3x + 2) = 0$

Example 2

With the coefficients of a quadratic equation, we can find the solutions.

If the quadratic equation $x^2 - 7x + 12 = 0$ has solutions $x = \alpha$ and $x = \beta$, then

$\alpha + \beta = -\dfrac{-7}{1} = 7$ and $\alpha \cdot \beta = \dfrac{12}{1} = 12$

$\therefore \ (\alpha - \beta)^2 = (\alpha + \beta)^2 - 4\alpha\beta = 7^2 - 4 \cdot 12 = 1 \quad \therefore \ \alpha - \beta = \pm 1$

$\therefore \quad \alpha + \beta = 7$

$+) \ \underline{\alpha - \beta = \pm 1}$

$\quad 2\alpha \quad = 7 \pm 1 \quad \therefore \ \alpha = 4 \ \text{ or } \ \alpha = 3$

Therefore, the solutions are $x = 4$ and $x = 3$.

(5) Signs of Solutions

For a quadratic function $ax^2 + bx + c = 0$ with real number coefficients,

if α and β are two real number solutions of the equation, then

① $\alpha > 0$ and $\beta > 0$ (Two positive solutions) $\iff$ $D \geq 0$, $\alpha + \beta > 0$, $\alpha\beta > 0$

② $\alpha < 0$ and $\beta < 0$ (Two negative solutions) $\iff$ $D \geq 0$, $\alpha + \beta < 0$, $\alpha\beta > 0$

③ $(\alpha > 0$ and $\beta < 0)$ or $(\alpha < 0$ and $\beta > 0)$ (Two different signs of solutions)

$\iff$ $\alpha\beta < 0$

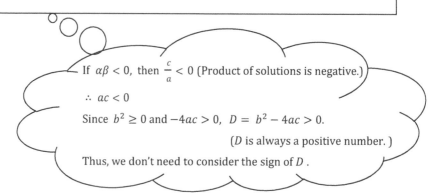

If $\alpha\beta < 0$, then $\dfrac{c}{a} < 0$ (Product of solutions is negative.)

$\therefore$ $ac < 0$

Since $b^2 \geq 0$ and $-4ac > 0$, $D = b^2 - 4ac > 0$.

(D is always a positive number.)

Thus, we don't need to consider the sign of D .

3-3 Polynomial Equations of Higher Degree

1. Solving Polynomial Equations

Use zero product property to solve factorable high-degree polynomial equations.

(1) Using Factorization Formulas

1) Factor by grouping pairs of terms that have a common factor.

For example,

$$x^3 - 3x^2 - 3x + 9 = 0 \implies x^2(x - 3) - 3(x - 3) = 0$$
$$\implies (x - 3)(x^2 - 3) = 0$$
$$\implies x - 3 = 0,\ x^2 = 3$$
$$\implies x = 3,\ x = \pm\sqrt{3}$$

$\therefore$ The equation has three solutions : 3 , $\sqrt{3}$, and $-\sqrt{3}$.

2) Factoring polynomials in quadratic form

For example,

$$2\,x^5 + 36x = 22\,x^3 \;\Rightarrow\; 2x^5 - 22\,x^3 + 36x = 0$$

$$\Rightarrow\; 2x(x^4 - 11x^2 + 18) = 0 \qquad \text{Factor common monomial}$$

$$\Rightarrow\; 2x(x^2 - 2)(x^2 - 9) = 0 \qquad \text{Factor trinomial}$$

$$\Rightarrow\; 2x(x^2 - 2)(x + 3)(x - 3) = 0 \qquad \text{Factorization}$$

$$\Rightarrow\; x = 0,\; x = \sqrt{2},\; x = -\sqrt{2},\; x = 3,\; x = -3$$
Use zero product property

∴ The equation has five solutions : $0,\ \sqrt{2},-\sqrt{2},\ 3,$ and -3.

(2) Using Factor Theorem and Synthetic Division

For a polynomial $P(x)$, if $P(\alpha) = 0$, then $p(x) = (x - \alpha)Q(x)$.

Using synthetic division, $Q(x)$ can be obtained.

Example When $1 + i$ is a solution of $x^3 + ax^2 - b = 0$,

find the values of real numbers a and b. Also, find the other solutions.

Since $1 + i$ is a solution of $x^3 + ax^2 - b = 0$, $(1 + i)^3 + a(1 + i)^2 - b = 0$

∴ $1^3 + 3(1^2)(i) + 3(1)(i^2) + i^3 + a(1 + 2i - 1) - b = 0$

∴ $1 + 3i - 3 - i + a + 2ai - a - b = 0$

∴ $(-2 - b) + 2(1 + a)i = 0$

Since a and b are real numbers, $-2 - b = 0$ and $1 + a = 0$

∴ $a = -1,\; b = -2$

Thus, the given equation is $x^3 - x^2 + 2 = 0$

Let $P(x) = x^3 - x^2 + 2$

Then, $P(-1) = (-1)^3 - (-1)^2 + 2 = -1 - 1 + 2 = 0$. By the synthetic division,

$$
\begin{array}{r|rrrr}
-1 & 1 & -1 & 0 & 2 \\
   &   & -1 & 2 & -2 \\
\hline
   & 1 & -2 & 2 & 0 \\
\end{array}
$$

∴ $P(x) = (x + 1)(x^2 - 2x + 2)$

By the quadratic formula, $x^2 - 2x + 2 = 0 \;\Rightarrow\; x = \dfrac{1 \pm \sqrt{1-2}}{1} = 1 \pm \sqrt{-1} = 1 \pm i$

∴ The solutions of the equation $x^3 - x^2 + 2 = 0$ are $x = -1,\; x = 1 + i,$ and $x = 1 - i$.

Therefore, the other solutions of the equation are $x = -1$ and $x = 1 - i$.

(3) Substituting the Common Terms with Other Variables and Factoring Polynomials in Quadratic Form

If polynomials have no common terms, factor by groups considering the newly combined terms have a common term.

For example,

① $(x+a)(x+b)(x+c)(x+d) = m$

$\Rightarrow (x^2 + \boxed{A}x + p)(x^2 + \boxed{A}x + q) = m$

Same coefficient of x

$\Rightarrow$ Substitute $X = x^2 + Ax$

② $ax^4 + bx^3 + cx^2 + bx + a = 0$

$\Rightarrow ax^2 + bx + c + \dfrac{b}{x} + \dfrac{a}{x^2} = 0$ Divide by x^2

$\Rightarrow a(x^2 + \dfrac{1}{x^2}) + b\left(x + \dfrac{1}{x}\right) + c = 0$

$\Rightarrow a((x + \dfrac{1}{x})^2 - 2) + b\left(x + \dfrac{1}{x}\right) + c = 0$

$\Rightarrow$ Substitute $X = x + \dfrac{1}{x}$

2. Properties of Imaginary Solutions of an Equation $x^3 = 1$

Note: $x^3 = 1 \Rightarrow x^3 - 1 = 0 \quad \therefore (x-1)(x^2 + x + 1) = 0$

(1) *Let ω be the solution of $x^3 = 1$. Then $\omega^3 = 1$.*

Since ω is an imaginary (complex number) solution, ω is a solution of $x^2 + x + 1 = 0$.

Thus, $\omega^2 + \omega + 1 = 0$.

That is, if ω is an imaginary solution of $x^3 = 1$, then $\omega^2 + \omega + 1 = 0$.

(2) *If ω is a solution of $x^2 + x + 1 = 0$, then $\omega^2 + \omega + 1 = 0$.*

By multiplying both sides by $\omega - 1$, $(\omega - 1)(\omega^2 + \omega + 1) = 0$

$\therefore \omega^3 - 1 = 0 \qquad \therefore \omega^3 = 1$

That is, if ω is one of the solutions of $x^2 + x + 1 = 0$, then $\omega^3 = 1$, $\omega^2 + \omega + 1 = 0$.

<u>Properties for a cube polynomial equation $x^3 = 1$</u>

If ω is an imaginary (complex number) solution of the equation $x^3 = 1$, then

① $\omega^3 = 1$, $\bar{\omega}^3 = 1$ ($\bar{\omega}$ is a conjugate of ω)

② $\omega^2 + \omega + 1 = 0$, $\bar{\omega}^2 + \bar{\omega} + 1 = 0$

③ $\omega + \bar{\omega} = -1$, $\omega\bar{\omega} = 1$

④ $\bar{\omega} = \dfrac{1}{\omega} = \omega^2$

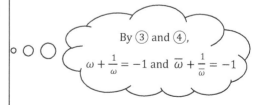

By ③ and ④,

$\omega + \dfrac{1}{\omega} = -1$ and $\bar{\omega} + \dfrac{1}{\bar{\omega}} = -1$

Note : (*Proof of* ③)

$$x^3 = 1 \quad \Rightarrow \quad x^3 - 1 = 0 \quad \Rightarrow \quad (x - 1)(x^2 + x + 1) = 0$$

Since ω is a solution of $x^2 + x + 1 = 0$, the conjugate $\overline{\omega}$ of ω is also a solution of the equation.

By the roots-coefficients relationship,

$\omega + \overline{\omega} = -1$ (The sum of two solutions) and $\omega\overline{\omega} = 1$ (The product of two solutions).

Note : (*Proof of* ④)

Since $\omega\overline{\omega} = 1$, $\overline{\omega} = \dfrac{1}{\omega}$

Since $\omega^3 = 1$, $\omega^3 = \omega\omega^2 = 1$ $\quad \therefore \omega^2 = \dfrac{1}{\omega}$

Therefore, $\overline{\omega} = \dfrac{1}{\omega} = \omega^2$

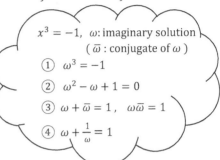

$x^3 = -1$, ω: imaginary solution
($\overline{\omega}$: conjugate of ω)

① $\omega^3 = -1$

② $\omega^2 - \omega + 1 = 0$

③ $\omega + \overline{\omega} = 1$, $\omega\overline{\omega} = 1$

④ $\omega + \dfrac{1}{\omega} = 1$

3. The Roots-Coefficients Relationship

Given the general cubic polynomial equation $ax^3 + bx^2 + cx + d = 0$, $a \neq 0$,

let α, β, and γ be the three roots. Then,

---

(1) $\alpha + \beta + \gamma = -\dfrac{b}{a}$

$\alpha\beta + \beta\gamma + \gamma\alpha = \dfrac{c}{a}$

$\alpha\beta\gamma = -\dfrac{d}{a}$

(2) The cubic polynomial equation with leading coefficient 1 is

$x^3 - (\alpha + \beta + \gamma)x^2 + (\alpha\beta + \beta\gamma + \gamma\alpha)x - \alpha\beta\gamma = 0$.

Note : (*Proof of the relationship between the roots and coefficients*)

If α, β, and γ are roots of a cubic polynomial equation $ax^3 + bx^2 + cx + d = 0$, $a \neq 0$, then

$ax^3 + bx^2 + cx + d = a(x - \alpha)(x - \beta)(x - \gamma)$

$\qquad\qquad\qquad\qquad = ax^3 - a(\alpha + \beta + \gamma)x^2 + a(\alpha\beta + \beta\gamma + \gamma\alpha)x - a\alpha\beta\gamma$

Since the expression in x is an identity,

$b = -a(\alpha + \beta + \gamma)$, $c = a(\alpha\beta + \beta\gamma + \gamma\alpha)$, and $d = -a\alpha\beta\gamma$ by comparing the coefficients of like

terms. Dividing each expression by a, we obtain $\alpha + \beta + \gamma = -\dfrac{b}{a}$,

$$\alpha\beta + \beta\gamma + \gamma\alpha = \dfrac{c}{a}, and$$

$$\alpha\beta\gamma = -\dfrac{d}{a}.$$

Note : **The Fundamental Theorem**

Every polynomial equation $P(x) = 0$ of degree n $(n \geq 1)$ has at least one root.

Number-of-Roots Theorem

A polynomial $P(x) = 0$ of degree n $(n \geq 1)$ has exactly n roots.

Note : $a_n x^n + a_{n-1} x^{n-1} + a_{n-2} x^{n-2} + \cdots\cdots + a_1 x + a_0 = 0, \quad a_n \neq 0$

$\Rightarrow$ The sum of all roots $= -\dfrac{a_{n-1}}{a_n}$

The sum of product of all two roots $= \dfrac{a_{n-2}}{a_n}$

The sum of product of all three roots $= -\dfrac{a_{n-3}}{a_n}$

$\vdots$

The product of all roots $= (-1)^n \dfrac{a_0}{a_n}$

4. Conjugate Pairs

In the complex number system,

for a cubic polynomial equation $ax^3 + b x^2 + cx + d = 0$,

Case 1. When a, b, c, and d are rational numbers,

if $\alpha + \beta\sqrt{m}$ is a solution, then the conjugate $\alpha - \beta\sqrt{m}$ is also a solution

where $\alpha, \beta \ (\beta \neq 0)$ are rational numbers and $\sqrt{m}$ is an irrational number.

Case 2. When a, b, c, and d are real numbers,

if $\alpha + \beta i$ is a solution, then the conjugate $\alpha - \beta i$ is also a solution

where $\alpha, \beta \ (\beta \neq 0)$ are real numbers and $i = \sqrt{-1}$.

Example   Find a fourth-degree polynomial equation, with real number coefficients,

that has $-2, -2$, and $3i$ as solutions.

Since $3i$ is a solution and the polynomial has real number coefficients,

the conjugate $-3i$ must also be a solution. From the linear factorization, the polynomial can be

written as : $a(x + 2)(x + 2)(x - 3i)(x + 3i)$

Letting $a = 1$ for simplicity, we have

$a(x + 2)(x + 2)(x - 3i)(x + 3i) = (x^2 + 4x + 4)(x^2 + 9)$

$$= x^4 + 4 x^3 + 13 x^2 + 36x + 36$$

3-4 Systems of Equations

1. Solving Systems of Linear Equations in Two Variables

A linear equation is an equation in the form of $ax + by = c$, where a, b, and c are constants and

$a \neq 0, b \neq 0$. For example, $2x + 3y - 5 = 0$ or $2x + 3y = 5$.

A system of equations is a pair of two or more linear equations with the same variables.

For example, $\begin{cases} x + y = 2 \\ 2x + y = 3 \end{cases}$

For the values of x and y which are satisfying the two linear equations, the ordered pair (x, y) is called the *solution* for the system of equations.

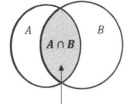

Note : A = Set of solutions for equation ①

B = Set of solutions for equation ②.

Solutions for the system of equations ① and ②.

Example

For any positive integers x and y, find the solution for the system $\begin{cases} x + y = 3 \cdots\cdots ① \\ 2x + y = 5 \cdots\cdots ② \end{cases}$

For the equation ①, the solutions are $(1, 2), (2, 1)$ and for the equation ②, the solutions are $(1, 3), (2, 1)$. Therefore, the solution for the system of equations ① and ② is $(2, 1)$.

(1) The Elimination Method

> The *elimination method* is a method for finding a solution by eliminating (removing) one variable by using addition or subtraction.

Note 1 : If the sign of the coefficient of one variable which is supposed to be removed is different from the sign of the coefficient of the other variable, use addition for the two equations to remove the variable. If the sign of the coefficient of one variable which is supposed to be removed is the same as the sign of the coefficient of the other variable, use subtraction for the two equations to remove the variable.

Note 2 : One or two of the given equations needs to be multiplied by a number in order to make the two equations with the same absolute value for the coefficient of the variable which is supposed to be removed.

Example 1 Solve a system $\begin{cases} 2x + y = 3 \cdots\cdots ① \\ x - y = 6 \cdots\cdots ② \end{cases}$.

Step 1. Add equations ① and ② to remove the variable y. Then solve for x.

$$2x + y = 3 \cdots\cdots ①$$
$$+)\ \underline{x - y = 6 \cdots\cdots ②}$$
$$3x\quad = 9\ ;\ x = 3$$

Step 2.   Substitute the value of x into any one of the given two equations and then solve for y.

$$x - y = 6 \;\Rightarrow\; 3 - y = 6 \;\Rightarrow\; y = -3$$

Step 3.   Find the solution.

$$(x, y) = (3, -3)$$

Example 2   Solve a system $\begin{cases} 2x + y = 4 & \cdots\cdots ① \\ 3x + y = 2 & \cdots\cdots ② \end{cases}$.

Step 1.   Subtract equation ② from equation ① to remove the variable y.

Then solve for x.

$$\begin{array}{r} 2x + y = 4 \quad\cdots\cdots ① \\ -)\ \underline{3x + y = 2 \quad\cdots\cdots ②} \\ -x \quad\ \ = 2 \quad;\ x = -2 \end{array}$$

Step 2.   Substitute the value of x into any one of the given two equations.

Then solve for y.

$$2x + y = 4 \;\Rightarrow\; -4 + y = 4 \;\Rightarrow\; y = 8$$

Step 3.   Find the solution.

$$(x, y) = (-2,\ 8)$$

Example 3   Solve a system $\begin{cases} 2x + 3y = 3 & \cdots\cdots ① \\ 3x + 4y = 1 & \cdots\cdots ② \end{cases}$.

Step 1.   Consider *Note* 1 and *Note* 2.

$$\begin{array}{r} 6x + 9y = 9 \quad\cdots\cdots ① \times 3 \\ -)\ \underline{6x + 8y = 2 \quad\cdots\cdots ② \times 2} \\ y = 7 \end{array}$$

Step 2.   Substitute the value of y into any one of the given two equations.

Then solve for x.

$$2x + 3y = 3 \;\Rightarrow\; 2x + 21 = 3 \;\Rightarrow\; 2x = -18 \;\Rightarrow\; x = -9$$

Step 3.   Find the solution.

$$(x, y) = (-9,\ 7)$$

(2) The Substitution Method

> The substitution method is a method for finding the solution by substituting the expression which is already solved for one variable into the other equation.

Note : If the coefficient of one variable is 1 or −1 or if one of the two equations is easily solved for one variable, – for example, (x = expression of y) or (y = expression of x) − this substitution method is the best way to find the solution.

Example 1   Solve a system $\begin{cases} 2x + y = 3 & \cdots\cdots ① \\ 3x + 2y = 5 & \cdots\cdots ② \end{cases}$.

Step 1.   Solve equation ① for one variable y in terms of the other ; $y = -2x + 3$

Step 2.   Substitute the expression $y = -2x + 3$ found in Step 1 into the other equation ②

to obtain an equation in one variable.

$$3x + 2(-2x + 3) = 5 \; ; \; -x = -1 \; ; \; x = 1$$

Step 3.   Back-Substitute the solution $x = 1$ into equation ① to find the value of the other

variable ; $2 \cdot 1 + y = 3 \; ; \; y = 1$

Step 4.   Find the solution.

$$(x, y) = (1, 1)$$

Step 5.   Check the solution to see that it satisfies each of the given equations.

Note : If the coefficients of variables are unknown and the solution for the system of equations is given, then do the following :

    Step 1.   Substitute the solution into the given system of equations.

    Step 2.   Find the coefficients of variables using the elimination method or substitution method.

Example 2   The solution for the system of equations $\begin{cases} ax + by = 3 & \cdots\cdots ① \\ 2bx - ay = -4 & \cdots\cdots ② \end{cases}$ is $(1, -2)$.

     Find the value of $a + b$.

Step 1. $\begin{cases} a - 2b = 3 & \cdots\cdots ③, \text{ from } ① \\ 2b + 2a = -4 & \cdots\cdots ④, \text{ from } ② \end{cases}$

Step 2. $\begin{aligned} a - 2b &= 3 \\ \underline{2a + 2b} &= \underline{-4} \\ 3a \quad\;\; &= -1 \; ; \; a = -\frac{1}{3} \end{aligned}$

Substituting $a = -\frac{1}{3}$ into ③, $b = -\frac{5}{3}$. Therefore, $a + b = -\frac{1}{3} - \frac{5}{3} = -2$

2. Solving Special Systems of Linear Equations in Two Variables

(1) If the coefficients of variables are fractions or decimals

$\Rightarrow$ Change the coefficients to integers. Then solve as usual.

Example   Solve a system $\begin{cases} \frac{1}{2}x - \frac{1}{3}y = 1 & \cdots\cdots ① \\ 0.3x + 0.2y = 0.2 & \cdots\cdots ② \end{cases}$.

Step 1. $\begin{cases} 3x - 2y = 6 & \cdots\cdots ① \times 6 \\ 3x + 2y = 2 & \cdots\cdots ② \times 10 \end{cases}$

Step 2. $3x - 2y = 6 \cdots\cdots ③$

$\qquad\quad -)\ \underline{3x + 2y = 2 \cdots\cdots ④}$

$\qquad\qquad\quad -4y = 4 \quad ; y = -1$

Substituting $y = -1$ into ③, $x = \frac{4}{3}$.   Therefore, $(x, y) = \left(\frac{4}{3}, -1\right)$

---

(2) $\begin{cases} px = qy + r \\ my = nx + t \end{cases}$ $\Rightarrow$ Rearrange or simplify in the form of $ax + by = c$

Example   Solve a system $\begin{cases} 2x = 3y + 5 \\ -y = 4x - 3 \end{cases}$. $\xrightarrow{\text{Rearrange}}$ $\begin{cases} 2x - 3y = 5 & \cdots\cdots ① \\ 4x + y = 3 & \cdots\cdots ② \end{cases}$

$\qquad\quad \begin{cases} 4x - 6y = 10 & \cdots\cdots ① \times 2 \\ 4x + y = 3 & \cdots\cdots ② \end{cases}$

$\qquad\qquad 4x - 6y = 10 \cdots\cdots ③$

$\qquad\quad -)\ \underline{4x + y = 3 \cdots\cdots ④}$

$\qquad\qquad\quad -7y = 7 \quad ; y = -1$

Substituting $y = -1$ into ④, $x = 1$

Therefore, $(x, y) = (1, -1)$

---

(3) $\begin{cases} \dfrac{a}{x} + \dfrac{b}{y} = c \\ \dfrac{m}{x} + \dfrac{n}{y} = t \end{cases}$ $\Rightarrow$ Let $\dfrac{1}{x} = A, \quad \dfrac{1}{y} = B$

Example   Solve a system $\begin{cases} \dfrac{2}{x} + \dfrac{3}{y} = 1 \\ \dfrac{1}{x} + \dfrac{2}{y} = 2 \end{cases}$. $\xrightarrow{\text{Replace}}$ $\begin{cases} 2A + 3B = 1 & \cdots\cdots ① \\ A + 2B = 2 & \cdots\cdots ② \end{cases}$

$$\begin{cases} 2A + 3B = 1 & \cdots\cdots ① \\ 2A + 4B = 4 & \cdots\cdots ② \times 2 \end{cases}$$

$$2A + 3B = 1 \quad\cdots\cdots ①$$
$$-)\ \underline{2A + 4B = 4} \quad\cdots\cdots ③$$
$$-B = -3 \quad ; B = 3 \; ; \frac{1}{y} = 3 \; ; y = \frac{1}{3}$$

Substituting $B = 3$ into ②, $A = -4$; $\frac{1}{x} = -4$; $x = -\frac{1}{4}$

Therefore, $(x, y) = \left(-\frac{1}{4}, \frac{1}{3}\right)$

---

(4) A system of equations in form of the $A = B = C$

⇒ Rewrite the equations as a system

$$\begin{cases} A = B \\ B = C \end{cases} \text{ or } \begin{cases} A = B \\ A = C \end{cases} \text{ or } \begin{cases} A = C \\ B = C \end{cases}$$

Note : If $A = B = k$, k is a constant, rewrite the equation as a system $\begin{cases} A = k \\ B = k \end{cases}$.

Example Solve a system $2x - y = x + 3y = y + 5$.

Step 1. (Rewrite)

$$\begin{cases} 2x - y = x + 3y & \cdots\cdots ① \\ 2x - y = y + 5 & \cdots\cdots ② \end{cases}$$

Step 2. (Rearranging ① and ②)

$$\begin{cases} x - 4y = 0 & \cdots\cdots ③ \\ 2x - 2y = 5 & \cdots\cdots ④ \end{cases}$$

Step 3. (Solve)

$$2x - 8y = 0 \quad\cdots\cdots ③ \times 2$$
$$-)\ \underline{2x - 2y = 5} \quad\cdots\cdots ④$$
$$-6y = -5 \quad ; y = \frac{5}{6}$$

Substituting $y = \frac{5}{6}$ into ③, $x = \frac{10}{3}$

Therefore, $(x, y) = \left(\frac{10}{3}, \frac{5}{6}\right)$

Note : A system of linear equations is consistent if it has at least one solution,

and it is inconsistent if it has no solution.

---

(5) If a system $\begin{cases} ax + by = c \\ mx + ny = t \end{cases}$ has the special condition $\dfrac{a}{m} = \dfrac{b}{n} = \dfrac{c}{t}$, then

the system has an unlimited number of solutions. (Consistent and dependent)

Example Solve a system $\begin{cases} 2x + 3y = 5 \\ 4x + 6y = 10 \end{cases}$.

If the two equations are the same (in that they have the same ratios), that is, form of $0 = 0$, then it's always true.

$\Rightarrow \dfrac{2}{4} = \dfrac{3}{6} = \dfrac{5}{10}$

$\therefore$ The system has unlimited solutions.

If the two equations have the same coefficients for the variables but constants, that is, form of $0 = a$ (a is a non-zero number), then it's impossible.

---

(6) If a system $\begin{cases} ax + by = c \\ mx + ny = t \end{cases}$ has the special condition $\dfrac{a}{m} = \dfrac{b}{n} \neq \dfrac{c}{t}$, then

the system does not have a solution. (Inconsistent)

Note : $ax + by + c = 0$ $\xrightarrow{Transform}$ $y = px + q$

(Equation $\xrightarrow{Transform}$ Function)

If $y_1 = px + q_1$ and $y_2 = px + q_2$, $q_1 \neq q_2$, then y_1 and y_2 are parallel.

So, the lines do not intersect at all.

Therefore, $\dfrac{a}{m} = \dfrac{b}{n} \neq \dfrac{c}{t}$

$\Rightarrow$ *The system of equations does not have a solution.*

Example Solve a system $\begin{cases} 3x + y = 2 \\ 6x + 2y = 3 \end{cases}$.

$\Rightarrow \dfrac{3}{6} = \dfrac{1}{2} \neq \dfrac{2}{3}$

$\therefore$ The system has no solution.

3. Solving Systems of Linear Equations in More than Two Variables

(1) The Back-Substitution Method

The system in triangular form is easily solved by back-substitution method. To solve such a system, solve the last equation first and then work backward, substituting the values we obtain into the previous equation.

Example   Solve a system
$$\begin{cases} 2x + y - z = -4 & \cdots\cdots \text{①} \\ -3y + 5z = -2 & \cdots\cdots \text{②} \\ 2z = 4 & \cdots\cdots \text{③} \end{cases}$$

———— Triangular form

Step 1. (Solve the last equation to obtain z)

From ③, $2z = 4$ $\therefore z = 2$

Step 2. (Solve for y)

Substitute $z = 2$ into ② to solve for y. Then, $-3y + 5(2) = -2$ $\therefore y = 4$

Step 3. (Solve for x)

Substitute $y = 4$ and $z = 2$ into ① to obtain the variable of x.

Then, $2x + 4 - 2 = -4$ $\therefore x = -3$

Step 4. $(x, y, z) = (-3, 4, 2)$

Step 5. Check this solution in the given system.

(2) The Elimination Method

The elimination method can be used to a system of linear equations in more than two variables. To have an efficient method for solving systems of three or more linear equations, rewrite the system in a form to which back-substitution can be used.

Steps for Solving a system of three equations with three variables

Step 1. Eliminate one of the variables to create a system of equations with two variables.

Step 2. Solve the new system for two variables.

Step 3. Substitute the two values to find the eliminated variable.

Step 4. Find the solution.

Step 5. Check this solution in the given system.

Example Find the solution (x, y, z) for a system $\begin{cases} 2x - y + z = 30 & \cdots\cdots ① \\ x + 2y + z = 2 & \cdots\cdots ② \\ 3x + 2y - z = -18 & \cdots\cdots ③ \end{cases}$.

Step 1. (To remove z)

$\begin{cases} 5x + y = 12 & \cdots\cdots ④ \ (\text{using } ① + ③) \\ 4x + 4y = -16 & \cdots\cdots ⑤ \ (\text{using } ② + ③) \end{cases}$

Step 2. (To find (x, y))

$$20x + 4y = 48 \ (\text{using } ④ \times 4)$$

$$-) \ \underline{4x + 4y = -16}$$

$$16x \qquad = 64 \ ; \ x = 4$$

Substituting $x = 4$ into ④, $y = 12 - 20 = -8$

So, $(x, y) = (4, -8)$

Step 3. (To find z)

Substituting $(x, y) = (4, -8)$ into ①, $z = 30 - 2x + y = 30 - 2 \cdot 4 - 8 = 14$

Step 4. $(x, y, z) = (4, -8, \ 14)$

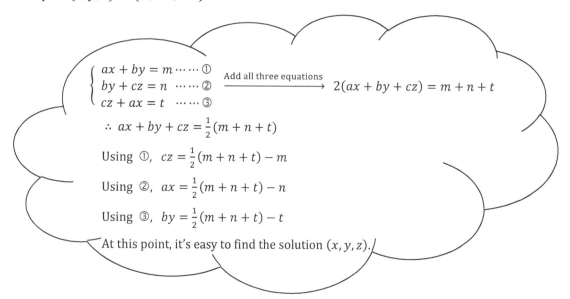

$\begin{cases} ax + by = m & \cdots\cdots ① \\ by + cz = n & \cdots\cdots ② \\ cz + ax = t & \cdots\cdots ③ \end{cases}$ $\xrightarrow[]{\text{Add all three equations}}$ $2(ax + by + cz) = m + n + t$

$\therefore \ ax + by + cz = \frac{1}{2}(m + n + t)$

Using ①, $cz = \frac{1}{2}(m + n + t) - m$

Using ②, $ax = \frac{1}{2}(m + n + t) - n$

Using ③, $by = \frac{1}{2}(m + n + t) - t$

At this point, it's easy to find the solution (x, y, z).

4. Solving Systems of Quadratic Equations in Two Variables

(1) System of a Linear Equation and Quadratic Equation

Step 1. Solve the linear equation for one variable.

Step 2. Substitute the result into the quadratic equation to get the values of the other variable.

Step 3. Finally, substitute the values we obtain from Step 2 into one of the given equations.

Example   Solve a system $\begin{cases} x - y = 2 & \cdots\cdots ① \\ x^2 - 2xy - y = 2 & \cdots\cdots ② \end{cases}$

Step 1. From ①, $y = x - 2$

Step 2. $x^2 - 2x(x-2) - (x-2) = 2 \Rightarrow x^2 - 2x^2 + 4x - x + 2 = 2$

$\Rightarrow x^2 - 3x = 0 \Rightarrow x(x-3) = 0$

∴ $x = 0$ or $x = 3$

Step 3. $x = 0 \Rightarrow y = 0 - 2 = -2$

$x = 3 \Rightarrow y = 3 - 2 = 1$

Therefore, $(x, y) = (0, -2)$; $(x, y) = (3, 1)$

(2) System of Two Quadratic Equations

Use factorization or eliminate the constant term or the term of degree 2 to rewrite the system as the system of a linear equation and quadratic equation.

Example   Solve a system $\begin{cases} x^2 - xy - 2y^2 = 0 & \cdots\cdots ① \\ 3x^2 - 2y^2 = 20 & \cdots\cdots ② \end{cases}$

$x^2 - xy - 2y^2 = 0 \Rightarrow (x - 2y)(x + y) = 0$ ∴ $x = 2y$ or $x = -y$

(i) Substitute $x = 2y$ into ②.

Then, $3(2y)^2 - 2y^2 = 20$ ∴ $10y^2 = 20$ ∴ $y^2 = 2$ ∴ $y = \pm\sqrt{2}$, $x = \pm 2\sqrt{2}$

(ii) Substitute $x = -y$ into ②.

Then, $3(-y)^2 - 2y^2 = 20$ ∴ $y^2 = 20$ ∴ $y = \pm 2\sqrt{5}$, $x = \mp 2\sqrt{5}$

Therefore, $(x, y) = \left(2\sqrt{2}, \sqrt{2}\right)$; $(x, y) = \left(-2\sqrt{2}, -\sqrt{2}\right)$;

$(x, y) = \left(-2\sqrt{5}, 2\sqrt{5}\right)$; $(x, y) = \left(2\sqrt{5}, -2\sqrt{5}\right)$

Note : $x^2 + y^2 = (x+y)^2 - 2xy$.

If a system has terms $x + y$ and xy, then

consider x and y as the solutions of $X^2 - uX + v = 0$ where $u = x + y$, $v = xy$.

For example, $\begin{cases} x + y = 4 \\ xy = 2 \end{cases}$ *⇒ Consider an equation $X^2 - uX + v = 0$ where $u = x + y$, $v = xy$.*

Then, we have $X^2 - 4X + 2 = 0$.

By quadratic formula, $X = \dfrac{2 \pm \sqrt{(-2)^2 - 1(2)}}{1} = 2 \pm \sqrt{2}$ That is, the solutions are $2 + \sqrt{2}$ and $2 - \sqrt{2}$.

Since x and y are solutions, $(x, y) = \left(2 + \sqrt{2}, 2 - \sqrt{2}\right)$; $(x, y) = \left(2 - \sqrt{2}, 2 + \sqrt{2}\right)$.

5. Solving a System with Fewer Equations than Variables

If the number of equations is equal to the number of variables, then the system is called *square system*. If the number of equations differs from the number of variables, then the system is called *non-square system*.

Example Solve a system of linear equations $\begin{cases} x + 2y - z = 3 & \cdots\cdots ① \\ 2x - y + 3z = 1 & \cdots\cdots ② \end{cases}$

$① \times 2 - ② :$

$$\begin{array}{r} 2x + 4y - 2z = 6 \\ -)\ 2x - y + 3z = 1 \\ \hline 5y - 5z = 5 \ ;\ y - z = 1 \end{array}$$

$\begin{cases} x + 2y - z = 3 \cdots\cdots ① \\ y - z = 1 \qquad \cdots\cdots ③ \end{cases}$

Solving for y in terms of z, we have $y = z + 1$

Back-substitution into $①$: $x + 2(z + 1) - z = 3$; $x = -z + 1$

By letting $z = \alpha$ (α is a real numner),

the solution of the system is $(x,\ y,\ z) = (-\alpha + 1,\ \alpha + 1,\ \alpha)$

(1) If the solutions are integers,

Transfer to the form of [(Linear Eqaution) × (Linear Equation) = Integer]

Example Find the values of integers x and y which satisfy the equation $xy - x - y - 3 = 0$.

$xy - x - y - 3 = 0 \ \Rightarrow \ (x - 1)(y - 1) = 4$

Since $x - 1$ and $y - 1$ are integers, we have the following :

$x - 1$	1	2	4	-1	-2	-4
$y - 1$	4	2	1	-4	-2	-1

$\therefore \ (x, y) = (2, 5) \,;\ (3, 3) \,;\ (5, 2) \,;\ (0, -3) \,;\ (-1, -1) \,;\ (-3,\ 0)$

(2) If the solutions are real numbers,

1) Transfer to the form of [$A^2 + B^2 = 0$].

By using the property, $A^2 + B^2 = 0 \ \Leftrightarrow \ A = 0,\ B = 0$

Or 2) Rearrange terms in descending order of equation from left to right and use the discriminant $D \geq 0$.

Example 1 Find the values of real numbers x and y

which satisfy the equation $x^2 + y^2 - 4x + 6y + 13 = 0$.

$x^2 + y^2 - 4x + 6y + 13 = 0 \Rightarrow (x-2)^2 + (y+3)^2 = 0$ Form of $[\ A^2 + B^2 = 0\ ]$

Since x and y are real numbers, $x - 2 = 0$, $y + 3 = 0$ $\therefore$ $x = 2$, $y = -3$

OR

$x^2 + y^2 - 4x + 6y + 13 = 0 \Rightarrow x^2 - 4x + (y^2 + 6y + 13) = 0$ Descending order in x

Since x is a real number, the equation has real number solutions.

Thus, the discriminant ≥ 0.

$\therefore D = (-4)^2 - 4(y^2 + 6y + 13) \geq 0$

$\therefore y^2 + 6y + 9 \leq 0$ $\therefore (y+3)^2 \leq 0$

Since $(\text{real number})^2 \geq 0$, $(y+3)^2 = 0$

$\therefore y + 3 = 0$ $\therefore y = -3$

Substituting $y = -3$ into the equation, we have $x^2 - 4x + (9 - 18 + 13) = 0$.

$\therefore x^2 - 4x + 4 = 0$ $\therefore (x-2)^2 = 0$ $\therefore x = 2$

Therefore, $(x, y) = (2, -3)$

Example 2 Find the values of integers x and y

which satisfy the equation $(x^2 + y^2 + x - 3y - 2)^2 + (xy - 2x + 2y - 7)^2 = 0$.

Since x and y are integers, $\quad x^2 + y^2 + x - 3y - 2 = 0$ ······①

$\qquad\qquad\qquad\qquad\qquad xy - 2x + 2y - 7 = 0 \qquad$ ······②

From ②, $xy - 2x + 2y - 4 = 7 - 4$ $\therefore (x+2)(y-2) = 3$

Since $x + 2$ and $y - 2$ are integers, we have the following :

$x + 2$	1	3	-1	-3
$y - 2$	3	1	-3	-1

$\therefore (x, y) = (-1, 5)\ ;\ (1, 3)\ ;\ (-3, -1)\ ;\ (-5, 1)$

But, only $(x, y) = (1, 3)$ satisfies ①.

Therefore $x = 1$, $y = 3$

3-5 Linear Inequalities

1. Linear Inequalities and Their Solutions

(1) Linear Inequalities

Inequality is an expression using the algebraic inequality symbols to show the relationship between the values of numbers or variables.

Example

1) $x > 2$ (x is greater than 2) means :

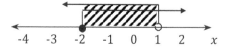

2 is not included (open circle) in the solution.

2) $-2 \le x < 1$ (x is greater than or equal to -2 and less than 1) means :

-2 is included (closed circle) in the solution.
1 is not included (open circle) in the solution.

(2) Solutions of Linear Inequalities

To solve a linear inequality with one variable (to identify the values of x which satisfy the inequality), isolate the variable on one side of the inequality and solve it exactly like a linear equation. The solution will consist of intervals or unions of intervals.

1) The linear inequality is formed by

$$ax < b\,(\,ax \le b) \quad \text{or} \quad ax > b\,(\,ax \ge b) \quad \text{for any } a \ne 0$$

2) The solution of the inequality $ax < b$ is

① $x < \dfrac{b}{a}$ when $a > 0$ (positive a)

② $x > \dfrac{b}{a}$ when $a < 0$ (negative a)

③ all real numbers when $a = 0$ and $b > 0$

④ not defined (no solution) when $a = 0$ and $b \le 0$

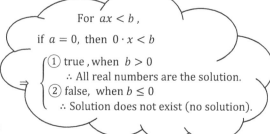

For $ax < b$,
if $a = 0$, then $0 \cdot x < b$
① true , when $b > 0$
∴ All real numbers are the solution.
② false, when $b \le 0$
∴ Solution does not exist (no solution).

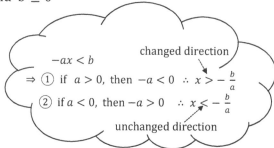

changed direction
$-ax < b$
⇒ ① if $a > 0$, then $-a < 0$ ∴ $x > -\dfrac{b}{a}$
② if $a < 0$, then $-a > 0$ ∴ $x < -\dfrac{b}{a}$
unchanged direction

Note : Find k.

    (1) *The solution of $ax < b$ is $x < k$.*

        $\Rightarrow$ *Since the direction of the inequality symbol of the solution is unchanged, $a > 0$.*

        From $ax < b$, $x < \dfrac{b}{a}$ $\therefore k = \dfrac{b}{a}$

    (2) *The solution of $ax < b$ is $x > k$.*

        $\Rightarrow$ *Since the direction of the inequality symbol of the solution is changed, $a < 0$.*

        From $ax < b$, $x > \dfrac{b}{a}$ $\therefore k = \dfrac{b}{a}$

2. Properties of Inequalities

For any real numbers a, b and c, the following properties apply:

(1) Transitive Property :

$$\boxed{a < b ,\ b < c \ \Rightarrow \ a < c}$$

(2) Adding or subtracting the same number to or from each side of an inequality does not change the direction of the inequality symbol.

$$\boxed{a < b \Rightarrow \ a + c < b + c \ \text{ and } \ a - c < b - c}$$

    Note : $a < b \Rightarrow \ a + c < b + c$ (Adding c both sides of the inequality)

               $c < d \Rightarrow \ b + c < b + d$ (Adding b both sides of the inequality)

               Since $a + c < b + c$ and $b + c < b + d$, $a + b < b + d$ (by the transitive property).

(3) Multiplying or dividing each side of an inequality by the same positive number does not change the direction of the inequality symbol.

$$\boxed{a < b ,\ c > 0 \ \Rightarrow \ a \cdot c < b \cdot c \ \text{ and } \ \dfrac{a}{c} < \dfrac{b}{c}}$$

    Note: $a < b ,\ c = 0 \ \Rightarrow \ a \cdot c = b \cdot c$

            $a < b$ $\nRightarrow a^2 < b^2$ ($\because -2 < -1$, but $(-2)^2 > (-1)^2$)

            $0 < a < b$ $\Rightarrow a^2 < b^2$

            $a < b$ $\Rightarrow a^3 < b^3$ ($\because -3 < -2$ and $(-3)^3 < (-2)^3$)

(4) Multiplying or dividing both sides of an inequality by the same negative number will change the direction by reversing the inequality symbol).

$$\boxed{a < b ,\ c < 0 \ \Rightarrow \ a \cdot c > b \cdot c \ \text{ and } \ \dfrac{a}{c} > \dfrac{b}{c}}$$

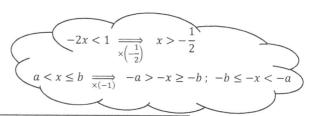

(5) $\boxed{\begin{array}{l} a < b \iff a - b < 0 \\ a > b \iff a - b > 0 \end{array}}$

(6) $\boxed{\begin{array}{l} \text{If } a \text{ and } b \text{ have the same sign} \Rightarrow a \cdot b > 0, \ \dfrac{a}{b} > 0, \text{ and } \dfrac{b}{a} > 0 \\[2ex] \text{If } a \text{ and } b \text{ have different signs} \Rightarrow a \cdot b < 0, \ \dfrac{a}{b} < 0, \text{ and } \dfrac{b}{a} < 0 \end{array}}$

(7) Expanded Properties :

$\boxed{\begin{array}{l} a < x < b \quad\quad\quad \Rightarrow \quad a + c < x + c < b + c, \quad a - c < x - c < b - c \\[2ex] a < x < b, \ c > 0 \ \Rightarrow \ ac < xc < bc, \quad \dfrac{a}{c} < \dfrac{x}{c} < \dfrac{b}{c} \\[2ex] a < x < b, \ c < 0 \ \Rightarrow \ ac > xc > bc, \quad \dfrac{a}{c} > \dfrac{x}{c} > \dfrac{b}{c} \end{array}}$

(8) $\boxed{\text{When } a > 0 \text{ and } b > 0, \quad a > b \iff a^2 > b^2}$

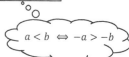

Note that : if $a > 0$ and $b > 0$, then $a + b > 0$

$(\Rightarrow)$: $a > b \ \Rightarrow \ a - b > 0$ and $a + b > 0$

$\quad\quad \therefore \ a^2 - b^2 = (a + b)(a - b) > 0$

$\quad\quad \therefore \ a^2 > b^2$

$(\Leftarrow)$: $a^2 > b^2 \ \Rightarrow \ a^2 - b^2 > 0$

$\quad\quad \therefore \ (a + b)(a - b) > 0$

$\quad\quad$ Since $a + b > 0$ and $a - b > 0$

$\quad\quad \therefore \ a > b$

Therefore, $a > b \iff a^2 > b^2$

(9) $\boxed{\text{When } a < 0 \text{ and } b < 0, \quad a < b \iff a^2 > b^2}$

Note that : if $a < 0$ and $b < 0$, then $a + b < 0$

$(\Rightarrow)$: $a < b \ \Rightarrow \ a - b < 0$ and $a + b < 0$

$\quad\quad \therefore \ (a + b)(a - b) > 0$

$\quad\quad \therefore \ a^2 - b^2 > 0$

$\quad\quad \therefore \ a^2 > b^2$

$(\Leftarrow)$: $a^2 > b^2 \ \Rightarrow \ a^2 - b^2 > 0$

$\quad\quad \therefore \ (a + b)(a - b) > 0$

$\quad\quad$ Since $a + b < 0$, $a - b < 0 \quad\quad \therefore \ a < b$

Therefore, $a < b \iff a^2 > b^2$

3. Operations of Inequalities

(1) When $a < x < b$ and $c < y < d$,

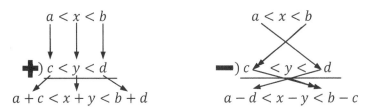

$$a + c < x + y < b + d \qquad a - d < x - y < b - c$$

(2) When $\boxed{0 <} a < x < b$ and $\boxed{0 <} c < y < d$,

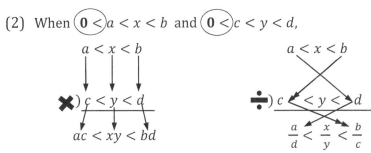

$$ac < xy < bd \qquad \frac{a}{d} < \frac{x}{y} < \frac{b}{c}$$

If "$<$" and "$\leq$" are involved in operations together, determine the result as "$<$".

Example

$$\begin{array}{c} -2 < x \leq 3 \\ +)\ \underline{1 < y \leq 5} \\ -1 < x + y \leq 8 \end{array} \qquad \begin{array}{c} 6 \leq x \leq 9 \\ \div)\ \underline{1 \leq y < 3} \\ \frac{6}{3} < \frac{x}{y} \leq \frac{9}{1} \end{array}$$

$$\begin{array}{c} -4 < x < -2 \\ \times)\ \underline{-3 < y < -1} \\ \text{Negative numbers} \end{array} \xRightarrow{\text{Transfer to positive numbers}} \begin{array}{c} 2 < -x < 4 \\ \times)\ \underline{1 < -y < 3} \\ 2 < xy < 12 \end{array}$$

4. Solving Linear Inequalities

(1) Solving Simple Inequalities

For the linear inequality $ax < b$ or $ax \leq b$ with $a > 0$, the solution is given by the simpler

inequality: $x < \dfrac{b}{a}$ or $x \leq \dfrac{b}{a}$.

This is because the solution has an infinite number of values and cannot be expressed in a

simpler form.

The usual representation of this solution is given by its graph on a number line, using open

points and closed points.

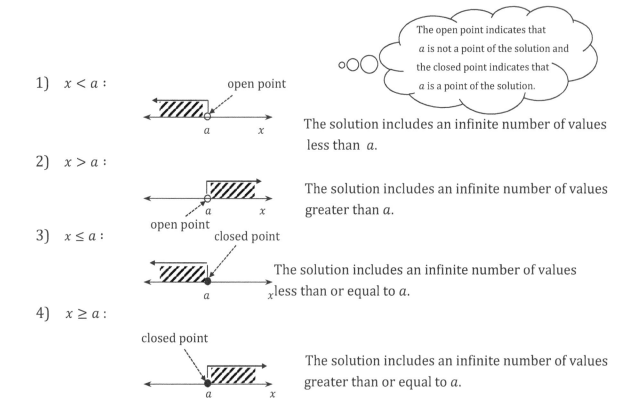

The open point indicates that a is not a point of the solution and the closed point indicates that a is a point of the solution.

1) $x < a$:

open point

The solution includes an infinite number of values less than a.

2) $x > a$:

open point

The solution includes an infinite number of values greater than a.

3) $x \leq a$:

closed point

The solution includes an infinite number of values less than or equal to a.

4) $x \geq a$:

closed point

The solution includes an infinite number of values greater than or equal to a.

(2) Solving Compound (Combined) Inequalities ; Double Inequalities

A compound inequality consists of two inequalities joined by "and" or "or".

Solve the two given inequalities together.

For $m < n$,

1) $m < x < n \implies m < x$ and $x < n$

2) $x > n$ or $x \leq m$

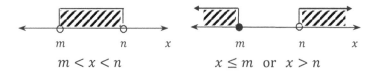

$m < x < n$ $x \leq m$ or $x > n$

To solve this compound inequality, isolate the variable between the inequality symbols, or isolate the variable in each inequality.

Examples

(1) Solve an "And" compound inequality :

$$1 < 2x - 3 \leq 5 \ \Rightarrow \ 1 + 3 < 2x - 3 + 3 \leq 5 + 3$$

$$\Rightarrow \ 4 < 2x \leq 8$$

$$\Rightarrow \ \frac{4}{2} < x \leq \frac{8}{2}$$

$$\Rightarrow \ 2 < x \leq 4 \ \Rightarrow \ x \leq 4 \ \text{and} \ x > 2$$

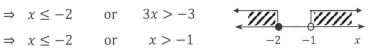 •- - - - - - 2 is not in the graph by placing an open circle above it .
4 is a point of the graph by filling in the circle above it.

(2) Solve an "Or " compound inequality :

$$3x + 4 \leq 2 + 2x \ \text{ or } \ -2x < x + 3$$

$$\Rightarrow \ x \leq -2 \quad \text{ or } \quad 3x > -3$$

$$\Rightarrow \ x \leq -2 \quad \text{ or } \quad x > -1$$

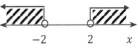

5. Solving Absolute Value Inequalities

To solve inequalities involving absolute values, consider the following comparisons.

$|x| = 2 \ \Rightarrow$ Solutions are $x = -2$ and $x = 2$.

$|x| < 2 \ \Rightarrow$ Solution is $-2 < x < 2$.

$|x| > 2 \ \Rightarrow$ Solution is $x > 2$ or $x < -2$.

(1) Transformations of Absolute Value Inequalities

For $a > 0, \ b > 0,$

(1) $\|x\| < a \ \Leftrightarrow \ -a < x < a$
(2) $\|x\| > a \ \Leftrightarrow \ x > a \ \text{ or } \ x < -a$
(3) $a < \|x\| < b \ \Leftrightarrow \ -b < x < -a \ \text{ or } \ a < x < b$

(2) Linear Inequalities with Absolute Values

$$|a| = \begin{cases} a, & a \geq 0 \\ -a, & a < 0 \end{cases}$$

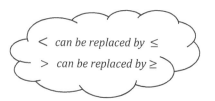

> $<$ can be replaced by $\leq$
> $>$ can be replaced by $\geq$

1) For $a > 0$, $|x| > a \Rightarrow x > a$ or $-x > a$

Since $-x > a$ is equivalent to $x < -a$,

$|x| > a$ is equivalent to $(x > a$ or $x < -a)$: x is beyond $-a$ and a.

Similarly, $|x + b| > a$ is equivalent to $(x + b > a$ or $x + b < -a)$.

2) $|x| < a$, $a > 0$ $\Leftrightarrow$ $-a < x < a$: x is between $-a$ and a.

$|x + b| < a$, $a > 0$ $\Leftrightarrow$ $-a < x + b < a$ or $-a - b < x < a - b$

$|cx + b| < a$, $a > 0$ $\Leftrightarrow$ $-a < cx + b < a$ or $-a - b < cx < a - b$

or $\dfrac{-a-b}{c} < x < \dfrac{a-b}{c}$, $c > 0$

Examples Solve each inequality involving absolute value.

(1) $2 < |x - 1| < 3$

  ① When $x - 1 \geq 0$ $(x \geq 1)$,

    $2 < |x - 1| < 3 \Rightarrow 2 < x - 1 < 3 \Rightarrow 2 + 1 < x < 3 + 1 \Rightarrow 3 < x < 4$

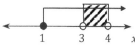

  So $3 < x < 4$

  ② When $x - 1 < 0$ $(x < 1)$,

    $2 < |x - 1| < 3 \Rightarrow 2 < -(x - 1) < 3 \Rightarrow 2 - 1 < -x < 3 - 1 \Rightarrow 1 < -x < 2$

      $\Rightarrow -2 < x < -1$

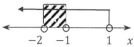

  So $-2 < x < -1$

  Therefore, $3 < x < 4$ or $-2 < x < -1$

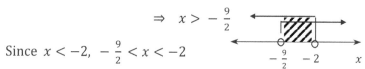

(2) $|x + 2| + |x - 3| < 10$

  Since $x + 2 = 0 \Rightarrow x = -2$ and $x - 3 = 0 \Rightarrow x = 3$,

  consider the three cases, $x < -2$, $-2 \leq x < 3$, and $x \geq 3$.

  ① Case 1 : When $x < -2$,

    $|x + 2| + |x - 3| < 10 \Rightarrow -(x + 2) - (x - 3) < 10 \Rightarrow -2x < 9$

      $\Rightarrow x > -\dfrac{9}{2}$

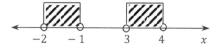

    Since $x < -2$, $-\dfrac{9}{2} < x < -2$

②   Case 2 : When $-2 \leq x < 3$,

$$|x + 2| + |x - 3| < 10 \Rightarrow (x + 2) - (x - 3) < 10$$

$$\Rightarrow 0 \cdot x < 5 \quad ; \text{ Always true}$$

$$\therefore \ -2 \leq x < 3$$

③   Case 3 : When $x \geq 3$,

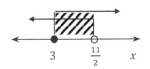

$$|x + 2| + |x - 3| < 10 \Rightarrow (x + 2) + (x - 3) < 10$$

$$\Rightarrow 2x < 11$$

$$\Rightarrow x < \frac{11}{2}$$

Since $x \geq 3, \quad 3 \leq x < \frac{11}{2}$

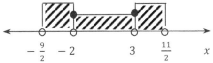

Therefore, the sum of all three intervals is $-\frac{9}{2} < x < \frac{11}{2}$.

3-6   Quadratic Inequalities in One Variable

To solve quadratic inequalities in one variable such as

$$ax^2 + bx + c > 0, \ ax^2 + bx + c \geq 0, \ ax^2 + bx + c < 0, \ ax^2 + bx + c \leq 0,$$

consider the discriminant $D = b^2 - 4ac$ of $ax^2 + bx + c = 0$.

1.   When $D > 0$,

(The quadratic equation $ax^2 + bx + c = 0, \ a > 0$, has two different real number solutions.)

Let α and β $(\alpha < \beta)$ be the two real number solutions.

Then, $ax^2 + bx + c = a(x - \alpha)(x - \beta)$

∴ The sign of $ax^2 + bx + c$ is equivalent to the sign of $(x - \alpha)(x - \beta)$.

x	$x < \alpha$	$x = \alpha$	$\alpha < x < \beta$	$x = \beta$	$x > \beta$
$x - \alpha$	$-$	0	$+$	$+$	$+$
$x - \beta$	$-$	$-$	$-$	0	$+$
$(x - \alpha)(x - \beta)$	$+$	0	$-$	0	$+$

Above ; Positive $+$

Below ; Negative

α $-$ β Number line

Therefore,

(1) The solutions of $ax^2 + bx + c > 0$ are

$x < \alpha$ or $x > \beta$.

(2) The solutions of $ax^2 + bx + c \geq 0$ are

$x \leq \alpha$ or $x \geq \beta$.

(3) The solutions of $ax^2 + bx + c < 0$ are

$\alpha < x < \beta$.

(4) The solutions of $ax^2 + bx + c \leq 0$ are

$\alpha \leq x \leq \beta$.

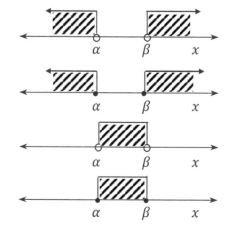

2. When $D = 0$,

(The quadratic equation $ax^2 + bx + c = 0$, $a > 0$, has only one real number solution.)

Let α be the double root.

Then, $ax^2 + bx + c = a(x - \alpha)^2$

Therefore,

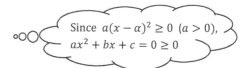

Since $a(x - \alpha)^2 \geq 0$ $(a > 0)$,
$ax^2 + bx + c = 0 \geq 0$

(1) The solutions of $ax^2 + bx + c > 0$ are all real numbers with $x \neq \alpha$.

(2) The solutions of $ax^2 + bx + c \geq 0$ are all real numbers.

(3) The solution of $ax^2 + bx + c < 0$ does not exist.

(4) The solution of $ax^2 + bx + c \leq 0$ is only $x = \alpha$.

3. When $D < 0$,

(The quadratic equation $ax^2 + bx + c = 0$, $a > 0$, has two imaginary (complex number) solutions.)

If $a > 0$ and $D = b^2 - 4ac < 0$, then for any real number x,

$$ax^2 + bx + c = a\left(x + \frac{b}{2a}\right)^2 - \frac{b^2 - 4ac}{4a} = a\left(x + \frac{b}{2a}\right)^2 + \frac{-D}{4a} > 0$$

$a > 0$,
$\left(x + \frac{b}{2a}\right)^2 \geq 0$,
$D < 0$

Therefore,

(1) The solutions of $ax^2 + bx + c > 0$ and $ax^2 + bx + c \geq 0$ are all real numbers.

(2) The solutions of $ax^2 + bx + c < 0$ and $ax^2 + bx + c \leq 0$ do not exist.

For a quadratic equation $ax^2 + bx + c = 0,\ a > 0$, with the discriminant $D = b^2 - 4ac$ and two real solutions $\alpha, \beta\ (\alpha \le \beta)$, we have the following results:

$a > 0$	$D > 0$	$D = 0$	$D < 0$
Solutions of $ax^2 + bx + c > 0$	$x < \alpha$ or $x > \beta$	All real #s but $x \ne \alpha$	All real #s
Solutions of $ax^2 + bx + c \ge 0$	$x \le \alpha$ or $x \ge \beta$	All real #s	All real #s
Solutions of $ax^2 + bx + c < 0$	$\alpha < x < \beta$	No solution	No solution
Solutions of $ax^2 + bx + c \le 0$	$\alpha \le x \le \beta$	$x = \alpha$	No solution

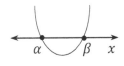 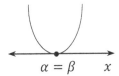 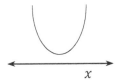

Note :

① *The quadratic inequality* $ax^2 + bx + c > 0$ *is always true for all real variables of* x

$$\underset{\text{if and only if}}{\Longleftrightarrow}\ a > 0,\ D = b^2 - 4ac < 0 \quad (or\quad a = 0,\ b = 0,\ c > 0)$$

② *The quadratic inequality* $ax^2 + bx + c \ge 0$ *is always true for all real variables of* x

$$\underset{\text{if and only if}}{\Longleftrightarrow}\ a > 0,\ D = b^2 - 4ac \le 0 \quad (or\quad a = 0,\ b = 0,\ c \ge 0)$$

③ *The quadratic inequality* $ax^2 + bx + c < 0$ *is always true for all real variables of* x

$$\underset{\text{if and only if}}{\Longleftrightarrow}\ a < 0,\ D = b^2 - 4ac < 0 \quad (or\quad a = 0,\ b = 0,\ c < 0)$$

④ *The quadratic inequality* $ax^2 + bx + c \le 0$ *is always true for all real variables of* x

$$\underset{\text{if and only if}}{\Longleftrightarrow}\ a < 0,\ D = b^2 - 4ac \le 0 \quad (or\quad a = 0,\ b = 0,\ c \le 0)$$

Example 1 Solve $2x^2 - x + 10 > 0$.

Since $D = b^2 - 4ac = 1 - 4 \cdot 2 \cdot 10 = -79 < 0$, the inequality is true for all x.

Example 2 Solve $-3x^2 + 2x - 5 > 0$.

Rewrite this inequality with $a > 0$ in the form $3x^2 - 2x + 5 < 0$.

Since $D = b^2 - 4ac = 4 - 4 \cdot 3 \cdot 5 = -56 < 0$, there are no x for which the inequality is true.

3-7 Systems of Inequalities

1. Definition

A *system of inequalities* is a pair of two or more linear inequalities.

The *solution* for a system of inequalities is the value or range which satisfies both inequalities at the same time.

2. Solving Systems of Inequalities

Step 1. Solve each inequality and find the solutions for each.

Step 2. Graph the solutions on a number line.

Step 3. Find the common range or value from the solutions of all inequalities.

Example Find the range of the solution for a system $\begin{cases} 2x - 3 \le 5 & \cdots\cdots ① \\ x^2 - 3x - 18 < 0 & \cdots\cdots ② \end{cases}$

$2x - 3 \le 5 \ \Rightarrow \ 2x \le 8 \quad \therefore \ x \le 4$

$x^2 - 3x - 18 < 0 \ \Rightarrow \ (x - 6)(x + 3) < 0 \quad \therefore \ -3 < x < 6$

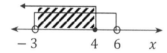

Therefore, the common interval of the solutions of inequalities ① and ② is $-3 < x \le 4$.

Note : When $a < b$,

(1) $\begin{cases} x < a \\ x < b \end{cases} \Rightarrow \ x < a$

(2) $\begin{cases} x > a \\ x > b \end{cases} \Rightarrow \ x > b$

(3) $\begin{cases} x > a \\ x < b \end{cases} \Rightarrow \ a < x < b$

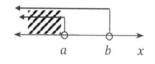

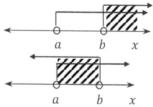

3. Solving Special Systems

(1) Solve parentheses by using the distributive property.

(2) Change fraction or decimal coefficients of variables to integers.

(3) Rewrite the system $A < B < C$ as $\begin{cases} A < B \\ B < C \end{cases}$.

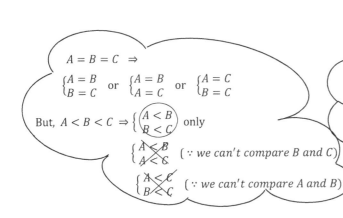

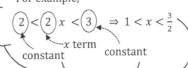

(4) $\begin{cases} x \leq a \\ x \geq a \end{cases} \Rightarrow x = a$ (This is the only solution.)

(5) If there is no common value or range,

$\begin{cases} x < a \\ x \geq a \end{cases}$ $\qquad$ $\begin{cases} x < a \\ x > a \end{cases}$ $\qquad$ $\begin{cases} x < a \\ x \geq b \end{cases}$, $a < b$

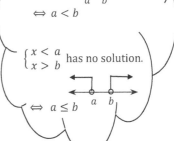

$\Rightarrow$ No solution $\qquad$ $\Rightarrow$ No solution $\qquad$ $\Rightarrow$ No solution

(6) Conditions for having solutions :

1) $\begin{cases} x \leq a \\ x \geq b \end{cases}$ 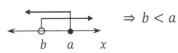 $\Rightarrow b \leq a$

$\qquad$ (∵ if $b = a$, then $\begin{cases} x \leq a \\ x \geq a \end{cases}$ ∴ $x = a$ is the solution.)

2) $\begin{cases} x \leq a \\ x > b \end{cases}$ 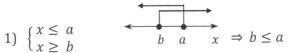 $\Rightarrow b < a$

$\qquad$ (∵ if $b = a$, then $\begin{cases} x \leq a \\ x > a \end{cases}$ ∴ There is no solution. Therefore, $b \nleqq a$ ($b < a$).)

3) $\begin{cases} x < a \\ x \geq b \end{cases}$ 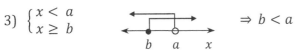 $\Rightarrow b < a$

$\qquad$ (∵ if $b = a$, then $\begin{cases} x < a \\ x \geq a \end{cases}$ ∴ There is no solution. Therefore, $b \nleqq a$ ($b < a$).)

4) $\begin{cases} x < a \\ x > b \end{cases}$ $\Rightarrow b < a$

$\left(\because \text{if } b = a, \text{ then } \begin{cases} x < a \\ x > a \end{cases} \quad \therefore \text{ There is no solution. Therefore, } b \lneqq a \ (b < a).\right)$

(7) For any $a > 0$,

$\begin{cases} |x| \leq a \Rightarrow -a \leq x \leq a \\ |x| > a \Rightarrow x > a \text{ or } x < -a \end{cases}$

$|x| \leq a$ $|x| > a$

3-8 Proofs of Equalities and Inequalities

The following inequalities become true for all real variables.

1. $\boxed{|a + b| \leq |a| + |b|}$

$\because$ Note that $|a|^2 = a^2$

Since $|a + b| \geq 0$ and $|a| + |b| \geq 0$, show that $|a + b|^2 \leq (|a| + |b|)^2$.

$|a + b|^2 - (|a| + |b|)^2 = (a + b)^2 - (|a|^2 + 2|a||b| + |b|^2)$

$= (a^2 + 2ab + b^2) - (a^2 + 2|ab| + b^2)$

$= 2(ab - |ab|)$

Since $|ab| \geq ab$, $2(ab - |ab|) \leq 0$

$\therefore \ |a + b|^2 \leq (|a| + |b|)^2$

$\therefore \ |a + b| \leq |a| + |b|$

When $ab \geq 0$, $|a + b| = |a| + |b|$

2. $\boxed{|a - b| \geq |a| - |b|}$

$\because$ Show that : $|a| \leq |a - b| + |b|$

Since $|a| = |(a - b) + b|$ and $|a + b| \leq |a| + |b|$, $|(a - b) + b| \leq |a - b| + |b|$.

That is, $|a| \leq |a - b| + |b|$. $\therefore \ |a| - |b| \ \leq |a - b|$

When $|(a - b)b| = (a - b)b$; i.e., $(a - b)b \geq 0$, $|a| - |b| = |a - b|$

3. More Inequalities

For real numbers $a, b,$ and $c,$

(1) $\boxed{a^2 \pm 2ab + b^2 \geq 0}$

$\because$ Since $a^2 \pm 2ab + b^2 = (a \pm b)^2,\ a^2 \pm 2ab + b^2 \geq 0$

When $a = b = 0,\ a^2 \pm 2ab + b^2 = 0$

(2) $\boxed{a^2 + b^2 + c^2 - ab - bc - ca \geq 0}$

$\because a^2 + b^2 + c^2 - ab - bc - ca = \dfrac{1}{2}(2a^2 + 2b^2 + 2c^2 - 2ab - 2bc - 2ca)$

$$= \frac{1}{2}\{(a^2 - 2ab + b^2) + (b^2 - 2bc + c^2)(c^2 - 2ca + a^2)\}$$

$$= \frac{1}{2}\{(a - b)^2 + (b - c)^2 + (c - a)^2\}$$

Since $a, b,$ and c are real numbers,

$(a - b)^2 \geq 0, (b - c)^2 \geq 0,$ and $(c - a)^2 \geq 0$

$\therefore\ a^2 + b^2 + c^2 - ab - bc - ca \geq 0$

When $a = b = c,\ a^2 + b^2 + c^2 - ab - bc - ca = 0$

(3) $\boxed{a > 0,\ b > 0,\ c > 0\ \Rightarrow\ a^3 + b^3 + c^3 \geq 3abc}$

$\because$ Note that $a^3 + b^3 = (a + b)^3 - 3ab(a + b)$

$a^3 + b^3 + c^3 - 3abc = (a^3 + b^3) + c^3 - 3abc$

$$= (a + b)^3 - 3ab(a + b) + c^3 - 3abc$$

$$= (a + b)^3 + c^3 - 3ab(a + b + c)$$

Since $A^3 + B^3 = (A + B)(A^2 - AB + B^2),$

$\quad (a + b)^3 + c^3 - 3ab(a + b + c)$

$= (a + b + c)\{(a + b)^2 - (a + b)c + c^2\} - 3ab(a + b + c)$

$= (a + b + c)\{(a + b)^2 - (a + b)c + c^2 - 3ab\}$

$= (a + b + c)(a^2 + b^2 + c^2 + 2ab - ac - bc - 3ab)$

$= (a + b + c)(a^2 + b^2 + c^2 - ab - bc - ca)$

$\therefore\ a^3 + b^3 + c^3 - 3abc = (a + b + c)(a^2 + b^2 + c^2 - ab - bc - ca)$

Since $a + b + c > 0$ and $a^2 + b^2 + c^2 - ab - bc - ca \geq 0,$

$(a + b + c)(a^2 + b^2 + c^2 - ab - bc - ca) \geq 0.$

$\therefore\ a^3 + b^3 + c^3 \geq 3abc$

When $a = b = c,\ a^3 + b^3 + c^3 = 3abc$

4. The Relationship between Geometric Mean and Arithmetic Mean

For positive numbers a and b, $\sqrt{ab}$ is called the *geometric mean* of a and b ;

$\frac{a+b}{2}$ is called the *arithmetic mean* of a and b.

(1) Geometric Mean $\leq$ Arithmetic Mean

> If a and b are positive, $\sqrt{ab} \leq \frac{a+b}{2}$ (When $a = b$, $\sqrt{ab} = \frac{a+b}{2}$)

$\because$ Since $a > 0$ and $b > 0$, $a + b > 0$ and $ab > 0$ $\therefore \frac{a+b}{2} > 0$ and $\sqrt{ab} > 0$

To show $\sqrt{ab} \leq \frac{a+b}{2}$ is equivalent to show $\left(\sqrt{ab}\right)^2 \leq \left(\frac{a+b}{2}\right)^2$

; i.e., $ab \leq \frac{(a+b)^2}{4}$; $4ab \leq (a + b)^2$

$(a + b)^2 - 4ab = a^2 + 2ab + b^2 - 4ab = (a - b)^2 \geq 0$

$\therefore (a + b)^2 \geq 4ab$ $\therefore \frac{a+b}{2} \geq \sqrt{ab}$

If $a = b$, then $(a + b)^2 - 4ab = 0$; $\frac{a+b}{2} = \sqrt{ab}$

> *Note :* $a > 0$, $b > 0$ $\Rightarrow$ $\frac{2ab}{a+b} \leq \sqrt{ab} \leq \frac{a+b}{2}$ (When $a = b$, $\frac{2ab}{a+b} = \sqrt{ab} = \frac{a+b}{2}$)

$\qquad \because a > 0, b > 0 \Rightarrow$ ① $\frac{a+b}{2} - \sqrt{ab} = \frac{a+b-2\sqrt{ab}}{2} = \frac{\left(\sqrt{a}\right)^2 + \left(\sqrt{b}\right)^2 - 2\sqrt{ab}}{2} = \frac{\left(\sqrt{a}-\sqrt{b}\right)^2}{2} \geq 0$

$\qquad\qquad \therefore \frac{a+b}{2} \geq \sqrt{ab}$

$\qquad\qquad$ ② $\sqrt{ab} - \frac{2ab}{a+b} = \frac{\sqrt{ab}(a+b-2\sqrt{ab})}{a+b} = \frac{\sqrt{ab}\left(\sqrt{a}-\sqrt{b}\right)^2}{a+b} \geq 0$

$\qquad\qquad\qquad \therefore \sqrt{ab} \geq \frac{2ab}{a+b}$

(2) Minimum and Maximum

1) If the value of ab is a constant, then $a + b$ has a minimum value when $a = b$.

2) If the value of $a + b$ is a constant, then ab has a maximum value when $a = b$.

Example   When $a > 0$ and $b > 0$, find minimum value of $(a + b)\left(\frac{1}{a} + \frac{4}{b}\right)$.

$(a + b)\left(\frac{1}{a} + \frac{4}{b}\right) = 1 + \frac{b}{a} + \frac{4a}{b} + 4 = 5 + \frac{b}{a} + \frac{4a}{b}$

Since $\frac{b}{a} + \frac{4a}{b} \geq 2\sqrt{\frac{b}{a} \cdot \frac{4a}{b}}$, $(a + b)\left(\frac{1}{a} + \frac{4}{b}\right) \geq 5 + 2\sqrt{\frac{b}{a} \cdot \frac{4a}{b}} = 5 + 2\sqrt{4} = 9$

$\therefore (a + b)\left(\frac{1}{a} + \frac{4}{b}\right)$ has minimum value 9.

When $\frac{b}{a} = \frac{4a}{b}$, $4a^2 = b^2$; $2a = b$

Wrong Approach : Since $a + b \geq 2\sqrt{ab}$ and $\frac{1}{a} + \frac{4}{b} \geq 2\sqrt{\frac{1}{a} \cdot \frac{4}{b}}$,

$$(a + b)\left(\frac{1}{a} + \frac{4}{b}\right) \geq 2\sqrt{ab} \cdot 2\sqrt{\frac{1}{a} \cdot \frac{4}{b}} = 4\sqrt{ab \cdot \frac{1}{a} \cdot \frac{4}{b}} = 8 \ (\times)$$

If $a = b$, then $a + b \geq 2\sqrt{ab}$ is true.

If $b = 4a$, then $\frac{1}{a} + \frac{4}{b} \geq 2\sqrt{\frac{4}{ab}}$ is true.

However, there are no positive numbers a and b so that both inequalities are true.

Be sure that the equality when you use more than two times the relationship between geometric mean and arithmetic mean.

5. Cauchy's Inequality

(1) For any real numbers a, b, x, and y,

$$\boxed{(ax + by)^2 \leq (a^2 + b^2)(x^2 + y^2)}$$

(when $\frac{x}{a} = \frac{y}{b}$, $(ax + by)^2 = (a^2 + b^2)(x^2 + y^2)$)

(2) For any real numbers a, b, c, x, y, and z,

$$\boxed{(ax + by + cz)^2 \leq (a^2 + b^2 + c^2)(x^2 + y^2 + z^2)}$$

(when $\frac{x}{a} = \frac{y}{b} = \frac{z}{c}$, $(ax + by + cz)^2 = (a^2 + b^2 + c^2)(x^2 + y^2 + z^2)$)

Note : Proof of Cauchy's Inequality

$$(ax + by)^2 - (a^2 + b^2)(x^2 + y^2)$$

$$= a^2x^2 + 2abxy + b^2y^2 - (a^2x^2 + b^2x^2 + a^2y^2 + b^2y^2)$$

$$= 2abxy - b^2x^2 - a^2y^2 = -(b^2x^2 - 2abxy + a^2y^2)$$

$$= -(bx - ay)^2 \leq 0$$

$$\therefore \ (ax + by)^2 \leq (a^2 + b^2)(x^2 + y^2)$$

Exercises

#1 Solve the following equations for x :

(1) $3x - 2 = 7$

(2) $2x + 3 = 3x - 2$

(3) $5x - 2 = \frac{1}{2}x - 1\frac{1}{4}$

(4) $0.2x - 0.3 = 0.4x - 0.5$

(5) $\frac{3}{4}\left(x - \frac{1}{3}\right) = \frac{1}{2}\left(\frac{1}{5} + 4x\right)$

(6) $\frac{x-3}{2} - 1 = \frac{x}{4} - 3$

(7) $3(1 - 2x) + 7 = -2x - 2$

(8) $3 - \frac{2x-1}{3} = 5x - \frac{x-2}{6}$

#2 Find the value.

(1) For any constants $a, b,$ the solution of the equation $3x - 2 = ax - 4$ is $x = -1$ and the solution of the equation $\frac{1}{2}x + b = ax + 3$ is $x = -2$. Find $a \cdot b$

(2) The solution of the equation $2ax + 5 = -3$ is half of the solution of the equation $x - 5 = 3x + 7$. Find the value of $3a - 4$.

(3) For any constants a and b, $\frac{1}{a} - \frac{1}{b} = 3$ $(ab \neq 0)$. Find the value of $\frac{5a - 3ab - 5b}{a - b}$.

(4) $\begin{cases} ① & \frac{a+3}{4} - \frac{2x-2}{3} = 1 \\ ② & \frac{3a-2}{2} - \frac{2a-x}{3} = 1 \end{cases}$

When the ratio of the solution of ① to the solution of ② is $1 : 4$, find the value of a.

(5) For any x, the equation $3x - 5a = 2bx + 6$, where a and b are constants, is always true. Find the value of $\frac{a}{2b}$.

(6) The solution of an equation $\frac{2x-5a}{3} + x + 4 = 8$ is a negative integer. Find the greatest value of a.

(7) $a@b = ab^2 + a^2b$ When $\frac{1}{a} = 2$, $\frac{1}{b} = -3$, find the value of $b@a$.

(8) A quadratic equation $x^2 - x + 1 = 0$ has two solutions α and β. Find the value of $(1 - \alpha)(1 - \beta) + (2 - \alpha)(2 - \beta)$.

#3 Solve the equation involving absolute value.

(1) $|x - 3| = 5x + 2$

(2) $|x - 4| + |x + 2| = 10$

(3) $|x - 2| - |5 - x| = 0$

#4 Solve each equation for x by using the factorization.

(1) $x^2 - 2x - 3 = 0$

(2) $2x^2 - 7x + 5 = 0$

(3) $-3x^2 + 6x = 0$

(4) $2x^2 + 2x - 4 = 0$

(5) $x(x + 5) = 6$

(6) $x^2 = \frac{x+1}{2}$

#5 Solve each equation for x by using the square root.

(1) $2x^2 = 8$

(2) $9x^2 - 5 = 0$

(3) $3(x - 1)^2 = 15$

(4) $(2x + 5)^2 - 3 = 0$

(5) $4(x - 2)^2 - 1 = 0$

#6 Solve each quadratic equation by using perfect squares.

(1) $x^2 - 3x - 3 = 0$

(2) $2x^2 + 5x = 7$

(3) $-x^2 - 3x + 5 = 0$

(4) $3x^2 - 4x + 1 = 0$

#7 Find the constant k for each quadratic equation with a double root.

(1) $(3x - 4)^2 - k^2 = 0$

(2) $x^2 - kx + 5 = 0$

(3) $x^2 + 2x + k^2 = 0$

(4) $kx^2 + 3x + 2 = 0$

(5) $2x^2 + 3x + k - 5 = 0$

(6) $x^2 + kx + (k - 1) = 0$

(7) $\frac{1}{3}x^2 + (k + 1)x + 8 = 0$

#8 Find the value of $p + q$ for each quadratic equation with the solution $x = p \pm \sqrt{q}$.

(1) $-2x^2 + 5x + 1 = 0$

(2) $3(x - 1)^2 = 4$

(3) $-(x + 1)^2 + 5 = 0$

#9 Find the constant a or the range of a for each quadratic equation with a condition.

(1) $(x + 1)^2 = a + 2$ has no real number solution.

(2) $x^2 + 3x + 3a = 0$ has two different real number solutions.

(3) $ax^2 + x + 2 = 0$ has one real number solution.

(4) $3x^2 - x + a = 0$ has no real number solution.

(5) $x^2 + (a + 1)x + \frac{a+3}{2} = 0$ has a double root.

(6) $x^2 + x + a = 0$ has two different real number solutions, $x = 2$ and $x = b$.

(7) $x^2 + 2ax + b = 0$ has a double root $x = 3$.

(8) $x^2 + ax + b = 0$ has a real number solution $x = 1 + \sqrt{2}$.

(9) $x^2 - ax - 2b^2 = 0$ has two different real number solutions, $x = 4 \pm \sqrt{2a}$.

#10 Solve each quadratic equation by using quadratic formulas.

(1) $x^2 - 2x - 4 = 0$

(2) $3x^2 + 5x - 1 = 0$

(3) $5x^2 - 2x - 1 = 0$

(4) $-2x^2 + 3x + 5 = 0$

(5) $\frac{1}{2}x^2 - 3x + 2 = 0$

(6) $\frac{1}{6}x^2 - 0.5x + \frac{1}{4} = 0$

(7) $(x + 1)^2 = 3(x + 2)$

(8) $(x + 2)^2 + 3(x + 2) - 2 = 0$

(9) $-\frac{(x-1)^2}{2} + x = 0.4(x + 1)$

(10) $(x + 3)(2x + 6) = 5$

#11 Find the value of the given expression.

(1) $a + b$ when $(2x + 1)^2 = 3$ has solutions $x = a \pm b\sqrt{3}$.

(2) $a - b$ when $2x^2 - 8x + 1 = 0$ has solutions $x = \frac{a \pm 3\sqrt{b}}{6}$.

(3) ab when $ax^2 + 5x + 2 = 0$ has solutions $x = \frac{-5 \pm 2\sqrt{b}}{4}$.

(4) $\frac{b}{a}$ when $3x^2 - 5x + 1 = 0$ has solutions $x = a \pm \sqrt{b}$.

(5) $\frac{a+b}{ab}$ when $x^2 - 3x + 1 = 0$ has solutions $x = \frac{a \pm 2\sqrt{b}}{2}$.

(6) $\frac{a-b}{a^2-b^2}$ when $ax^2 + 3x - 3b = 0$ has solutions $x = -1 \pm \sqrt{5}$.

#12 State the number and type of solutions for each quadratic equation.

(1) $x^2 + 2x - 3 = 0$

(2) $-x^2 + x - 5 = 0$

(3) $4x^2 - 4x + 1 = 0$

(4) $kx^2 - (k + 5)x + 1 = 0$

(5) $3x^2 - x - k^2 = 0$

(6) $x^2 - 4kx + 5k^2 + 1 = 0$

#13 Determine the value of a or the range of the values of a at which the following quadratic equation will have a given condition.

(1) $x^2 + 5x + a = x + 2$ will have no real number solution.

(2) $(a + 3)x^2 - 2ax + a - 1 = 0$ will have two different real number solutions.

(3) $x^2 + 3ax - 2a + 3 = 0$ will have only one real number solution.

(4) $x^2 + ax + a + 2 = 0$ will have a double root and $x^2 + 4ax + (2a - 1)^2 = 0$ will have two different real number solutions.

(5) $2x^2 + (2a - 1)x + a^2 + \frac{1}{4} = 0$ will have real number solutions.

(6) $(a - 1)x^2 + 2(a - 1)x + (a + 1) = 0$ will have real number solutions.

(7) The quadratic equation $x^2 - 3x + 2a = 0$ will have two different positive real number solutions.

(8) The quadratic equation $ax^2 + 2x + 3 = 0$ will have two different negative real number solutions.

(9) The quadratic equation $x^2 - 4x + 3a = 0$ will have two different real number solutions α and β with opposite signs.

(10) The quadratic equation $(a + 3) x^2 - 2ax + a - 1 = 0$ will have real number solutions.

(11) The quadratic equation $x^2 + (a^2 + a - 6)x - a - 1 = 0$ will have two different real number solutions α and β with $|\alpha| = |\beta|$.

(12) The quadratic equation $x^2 + 2(3 - a)x + a^2 = 0$ will have two different real number solutions and $2x^2 - x + a = 0$ will have no real number solution.

(13) The quadratic equation $x^2 - (a + 5)x - a - 1 = 0$ will have roots α and β which are integers.

#14 Find the value of the given expression.

   (1) A quadratic equation $3x^2 + 5x - 2 = 0$ has two solutions, $x = \alpha$ and $x = \beta$.

   1) $\alpha + \beta$ 4) $\alpha^2 - \beta^2$

   2) $\alpha^2 + \beta^2$ 5) $\frac{1}{\alpha} + \frac{1}{\beta}$

   3) $\alpha - \beta$

   (2) A quadratic equation $x^2 + 3kx + 2k^2 - 4k - 1 = 0$ has two solutions α and β.
       Find the value of the given expressions in terms of k

   1) $\alpha + \beta$ 4) $(\alpha - \beta)^2$

   2) $\alpha\beta$ 5) $\frac{\beta}{\alpha} + \frac{\alpha}{\beta}$

   3) $\alpha^2 + \beta^2$

   (3) $x^2 + ax + b = 0$ has two solutions -2 and -3.
       Find the value of $\alpha^2 + \beta^2$ for $x^2 - bx - a = 0$ which has solutions α and β .

   (4) $x^2 - x - 1 = 0$ has two solutions α and β. Find the value of $(\alpha^4 + 1) + (\beta^4 + 1)$.

(5) $(1-i)x^2 + 2(1+i)x + 3(1-i) = 0$ has two solutions α and β.

Find the value of $(\alpha - \beta)^2$.

(6) When $x^2 - |x-1| - 1 = 0$ and $x^2 + ax + b = 0$ have the same solution,

find the value of $a^2 + b^2$.

(7) When $x^2 - 4xy + ay^2 + 2y - 1$ is factorized as a product of two linear factors,

find the value of a.

(8) For real numbers a and b, $x^2 + ax + b = 0$ has a solution $\dfrac{1}{1+i}$. Find the value of ab.

(9) $x^2 + 2(3+i)x + (a-4i) = 0$ has real number solutions. Find the value of a.

(10) $x^2 - 2x + 2 = 0$ has two roots α and β. Find the value of $\dfrac{\beta}{\alpha^2 - 3\alpha + 2} + \dfrac{\alpha}{\beta^2 - 3\beta + 2}$.

(11) $ax^2 - 2x + b = 0$, $a \neq 0$, has a solution $1 + 2i$. Find the value of $a + b$.

(12) For rational numbers a and b, $x^2 - ax + b = 0$ has a solution $\sqrt{4 - 2\sqrt{3}}$.

Find the value of $a^2 - b^2$.

#15 Find the solution for the quadratic equation $ax^2 + (b-1)x + 4 = 0$:

(1) When the quadratic equation $2x^2 + (a-1)x + b = 0$ has two solutions $\dfrac{1}{2}$ and $\dfrac{1}{3}$.

(2) When the quadratic equation $3ax^2 + 8bx + 3 = 0$ has a double root -2.

(3) When the quadratic equation $ax^2 + 3ax - 4 = 0$ has two solutions, b and $b + 1$.

(4) When the quadratic equation $ax^2 + 3x + b = 0$ has two solutions, α and β, which

satisfy the conditions $\alpha + \beta = -2$ and $\alpha\beta = 4$.

(5) When the quadratic equation $x^2 + ax + 3 = 0$ has two different solutions. The one of the

solutions is $x = -2 + 3\sqrt{b}$.

#16 The following quadratic equations have only one solution.

Find the solution (a double root) for each equation.

(1) $x^2 + kx + 2k - 3 = 0$

(2) $(k+2)x^2 - 2kx + k + 1 = 0$

(3) $x^2 + (k+2)x + k^2 - k + 2 = 0$

#17 For $a \neq b$, a quadratic equation $x^2 + ax + b + 5 = 0$ has two solutions α, β and a quadratic

equation $x^2 + bx + a + 5 = 0$ has two solutions α, γ. Find the value of $\alpha + \beta + \gamma$.

#18 Create a quadratic equation with leading coefficient 1.

(1) $x^2 + 2x - 3 = 0$ has two roots α and β.

Find a quadratic equation that has the roots $\alpha + \beta$ and $\alpha\beta$.

(2) $x^2 + 2x + 3 = 0$ has two roots α and β.

Find a quadratic equation that has the roots $\dfrac{1}{\alpha}$ and $\dfrac{1}{\beta}$.

(3) Find a quadratic equation that has the roots $\sqrt{2 + \sqrt{3}}$ and $\sqrt{2 - \sqrt{3}}$.

#19 Solve each quadratic equation.

(1) $x^2 + |x| + x - 1 = 0$

(2) $|x| - 1 = \sqrt{(2x - 5)^2}$

#20 For rational numbers a, b, and c, $ax^2 + bx + c = 0$ has a solution which is the decimal part of $\sqrt{4 + \sqrt{12}}$. When α and β are solutions of $cx^2 + bx + a = 0$, find the value of $|\alpha - \beta|$.

#21 Solve each equation.

(1) $x^3 - x^2 + x - 1 = 0$

(2) $2x^3 - 5x^2 + 5x - 2 = 0$

(3) $x^3 - 4x^2 - 4x - 5 = 0$

(4) $x^4 + x^3 - 2x^2 - x + 1 = 0$

(5) $2x^4 - 3x^3 - x^2 - 3x + 2 = 0$

(6) $(x + 1)(x + 2)(x + 3)(x + 4) - 3 = 0$

(7) $x^3 + ax^2 - b = 0$ with a root $1 + i$

#22 Find the value of each expression.

(1) A fourth-degree polynomial equation $x^4 - 2x^2 + 3x - 2 = 0$ has a complex number solution α. Find the value of $\alpha + \dfrac{1}{\alpha}$.

(2) A fourth-degree polynomial equation $x^4 - x^3 - 4x^2 - x + 1 = 0$ has a positive solution α. Find the value of $\alpha^3 + \dfrac{1}{\alpha^3}$.

#23 For a cube polynomial equation $x^3 = 1$, ω is a complex number solution. Find the value of each expression.

(1) $\omega^2 + \omega + 1$

(2) $\omega + \dfrac{1}{\omega}$

(3) $\omega^2 + \dfrac{1}{\omega^2}$

(4) $\omega + \dfrac{1}{\omega} + \bar{\omega} + \dfrac{1}{\bar{\omega}}$

(5) $\dfrac{1}{1+\omega} + \dfrac{1}{1+\bar{\omega}}$

#24 Find the value of given expression:

(1) For a cube polynomial equation $x^3 - x^2 - 6x + 7 = 0$ with roots α, β, and γ,

find the value of $(\alpha + \beta)(\beta + \gamma)(\gamma + \alpha)$.

(2) For a cube polynomial equation $x^3 + ax^2 + 4x + b = 0$ with a root $1 + i$,

find the value of $a + b$ (a, b are real numbers).

(3) For a cube polynomial equation $x^3 + ax + b = 0$ with a root α ($\alpha < 0$) such that $\alpha^2 = 6 - 4\sqrt{2}$, find the value of $a - b$ (where a and b are rational numbers).

#25 Find the range of a real number a so that a fourth-degree quadratic equation

$x^4 - 2(a + 1)x^2 + a^2 - 2a - 3 = 0$ will have four different real number solutions.

#26 Find the value of ab for each system.

(1) $\begin{cases} ax - by = -2 \\ bx + 2y = a \end{cases}$ with solution $(3,2)$

(2) $\begin{cases} x + 5y = -3 \\ 2x - by = 5 \end{cases}$ with solution $(a, -1)$

(3) $\begin{cases} -2x + y = 5 \\ x - 2y = -1 \end{cases}$ with solution (a, b)

(4) $\begin{cases} 3x - by = 2 \\ ax + y = -2 \end{cases}$ with solution $(b - 1, 2)$

#27 The system $\begin{cases} 2x + 3y = 5 \\ -x - 2y = -3 \end{cases}$ has a solution (a, b).

Find the solution for the system $\begin{cases} (3 - a)x + 2y = -2 \\ 2x + 3y = 2b + 1 \end{cases}$.

#28 Find the value for the following :

(1) The system $\begin{cases} 2x - y = 3 \\ x + 3y = 5 \end{cases}$ has a solution $(a + 1, b - 1)$.

Find the value of $(a + b)^2 - (a - b)^2$.

(2) The solution of the system $\begin{cases} 3x - 2y = -2 \\ (k - 1)x + y = -3 \end{cases}$

is the same as the solution of the equation $2x - y = 3$. Find the constant k.

(3) Two systems $\begin{cases} 2x + by = 4 \\ x + 2y = -3 \end{cases}$ and $\begin{cases} x - 3y = 2 \\ ax + 2y = -1 \end{cases}$ have the same solution.

Find the value of $a + b$.

(4) The system $\begin{cases} \frac{a+1}{2}x - \frac{3}{4}y = -2 \\ 5x + \frac{b-1}{2}y = 4 \end{cases}$ has all real number solutions.

Find the value of $a + b$.

(5) The system $\begin{cases} a(x-y) + \frac{y}{2} = -1 \\ -\frac{x}{2} - \frac{1}{a}y = 3 \end{cases}$ has no solution. Find the value of a.

(6) The system $\begin{cases} 2kx - (3x+y) = 2y \\ -(k-1)x + 2y = kx \end{cases}$ has a solution other than $(0,0)$.

Find the value of the constant k.

(7) The system $\begin{cases} 3x - 2y = k \\ -2x + y = 3 \end{cases}$ has the solution (a, b) with the condition $a : b = 1 : 3$.

Find the constant k.

(8) Find the value of $\frac{1}{x} - \frac{1}{y}$ for variables x and y that satisfy the system

$\begin{cases} 2x - xy - 2y - 3 = 0 \\ 3x + 2xy - 3y + 1 = 0 \end{cases}$.

#29 Find the value of $x + y$ for variables x and y that satisfy the equations $2^x \cdot 8^y = 32$ and $3^{x+1} \cdot 9^{y-1} = 3^3$.

#30 Solve the following systems:

(1) $\begin{cases} x^2 + xy + y^2 = 1 \\ x^2 + y^2 + x + y = 2 \end{cases}$

(2) $\begin{cases} x^2 + y^2 = 10 \\ xy = 3 \end{cases}$

#31 Determine the range.

(1) Find the range of x for each expression when $-1 \le x \le 1$.

① $2x + 1$ ② $-3x - 2$ ③ $\frac{1}{4}x - 3$

(2) When $y = \frac{4-2x}{3}$

① Find the range of y when $1 < x < 5$.

② Find the range of y when $-3 < x < -1$.

③ Find the range of x when $2 \le y \le 4$.

#32 Find the constant k if :

(1) The inequality $\frac{1}{2}x - \frac{k}{3} < -1$ has the solution $x < 2$.

(2) The inequality $\frac{kx}{4} - \frac{1}{2} > 1$ has the solution $x < -1$.

(3) The inequality $\frac{2-kx}{5} - 2 \le \frac{x}{2} + 1$ has the solution $x \le -4$.

(4) Two inequalities $2(1 - 2x) - 3 \le x - 5$ and $\frac{3k-2x}{3} \le x + 2k$ have the same solution.

(5) The inequality $2 - kx < 2x + k$ has no solution.

(6) The inequality $-2kx + 5 > 6$ has the solution $x > 2$.

(7) $1 - 5x \le 2x - 5k$ has -2 as minimum value of the solution.

(8) The inequality $x - \left(3 + \frac{k}{2}\right) > 2x + k$ has no positive solution.

#33 Solve each inequality.

(1) The inequality $(-a + 2b)x + b - 3a \le 0$ has the solution $x \le -1$.

   Find the solution for the inequality $(a - b)x + a - 2b > 0$, where $b > 0$.

(2) The inequality $(a + b)x \ge 3a - b$ has the solution $x \le \frac{3}{2}$.

   Find the solution for the inequality $(a - 2b)x < 2a + 3b$.

#34 Solve each inequality for x and graph the solution :

(1) $|x - 2| \le 0$

(2) $|3x + 9| > 0$

(3) $|x + 4| < 0$

(4) $|-2x + 1| + 3 \le 6$

(5) $0 < |2 - 4x| < 8$

(6) $2 < |x + 1| < 3$

(7) $2|x + 1| + |x - 2| \le 5$

(8) $|x + 1| + \sqrt{x^2 - 4x + 4} < 4$

(9) $|x| - 2|x - 3| > 1 - x$

(10) $2|x + 1| - 3|x - 3| \ge 1$

(11) $x^2 - x - 5 > |2x - 1|$

(12) $|x - 3| < 4$

(13) $|x + 2| < 3x - 4$

(14) $|x + 2| + |3 - x| > 10$

#35 When $1 \le x \le 2$, find maximum and minimum values of $\frac{2x}{x+1}$.

#36 Find the value.

(1) When the inequality $|2x - a| > 4$ has solution: $x < b$ or $x > 3$, find the values of real

   numbers a and b.

(2) When the inequality $|4x + 2| - 1 \leq a$ has solution: $-3 \leq x \leq 2$, find the value of a.

#37 Find the value of k for the following conditions :

(1) The system $\begin{cases} x + 5 < 2k \\ 3x - 2 \geq 4 \end{cases}$ has the solution $2 \leq x < 5$.

(2) The system $\begin{cases} \frac{2x+1}{3} > \frac{x-3}{5} \\ 0.6x - 2.4 < kx - 0.8 \end{cases}$ has the solution $-2 < x < 2$.

(3) The system $\begin{cases} -x + 2 \leq 0 \\ \frac{x}{2} + 3 \leq -k + 5 \end{cases}$ has the solution $x = 2$.

(4) The system $\begin{cases} \frac{k-x}{2} \leq x + 5 \\ 3 - 2x < 3x - 2 \end{cases}$ has the solution $x \geq 3$.

(5) The system $\begin{cases} 2x + 3 \leq 4x - 5 \\ 3(x - k) \leq x + 3 \end{cases}$ has only one solution.

(6) The system $3 < \frac{k-4x}{-2} < 5$ has the solution $1 < x < 2$.

#38 Find the range of k for the following conditions :

(1) The system $\begin{cases} 2x \leq 5 - k \\ 3x - 3 \geq 2x - 1 \end{cases}$ has no solution.

(2) The system $\begin{cases} x - 3 \leq 2x - 6 \\ 5x + k < 3x + 1 \end{cases}$ has no solution.

(3) The system $\begin{cases} 2x + 3 \leq -5 \\ x + k > 1 \end{cases}$ has only one integer in the solution.

(4) The system $\begin{cases} x - 3 \geq 0 \\ 3x + k \leq 2x + 3 \end{cases}$ has solutions.

#39 Find the sum of all integers that satisfy the following systems :

(1) $\begin{cases} 3x - 5 \leq 7 \\ \frac{x-1}{2} < x + 3 \\ 2x - 5 < 5x + 4 \end{cases}$

(2) $\begin{cases} |x| \leq 5 \\ |x| > 2 \end{cases}$

#40 The system $\begin{cases} x - 1 \geq 2x - 4 \\ \frac{x+k}{2} < 3x - 2 \end{cases}$ has 5 integers in the solution. What is minimum value for k?

#41 Find the value.

(1) Find the real numbers a and b so that a quadratic inequality $ax^2 + bx - 12 > 0$ will have solution $2 < x < 3$.

(2) When the solution of a quadratic inequality $x^2 - ax + 12 \leq 0$ is $\alpha \leq x \leq \beta$ and the solution of a quadratic inequality $x^2 - 5x + b \geq 0$ is $x \leq \alpha - 1$ or $x \geq \beta - 1$, find the value of $a + b$.

(3) When the solution of a quadratic inequality $x^2 + |x| - 6 \leq 0$ is $a \leq x \leq b$, find the value of $a - b$.

#42 Find the range of a real number a so that:

(1) A quadratic equation $x^2 - 2ax + 2a = 0$ will have two different real number solutions.

(2) A quadratic inequality $ax^2 - x + a > 0$ is always true for all real number x.

(3) A quadratic equation $x^2 + 2ax + 3 - 2a = 0$ will have real number solutions and a quadratic equation $x^2 - ax + a = 0$ will have two different complex number solutions.

(4) A quadratic equation $x^2 + 2(a - 2)x + 2a - 1 = 0$ will have two different positive real number solutions.

(5) A quadratic inequality $ax^2 + 4x + a > 3$ will have all real number solutions.

(6) A quadratic inequality $a(2x^2 + 1) \leq (x - 1)^2$ will have no real number solution.

(7) A system $\begin{cases} x^2 - 5x < 0 \\ (x - 4)(x - a) < 0 \end{cases}$ will have only one positive integer.

#43 Solve each inequality.

Let $x = n + \alpha$ (n; integer, $0 \leq \alpha < 1$)

($[x]$ = Maximum value of an integer which is not greater than x .)

(1) $2[x]^2 - 5[x] - 3 < 0$

(2) $[x - 1]^2 + 3[x] - 3 < 0$

#44 Prove each inequality.

    (1) $a^2 + b^2 > ab$ when $a + b = 1$

    (2) $a + b > ab$ when $a + b = 1$

    (3) $(a + b)(b + c)(c + a) \geq 8abc$ when $a > 0$, $b > 0$, and $c > 0$

    (4) $a^3 + b^3 \geq ab(a + b)$ when $a > 0$ and $b > 0$

#45 Prove each inequality.

    Then state for which values the *LHS* (Left Hand Side) equals the *RHS* (Right Hand Side).

    (1) $a^2 + b^2 \geq ab$

    (2) $a + \dfrac{9}{a} \geq 6$

    (3) $\left(a + \dfrac{1}{b}\right)\left(b + \dfrac{1}{a}\right) \geq 4$ when $a > 0$ and $b > 0$

#46 Minimum value.

    (1) When $x > 0$, find minimum value of $x + 2 + \dfrac{4}{x+2}$.

    (2) When $x > 0$ and $y > 0$, find minimum value of $(2x + 8y)\left(\dfrac{2}{x} + \dfrac{2}{y}\right)$.

    (3) When $x^2 + ax - b = 0$ has two real number roots α, β and $x^2 + bx - a = 0$ has two real

        number roots γ, δ for negative numbers a, b, find minimum value of $\dfrac{1}{\alpha} + \dfrac{1}{\beta} + \dfrac{3}{\gamma} + \dfrac{3}{\delta}$.

    (4) When $\sqrt{2x} + \sqrt{3y} = 10$ for positive numbers x and y, find minimum value of $x + y$.

    (5) When $x^2 + y^2 + z^2 + a \geq 6(x + y + z)$ for real numbers $x, y,$ and z,

        find minimum value of a (a is a real number).

    (6) When $ax^2 + 3xy + by^2 \geq 0$ for any real numbers x and y,

        find minimum value of ab.

    (7) When a quadratic equation $x^2 - 3ax + a - 1 = 0$ has two different positive real number

        solutions for real number a, find minimum value of $2a - 3 + \dfrac{2}{a-1}$.

    (8) When $a, b,$ and c are positive real numbers,

        find minimum value of $\dfrac{a+b+c}{a} + \dfrac{a+b+c}{b} + \dfrac{a+b+c}{c}$.

    (9) When a and b are positive numbers,

        find minimum value of $a^2 - 4a + \dfrac{a}{b} + \dfrac{4b}{a}$.

#47 Maximum value.

(1) When $x > 0$, find maximum value of $\dfrac{x}{x^2+x+4}$.

(2) When $ab > 0$ and $a + 2b = 2$ for any real numbers a and b,

find maximum value of c such that $\dfrac{2}{a} + \dfrac{1}{b} \geq c$.

(3) When $\dfrac{x^2}{2} + \dfrac{y^2}{4} + \dfrac{z^2}{9} = 1$ for real numbers $x, y,$ and z,

find maximum value of $x + y + z$.

(4) When $a + 2\sqrt{2}b = 3 - x$ and $a^2 + 4b^2 = 9 - x^2$ for real numbers $a, b,$ and x,

find maximum value of x .

Chapter 4. Elements of Coordinate Geometry and Transformations

4-1 The Coordinate System

1. The Distance Between Two Points

(1) Coordinates

1) Coordinates On a Number Line

A *coordinate* is the number of a point assigned on a number line.

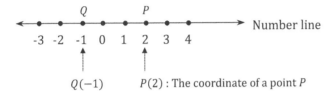

Note : *A number line is one-dimensional, and a coordinate plane is two-dimensional. Thus, an equation with 1 variable is graphed on a number line and an equation with 2 different variables is graphed on a coordinate plane.*

2) Coordinates On a Coordinate Plane

① Coordinate Planes

When two number lines are perpendicular to each other at the zero points of the two lines,

1) their intersection is called the *origin*,

2) the horizontal number line is called the *x-axis*,

3) the vertical number line is called the *y-axis*,

4) the plane formed by the *x*-axis and *y*-axis is called the *coordinate plane*.

The coordinate of origin is (0,0).

Note. *Each point P in a coordinate plane is assigned a pair of numbers.*

② 4 Quadrants

4 *Quadrants* are four rectangular regions in a coordinate plane which are labeled with Roman numerals I, II, III, and IV. The coordinate plane is divided by the two axes into 4 quadrants, which are named in counterclockwise order.

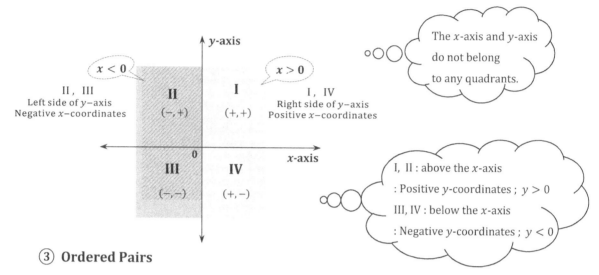

③ Ordered Pairs

Coordinates, which are associated with points on the real line, can be introduced into the plane by ordered pairs or real numbers.

An ordered pair (a, b) of real numbers is the coordinate of the point P of which

   a. a is the number of horizontal units moved from 0.

   b. b is the number of vertical units moved from 0.

Note: $(a, b) = (c, d)$ *if and only if* $a = c$ *and* $b = d$

        $(a, b) \neq (b, a)$ *if* $a \neq b$

④ Symmetric Transformation

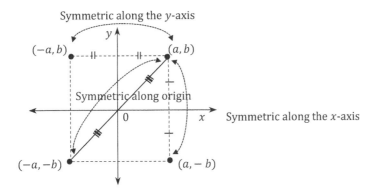

a. Symmetry of a point (a, b) along the x-axis ; $(a, -b)$: Opposite sign of y

b. Symmetry of a point (a, b) along the y-axis ; $(-a,\ b)$: Opposite sign of x

c. Symmetry of a point (a, b) along the origin ; $(-a, -b)$: Opposite signs of x and y

(2) The Pythagorean Theorem

In a right triangle, the square of the length of the hypotenuse is equal to the sum of the squares of the lengths of the legs.

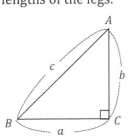

$$a^2 + b^2 = c^2$$

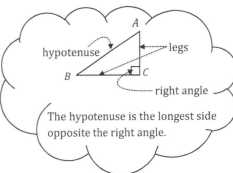

hypotenuse ⟶ legs

right angle

The hypotenuse is the longest side opposite the right angle.

If the lengths of any two sides of a right triangle are given, we can find the length of the other side by the Pythagorean Theorem.

$x^2 = a$, $a \geq 0$

$\Rightarrow x = \pm\sqrt{a}$

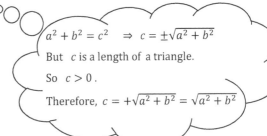

$a^2 + b^2 = c^2 \quad \Rightarrow c = \pm\sqrt{a^2 + b^2}$

But c is a length of a triangle.

So $c > 0$.

Therefore, $c = +\sqrt{a^2 + b^2} = \sqrt{a^2 + b^2}$

(3) The Distance Between Two Points

1) On a Number Line

The distance between two points $A(x_1)$ and $B(x_2)$ is $d(A, B) = |x_1 - x_2| = |x_2 - x_1|$

If $x_1 \leq x_2$, then $d(A, B) = x_2 - x_1$

If $x_1 > x_2$, then $d(A, B) = x_1 - x_2$

$\therefore \quad d(A, B) = |x_1 - x_2| = |x_2 - x_1|$

2) On a Coordinate Plane

① Lengths of Horizontal Segments

If $A(x_1, y)$ and $B(x_2, y)$ are on the same line parallel to the x-axis, then the distance between two points on a plane is $d(A, B) = |x_1 - x_2| = |x_2 - x_1|$.

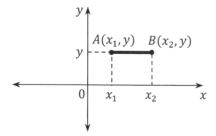

② Lengths of Vertical Segments

If $A(x, y_1)$ and $B(x, y_2)$ are on the same line parallel to the y-axis, then the distance between two points on a plane is $d(A, B) = |y_1 - y_2| = |y_2 - y_1|$.

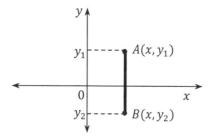

③ In general,

The distance between two points $A(x_1, y_1)$ and $B(x_2, y_2)$ is defined by the Pythagorean theorem. It gives the following formula:

$$d(A, B) = \sqrt{(x_2 - x_1)^2 + (y_2 - y_1)^2}$$

Since $(x_1 - x_2)^2 = (x_2 - x_1)^2$ and $(y_1 - y_2)^2 = (y_2 - y_1)^2$, $d(A, B) = d(B, A)$

Also, the distance between the origin $O(0,0)$ and $A(x_1, y_1)$ is $d(O, A) = \sqrt{x_1^2 + x_2^2}$

Note: (Proof of ③ *)*

Let C be the point (x_2, y_1) *and consider a right triangle* $\triangle ABC$.

Then, $d(A, C) = |x_2 - x_1|$ *and* $d(B, C) = |y_2 - y_1|$.

By the Pythagorean Theorem, $d(A, B)^2 = d(A, C)^2 + d(B, C)^2 = |x_2 - x_1|^2 + |y_2 - y_1|^2$

Since $|x|^2 = x^2$ *for any real number x,* $d(A, B)^2 = (x_2 - x_1)^2 + (y_2 - y_1)^2$

Therefore, $d(A, B) = \sqrt{(x_2 - x_1)^2 + (y_2 - y_1)^2}$

2. Centroid (Center of Gravity)

(1) Midpoint

For any two points $A(x_1, y_1)$ and $B(x_2, y_2)$ in a plane, the *midpoint* of the segment $\overline{AB}$ between

the points is $\boxed{M = \left(\dfrac{x_1 + x_2}{2}, \dfrac{y_1 + y_2}{2} \right)}$

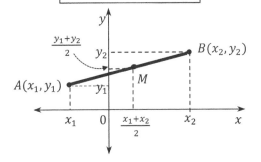
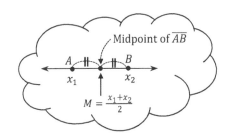

(2) The Median Theorem (Apollonius' Theorem)

The *median* of a triangle is a segment that connecting a vertex and the midpoint of the opposite side. For a triangle $\triangle ABC$, if a point M is the midpoint of $\overline{BC}$, the opposite side of $\angle A$, then the segment $\overline{AM}$ is called the median from a point A to $\overline{BC}$.

$$\boxed{\overline{AB}^2 + \overline{AC}^2 = 2(\overline{AM}^2 + \overline{BM}^2)}$$

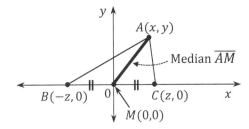

For three points $A(x, y), B(-z, 0),$ and $C(z, 0)$, let M be the midpoint of $\overline{BC}$.

Then, M is the origin $(0, 0)$.

$\overline{AB}^2 + \overline{AC}^2 = \{(x + z)^2 + y^2\} + \{(x - z)^2 + y^2\} = 2(x^2 + y^2 + z^2)$

$\overline{AM}^2 + \overline{BM}^2 = (x^2 + y^2) + z^2 = x^2 + y^2 + z^2$

$\therefore \overline{AB}^2 + \overline{AC}^2 = 2(\overline{AM}^2 + \overline{BM}^2)$

Note: If a rectangle □ABCD and a point P are on the same plane, $\overline{AP}^2 + \overline{CP}^2 = \overline{BP}^2 + \overline{DP}^2$.

∵ *Place the point B on the origin in a coordinate plane. Then we have the following figure:*

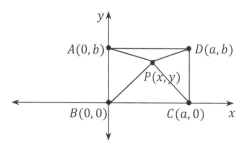

∴ $\overline{AP}^2 + \overline{CP}^2 = \{x^2 + (y-b)^2\} + \{(a-x)^2 + y^2\} = 2x^2 + 2y^2 - 2ax - 2by + a^2 + b^2$

$\overline{BP}^2 + \overline{DP}^2 = \{x^2 + y^2\} + \{(a-x)^2 + (b-y)^2\} = 2x^2 + 2y^2 - 2ax - 2by + a^2 + b^2$

Therefore, $\overline{AP}^2 + \overline{CP}^2 = \overline{BP}^2 + \overline{DP}^2$

(3) Section Formulas for Internal and External Divisions

1) On a Number Line

For two points $A(x_1)$ and $B(x_2)$ on a number line,

the point $P(x)$ which divides the segment $\overline{AB}$ internally in the ratio $\overline{AP} : \overline{PB} = m : n$

$(m > 0, \ n > 0)$ is given by

$$P(x) = \frac{mx_2 + nx_1}{m+n} \quad \text{and}$$

the point $Q(x)$ which divides the segment $\overline{AB}$ externally in the ratio $\overline{AQ} : \overline{QB} = m : n$

$(m > 0, \ n > 0, \ m \neq n)$ is given by

$$Q(x) = \frac{mx_2 - nx_1}{m-n}, \ m \neq n$$

① Section Formula for Internal Division

For two points $A(x_1)$ and $B(x_2)$, consider an interior point $P(x)$ which lies on the segment

$\overline{AB}$ such that $\overline{AP} : \overline{PB} = m : n \ (m > 0, \ n > 0)$

;i.e., $P(x)$ divides $\overline{AB}$ internally in the ratio $m : n$

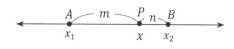   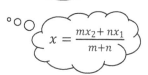

Interior points lie inside of the line segment.

Case 1. When $x_1 < x_2$

Since $x_1 < x < x_2$, $\overline{AP} = x - x_1$ and $\overline{PB} = x_2 - x$

Since $\overline{AP} : \overline{PB} = m : n$, $(x - x_1) : (x_2 - x) = m : n$

$x = \frac{mx_2 + nx_1}{m+n}$

$\therefore\ n(x - x_1) = m(x_2 - x)$

$\therefore\ (m + n)x = mx_2 + nx_1 \qquad \therefore\ x = \dfrac{mx_2 + nx_1}{m+n}$

Case 2. When $x_1 > x_2$

Similarly, we have the same result.

If $m = n$, then $P(x)$ is the midpoint of $\overline{AB}$. $\qquad \therefore\ x = \dfrac{x_1 + x_2}{2}$

Therefore, $P(x) = \dfrac{mx_2 + nx_1}{m+n}$

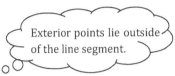
Exterior points lie outside of the line segment.

② **Section Formula for External Division**

For two points $A(x_1)$ and $B(x_2)$, consider an exterior point $Q(x)$ which lies outside of the segment $\overline{AB}$ such that $\overline{AQ} : \overline{QB} = m : n$ $(m > 0,\ n > 0,\ m \neq n)$

; i.e., $Q(x)$ divides $\overline{AB}$ externally in the ratio $m : n$

When $m > n$, When $m < n$,

Case 1. When $m > n,\ m \neq n$

Since $x_1 < x_2 < x$, $\overline{AQ} = x - x_1$ and $\overline{QB} = x - x_2$

Since $\overline{AQ} : \overline{QB} = m : n$, $(x - x_1) : (x - x_2) = m : n$

$x = \dfrac{mx_2 - nx_1}{m-n}$

$\therefore\ n(x - x_1) = m(x - x_2)$

$\therefore\ (m - n)x = mx_2 - nx_1 \qquad \therefore\ x = \dfrac{mx_2 - nx_1}{m-n}$

Case 2. When $m < n,\ m \neq n$

Similarly, we have the same result.

Therefore, $Q(x) = \dfrac{mx_2 - nx_1}{m-n},\ m \neq n$

2) On a Coordinate Plane

For two points $A(x_1, y_1)$ and $B(x_2, y_2)$ on a coordinate plane,

the point $P(x, y)$ which divides the segment $\overline{AB}$ internally in the ratio $\overline{AP} : \overline{PB} = m : n$

$(m > 0,\ n > 0)$ is given by

$$P(x, y) = \left(\frac{mx_2 + nx_1}{m+n},\ \frac{my_2 + ny_1}{m+n} \right) \text{ and}$$

the point $Q(x, y)$ which divides the segment $\overline{AB}$ externally in the ratio $\overline{AQ} : \overline{QB} = m : n$

$(m > 0,\ n > 0,\ m \neq n)$ is given by

$$Q(x, y) = \left(\frac{mx_2 - nx_1}{m-n},\ \frac{my_2 - ny_1}{m-n} \right),\ m \neq n$$

For two points $A(x_1, y_1)$ and $B(x_2, y_2)$, consider an interior point $P(x, y)$ such that $\overline{AP}$: $\overline{PB} = m : n$ $(m > 0,\ n > 0)$;i.e., $P(x, y)$ divides $\overline{AB}$ internally in the ratio $m : n$

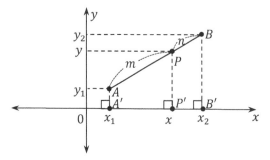

Let A', B', and P' be the three points on x-axis so that $\overline{AA'}$, $\overline{BB'}$, and $\overline{PP'}$ are perpendicular to the x-axis. Then, $\overline{A'P'} : \overline{P'B'} = m : n$

$\therefore$ P' divides $\overline{A'B'}$ in the ratio $\overline{A'P'} : \overline{P'B'} = m : n$

$\therefore$ $x = \dfrac{mx_2 + nx_1}{m+n}$

Similarly, $y = \dfrac{my_2 + ny_1}{m+n}$

Therefore, $P(x, y) = \left(\dfrac{mx_2 + nx_1}{m+n},\ \dfrac{my_2 + ny_1}{m+n} \right)$

If $m = n$, then the midpoint $M(x, y)$ of $\overline{AB}$ is $M(x, y) = \left(\dfrac{x_1 + x_2}{2},\ \dfrac{y_1 + y_2}{2} \right)$

In the same method,

the exterior point $Q(x, y)$ such that $\overline{AQ} : \overline{QB} = m : n$ $(m > 0,\ n > 0,\ m \neq n)$ is $Q(x, y) = \left(\dfrac{mx_2 - nx_1}{m-n},\ \dfrac{my_2 - ny_1}{m-n} \right)$, $m \neq n$

(4) Centroid (Center of gravity)

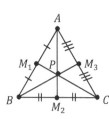

Let M_1, M_2, and M_3 be the three midpoints of $\overline{AB}$, $\overline{BC}$, and $\overline{AC}$, respectively. The medians $\overline{AM_2}$, $\overline{BM_3}$, and $\overline{CM_1}$ are formed by connecting three vertices and the midpoints of the opposite sides. The intersection point P of the three medians is called the *centroid* or *center of gravity* of the triangle.

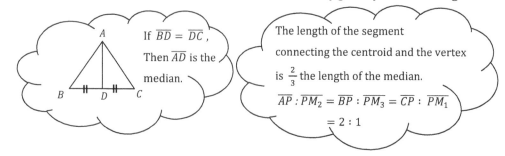

If $\overline{BD} = \overline{DC}$, Then $\overline{AD}$ is the median.

The length of the segment connecting the centroid and the vertex is $\dfrac{2}{3}$ the length of the median. $\overline{AP} : \overline{PM_2} = \overline{BP} : \overline{PM_3} = \overline{CP} : \overline{PM_1}$ $= 2 : 1$

For three points $A(x_1, y_1), B(x_2, y_2),$ and $C(x_3, y_3)$, the centroid $P(x, y)$ of a triangle $\triangle ABC$ is given by

$$P(x, y) = \left(\frac{x_1+x_2+x_3}{3} , \frac{y_1+y_2+y_3}{3} \right)$$

$\because$ Let M be the midpoint of $\overline{BC}$. Then $M = \left(\frac{x_2+x_3}{2} , \frac{y_2+y_3}{2} \right)$

Since $\overline{AP} : \overline{PM} = 2 : 1$,

$$x = \frac{2 \cdot \left(\frac{x_2+x_3}{2} \right) + 1 \cdot x_1}{2+1} = \frac{x_1+x_2+x_3}{3} \quad \text{and} \quad y = \frac{2 \cdot \left(\frac{y_2+y_3}{2} \right) + 1 \cdot y_1}{2+1} = \frac{y_1+y_2+y_3}{3}$$

$\therefore$ The coordinates of the centroid $P(x, y)$ are given by $x = \frac{x_1+x_2+x_3}{3}$, $y = \frac{y_1+y_2+y_3}{3}$

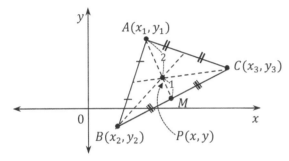

Note:

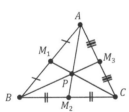

The intersection point P of all three medians is two-thirds of each median from the vertex to the opposite side.

$$\overline{AP} = \frac{2}{3}\overline{AM_2} , \ \overline{BP} = \frac{2}{3}\overline{BM_3}, \text{ and } \overline{CP} = \frac{2}{3}\overline{CM_1}$$

Example   For a triangle $\triangle ABC$ with $\overline{AB} = 10$, $\overline{BC} = 8$, and $\overline{AC} = 6$, let M be the midpoint of $\overline{BC}$ and P be the centroid of $\triangle ABC$. Find the length of $\overline{PM}$.

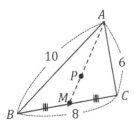

$\because$ Since M is the midpoint of $\overline{BC}$, $\overline{BM} = \frac{1}{2}\overline{BC} = 4$

By the median theorem, $\overline{AB}^2 + \overline{AC}^2 = 2(\overline{AM}^2 + \overline{BM}^2)$

$\therefore 100 + 36 = 2(\overline{AM}^2 + 16)$ $\therefore \overline{AM}^2 = 52$ $\therefore \overline{AM} = 2\sqrt{13}$

Since $\overline{AP} : \overline{PM} = 2 : 1$, $\overline{PM} = \frac{1}{3}\overline{AM} = \frac{1}{3} \cdot 2\sqrt{13} = \frac{2\sqrt{13}}{3}$

3. Equations of Lines

(1) The Slope of a Line

To find an equation of a line, consider the slope of the line.

The *slope* of a line measures the steepness and direction of the line. The absolute value of the slope determines a line's steepness. The sign of the slope determines a line's direction.

1) The Concept of Slope

A line which is not parallel to a coordinate axis may rise from lower left to upper right, or it may fall from upper left to lower right.

For a line passing through the given points (x_1, y_1) and (x_2, y_2), where $x_1 \neq x_2$, the positive number $y_2 - y_1$ is called the *rise*, and the positive number $x_2 - x_1$ is called the *run*. We define the *slope m* of the line by

$$m = \frac{y_2 - y_1}{x_2 - x_1} \quad \text{or} \quad m = \frac{y_1 - y_2}{x_1 - x_2}$$

Subtract the coordinates in the same order in both the numerator and denominator.

Note : $\quad Slope = \dfrac{rise}{run} = \dfrac{change\ in\ y}{change\ in\ x}$

$= Ratio\ of\ the\ vertical\ change\ of\ a\ line\ and\ the\ horizontal\ change\ of\ a\ line.$

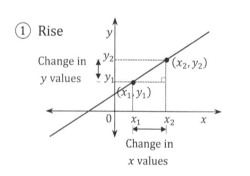 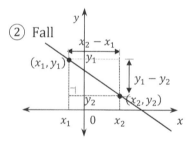

① Rise

Change in y values

Change in x values

② Fall

Note 1: For any horizontal line $y = k$, the slope is 0.

$(\because All\ points\ (coordinates)\ on\ the\ line\ y = k\ have\ the\ same\ y\ value,\ k.$

$That\ is,\ there\ is\ no\ change\ in\ y. \quad \therefore \quad slope = \dfrac{0}{change\ in\ x} = 0\ )$

Note 2: For any vertical line $x = k$, the slope does not exist.

$(\because All\ points\ (coordinates)\ on\ the\ line\ x = k\ have\ the\ same\ x\ value,\ k.$

$That\ is,\ there\ is\ no\ change\ in\ x. \quad \therefore \quad slope = \dfrac{change\ in\ y}{0}$

$This\ equation\ is\ undefined\ because\ a\ denominator\ can't\ be\ divided\ by\ zero.$

$Therefore, vertical\ lines\ have\ no\ slope\ ;i.e.,\ slope\ is\ undefined.\ )$

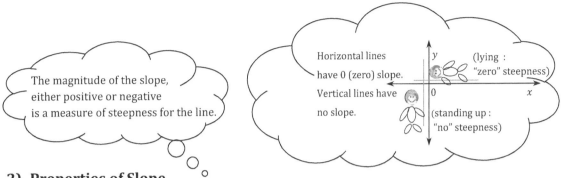

The magnitude of the slope, either positive or negative is a measure of steepness for the line.

Horizontal lines have 0 (zero) slope. (lying : "zero" steepness)

Vertical lines have no slope. (standing up : "no" steepness)

2) Properties of Slope

① For two lines $y = ax + b, a \neq 0$ and $y = cx + d, c \neq 0$,

the line $y = ax + b$ is steeper than the line $y = cx + d$ if $|a| > |c|$.

The larger the absolute value of the slope, the steeper the line will be.

Example

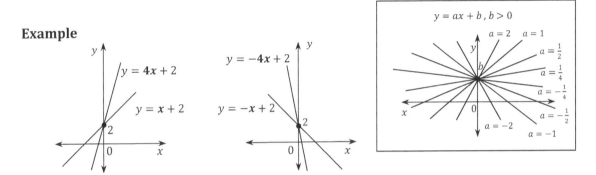

$y = 4x + 2$

$y = x + 2$

$y = -4x + 2$

$y = -x + 2$

$y = ax + b, b > 0$

$a = 2$ $a = 1$

$a = \frac{1}{2}$

$a = \frac{1}{4}$

$a = -\frac{1}{4}$

$a = -\frac{1}{2}$

$a = -2$

$a = -1$

② For a linear function $y = ax + b, a \neq 0$, the slope is a.

i) $a > 0$ (Positive slope)

: The line goes uphill from left to right. (The line rises to the right.)

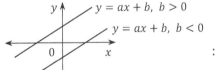

$y = ax + b, \ b > 0$

$y = ax + b, \ b < 0$

: x values increase $\Rightarrow$ y values increase

ii) $a < 0$ (Negative slope)

: The line goes downhill from left to right. (The line falls to the right.)

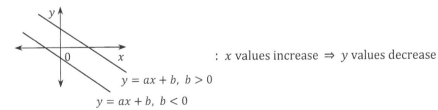

: x values increase $\Rightarrow$ y values decrease

$y = ax + b, \ b > 0$

$y = ax + b, \ b < 0$

(2) Methods of Creating the Equations of Lines

Note: **_Intercepts_**

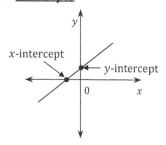

① x-intercept is the x-coordinate of the point where the line crosses (intersects) the x-axis.

x-intercept is the value of x when $y = 0$.

For example, $y = 2x + 3 \underset{y=0}{\Longrightarrow} 0 = 2x + 3 \Rightarrow x = -\frac{3}{2}$

$\therefore$ x-intercept is $-\frac{3}{2}$.

② y-intercept is the y-coordinate of the point where the line crosses (intersects) the y-axis.

y-intercept is the value of y when $x = 0$.

For example, $y = 2x + 3 \underset{x=0}{\Longrightarrow} y = 3$

$\therefore$ y-intercept is 3.

x-intercept is on the x-axis and y-intercept is on the y-axis.

1) A line which has slope m and y-intercept b (Slope-Intercept Form)

The equation of the line is $\boxed{y = mx + b}$

x-coefficient is slope m.

2) A line which has slope m and passes through a point (x_1, y_1) (Point-Slope Form)

The equation of the line is $\boxed{y - y_1 = m(x - x_1)}$

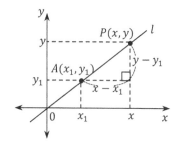

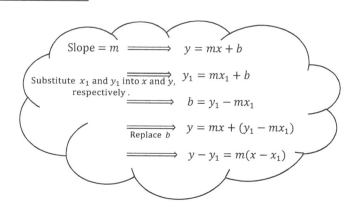

Slope $= m \Longrightarrow y = mx + b$

Substitute x_1 and y_1 into x and y, respectively. $\Longrightarrow y_1 = mx_1 + b$

$\Longrightarrow b = y_1 - mx_1$

Replace b $\Longrightarrow y = mx + (y_1 - mx_1)$

$\Longrightarrow y - y_1 = m(x - x_1)$

Let $P(x, y)$ be a point on a line l.

Then, the slope of $\overline{AP}$ is $m = \frac{y - y_1}{x - x_1}$.

Thus, the equation of the line l is $y - y_1 = m(x - x_1)$.

Therefore, any point $P(x, y)$ on the line l satisfies the equation.

Conversely, any point $P(x, y)$ that satisfies the equation is on the line l.

The equation of the line which passes through a point $A(x_1, y_1)$ and parallels to y-axis is $x = x_1$.

Hence, a line which has slope m and passes through a point $A(x_1, y_1)$ is $y - y_1 = m(x - x_1)$.

3) A line which passes through two different points (x_1, y_1) and (x_2, y_2), $x_1 \neq x_2$

⇒ ① Find slope m :

$$m = \frac{y_2 - y_1}{x_2 - x_1} \quad \text{or} \quad m = \frac{y_1 - y_2}{x_1 - x_2}$$

② Use the slope m and one of the two points. Using point-slope form,

$$y - y_1 = m(x - x_1) \quad \text{or} \quad y - y_2 = m(x - x_2)$$

Therefore, the equation of the line is given by

> When $x_1 \neq x_2$, $y - y_1 = \frac{y_2 - y_1}{x_2 - x_1}(x - x_1)$
>
> When $x_1 = x_2$, $x = x_1$

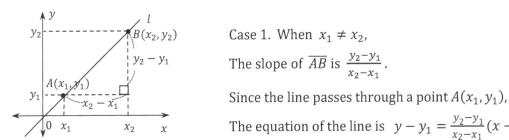

Case 1. When $x_1 \neq x_2$,

The slope of $\overline{AB}$ is $\frac{y_2 - y_1}{x_2 - x_1}$.

Since the line passes through a point $A(x_1, y_1)$,

The equation of the line is $y - y_1 = \frac{y_2 - y_1}{x_2 - x_1}(x - x_1)$.

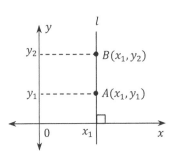

Case 2. When $x_1 = x_2$,

$\overline{AB}$ is parallel to y-axis and

passes through a point $A(x_1, y_1)$.

The equation of the line is $x = x_1$.

4) A line which has an x-intercept p and y-intercept q

: The same as a line which passes through two points $(p, 0)$ and $(0, q)$

⇒ ① Find slope m :

$$m = \frac{q - 0}{0 - p} \quad \text{or} \quad m = \frac{0 - q}{p - 0}$$

② Use the line which has slope m and y-intercept q :

$$y = mx + q$$

Therefore, the equation of the line is given by

> $\frac{x}{p} + \frac{y}{q} = 1 \quad (pq \neq 0)$

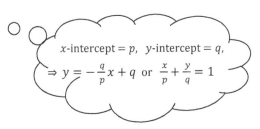

x-intercept $= p$, y-intercept $= q$,

⇒ $y = -\frac{q}{p}x + q$ or $\frac{x}{p} + \frac{y}{q} = 1$

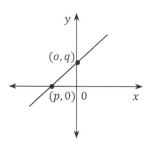

Using the equation of the line which has slope m and y-intercept q, we have

$$y = mx + q = -\frac{q}{p}x + q$$

$$\therefore \ \frac{q}{p}x + y = q$$

$$\therefore \ \frac{x}{p} + \frac{y}{q} = 1 \quad (pq \neq 0)$$

Example   Two lines $\frac{x}{3} + \frac{y}{4} = 1$ and $y = mx$ intersect at a point P.

When the ratio of the areas of $\triangle OAP$ and $\triangle OBP$ is $2:3$, find the value of m.

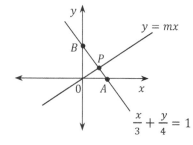

$\because$ The x-axis and y-axis of the line $\frac{x}{3} + \frac{y}{4} = 1$

are 3 and 4, respectively.

$$\therefore \ A = A(3,0) \ \text{and} \ B = B(0,4)$$

Since (The area of $\triangle OAP$) : (The area of $\triangle OBP$) = $2:3$,

the point P divides the segment $\overline{AB}$ internally in the ratio $2:3$ (i.e., $\overline{AP} : \overline{PB} = 2:3$)

$\because$ The coordinates of point P are

$$x = \frac{2 \cdot 0 + 3 \cdot 3}{2+3} = \frac{9}{5}, \quad y = \frac{2 \cdot 4 + 3 \cdot 0}{2+3} = \frac{8}{5}$$

Since the slope of the line which passes through the origin $(0,0)$ and $(\frac{9}{5}, \frac{8}{5})$ is

$$\frac{\frac{8}{5} - 0}{\frac{9}{5} - 0} = \frac{8}{9}, \ \text{we have the slope} \ m = \frac{8}{9}.$$

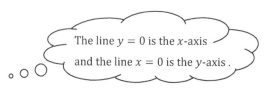

The line $y = 0$ is the x-axis and the line $x = 0$ is the y-axis .

(3) Horizontal and Vertical Lines

1) A *horizontal line* passes through a point on the y-axis and is parallel to the x-axis.

For any constant k, $y = k$ intersects the y-axis at k and the value of x doesn't matter at all.

2) A *vertical line* passes through a point on the x-axis and is parallel to the y-axis.

For any constant k, $x = k$ intersects the x-axis at k and the value of y doesn't matter at all.

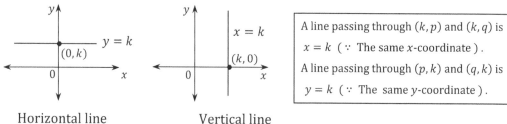

	A line passing through (k,p) and (k,q) is
	$x = k$ ($\because$ The same x-coordinate) .
	A line passing through (p,k) and (q,k) is
	$y = k$ ($\because$ The same y-coordinate) .

Horizontal line   Vertical line

(4) Lines and Their Equations

The two equations of lines $y - y_1 = m(x - x_1)$ and $x = x_1$ are represented by

$mx - y - mx_1 + y_1 = 0$ and $x - x_1 = 0$, respectively.

Therefore, an equation of a line is expressed as a linear equation for x and y :

$ax + by + c = 0 \ (a \neq 0 \text{ or } b \neq 0)$

Conversely, a line equation for x and y : $ax + by + c = 0 \ (a \neq 0 \text{ or } b \neq 0)$ is expressed as

equations of lines :   i) When $b \neq 0$, $y = -\dfrac{a}{b}x - \dfrac{c}{b}$

                       ii) When $b = 0$, $x = -\dfrac{c}{a}$

Equations of Lines

> ❶ Vertical line : $x = k$
>
> ❷ Horizontal line : $y = k$
>
> ❸ General linear equation : $ax + by + c = 0$
>
> ❹ Slope-Intercept form : $y = mx + b$
>
> ❺ Point-Slope form : $y - y_1 = m(x - x_1)$

4. Relations of Two Lines

For any two different lines $y = m_1x + a$ and $y = m_2x + b$, parallel and perpendicular lines

are identified by comparing their slopes, m_1 and m_2.

(1) Parallel Lines

For any two different lines $l = mx + n$ and $l' = m'x + n'$,

if the two lines are parallel ($l /\!/ l'$), then the slopes, m and m', are the same and

y-intercepts, n and n', are different ; i.e., $m = m'$ and $n \neq n'$

Conversely, if $m = m'$ and $n \neq n'$, then the two lines are parallel ($l /\!/ l'$).

Therefore, | two lines $l = mx + n$ and $l' = m'x + n'$ are parallel ($l /\!/ l'$)

$\underset{\text{if and only if}}{\Longleftrightarrow}$ $m = m'$ and $n \neq n'$

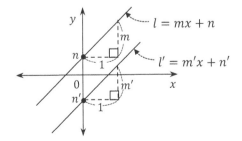

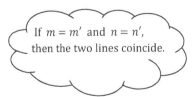

If $m = m'$ and $n = n'$,
then the two lines coincide.

(2) Perpendicular Lines

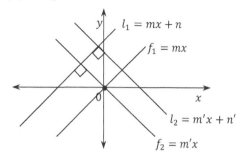

For two lines $l_1 = mx + n$ and $l_2 = m'x + n'$, if l_1 and l_2 are perpendicular ($l_1 \perp l_2$),

then the lines f_1 and f_2 such that $l_1 /\!/ f_1$, $l_2 /\!/ f_2$, respectively, and pass through the origin

$(0, 0)$ are $f_1 = mx$, $f_2 = m'x$, and $f_1 \perp f_2$.

Since the y-intercepts n and n' of l_1 and l_2 don't matter whether the lines are perpendicular,

we consider f_1 and f_2 instead of l_1 and l_2.

Now, consider lines $f_1 = mx$, $f_2 = m'x$, and $x = 1$.

For the intersection points $A(1, m)$ and $B(1, m')$, $\triangle OAC$ and $\triangle OBD$ are similar.

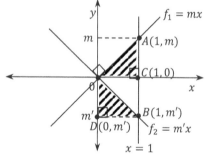

$\because \ m\angle OCA = m\angle ODB = 90°$

Since $m\angle DOB = 90° - m\angle BOC = m\angle AOC$,

$\triangle OAC \sim \triangle OBD$

Since $\overline{OC} : \overline{AC} = \overline{OD} : \overline{BD}$, $1 : m = -m' : 1$

$\therefore \ mm' = -1$

> Since $m' < 0$,
> the length of $\overline{OD}$ is $-m'$.

Conversely, if $mm' = -1$, then $1 : m = -m' : 1$ and $\overline{OC} : \overline{AC} = \overline{OD} : \overline{BD}$

$\therefore \ \triangle OAC \sim \triangle OBD$

$\therefore$ The two lines $f_1 = mx$ and $f_2 = m'x$ are perpendicular.

Therefore, two lines $l_1 = mx + n$ and $l_2 = m'x + n'$ are perpendicular

$\xleftrightarrow[\text{if and only if}]{}$ $mm' = -1$ (Slopes are negative reciprocals of each other.)

Example Find an equation of a line l_1 which passes through a point $(-1, 2)$ and is

perpendicular to a line $l_2 = -\dfrac{1}{2}x + 1$.

Let m be the slope of l_1. Since the slope of l_2 is $-\dfrac{1}{2}$, $m \cdot \left(-\dfrac{1}{2}\right) = -1$ $\therefore \ m = 2$

Since l_1 passes through $(-1, 2)$, $l_1 - 2 = 2(x + 1)$.

Therefore, the equation of the line is $l_1 = 2x + 4$.

(3) Relations of Two Lines

The relation of two lines or two linear equations in a coordinate plane is one of the following:

Relations	Two Lines $l_1 = mx + n,\ l_2 = m'x + n'$	Two Linear Equations $ax + by + c = 0,\ a'x + b'y + c' = 0$
Intersect at one point	$m \neq m'$ (Different slopes)	$\dfrac{a}{b} \neq \dfrac{a\prime}{b\prime}$
Parallel	$m = m',\ \ n \neq n'$ (The same slope, Different y-intercepts)	$\dfrac{a}{a\prime} = \dfrac{b}{b\prime} \neq \dfrac{c}{c\prime}$
Perpendicular	$mm' = -1$	$aa' + bb' = 0$
Coincide	$m = m',\ \ n = n'$ (The same slope, The same y-intercept)	$\dfrac{a}{a\prime} = \dfrac{b}{b\prime} = \dfrac{c}{c\prime}$

5. A Line Which Passes Through the Intersection Point of Other Two Lines

For two lines : $x + y - 2 = 0$ $\cdots\cdots$ ①

$\qquad\qquad x - 2y + 3 = 0$ $\cdots\cdots$ ②,

and any real number k, consider a line $(x + y - 2) + k(x - 2y + 3) = 0$ $\cdots\cdots$ ③

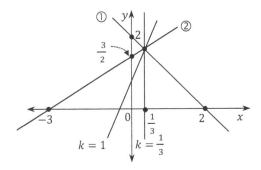

If a point (x_1, y_1) is the intersection point of ① and ②,

then $x_1 + y_1 - 2 = 0$ and $x_1 - 2y_1 + 3 = 0$

∴ ③ passes through the point (x_1, y_1) for any k.

Conversely, all lines (except ②) which pass through

the intersection point of ① and ② can be expressed in

the form of ③.

Therefore,

> the line $(ax + by + c) + k(a'x + b'y + c') = 0$
>
> passes through the intersection point of the two lines
>
> $ax + by + c = 0$ and $a'x + b'y + c' = 0$ for any real number k.

Example   Find the equation of a line which passes through a point $A(1, 0)$ and

the intersection point of two lines : $3x + 2y + 1 = 0$ and $x - y + 3 = 0$.

For any real number k, $(3x + 2y + 1) + k(x - y + 3) = 0 \cdots\cdots$ ①

Since ① passes through $A(1, 0)$, $3 + 1 + k(1 - 0 + 3) = 0$ $\quad$ ∴ $k = -1$

Substituting $k = -1$ into ①, $2x + 3y - 2 = 0$

Therefore, the equation of a line is $2x + 3y - 2 = 0$.

6. Distance Between a Point and a Line

The distance between a point $A(x_1, y_1)$ and a line $ax + by + c = 0$ is defined by $\frac{|ax_1 + by_1 + c|}{\sqrt{a^2 + b^2}}$.

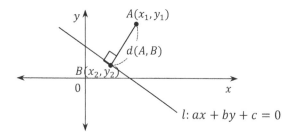

For a point $B(x_2, y_2)$ on the line $ax + by + c = 0$ so that the line segment $\overline{AB}$ and the line are perpendicular, $d(A, B)$ is the distance between a point A and a line $ax + by + c = 0$ $\cdots\cdots$ ①

(1) Case 1. $a \neq 0,\ b \neq 0$

Since the slope of the line is $-\frac{a}{b}$, the slope of line $\overleftrightarrow{AB}$ is $\frac{b}{a}$.

∴ The equation of the line $\overleftrightarrow{AB}$ is $y - y_1 = \frac{b}{a}(x - x_1)$ $\cdots\cdots$ ②

Since both ① and ② pass through $B(x_2, y_2)$, we have

$ax_2 + by_2 + c = 0$ $\cdots\cdots$ ③ and $y_2 - y_1 = \frac{b}{a}(x_2 - x_1)$ $\cdots\cdots$ ④

Putting $-ax_1 - by_1$ on both sides of ③, $a(x_2 - x_1) + b(y_2 - y_1) = -(ax_1 + by_1 + c) = 0$
$\cdots\cdots$ ⑤

From ④, $b(x_2 - x_1) = a(y_2 - y_1)$ $\cdots\cdots$ ⑥

⑤ $\times\ ab$: $a^2 b(x_2 - x_1) + ab^2(y_2 - y_1) = -ab(ax_1 + by_1 + c) = 0$

By ⑥, $a^2\{a(y_2 - y_1)\} + ab^2(y_2 - y_1) = -ab(ax_1 + by_1 + c) = 0$

$a(y_2 - y_1)(a^2 + b^2) = -ab(ax_1 + by_1 + c) = 0$

∴ $y_2 - y_1 = \frac{-b(ax_1 + by_1 + c)}{a^2 + b^2}$ $\quad (\because a \neq 0)$

Similarly, $x_2 - x_1 = \frac{-a(ax_1 + by_1 + c)}{a^2 + b^2}$

Since $d(A, B) = \sqrt{(x_2 - x_1)^2 + (y_2 - y_1)^2}$,

$$d(A, B) = \sqrt{\left(\frac{-a(ax_1 + by_1 + c)}{a^2 + b^2}\right)^2 + \left(\frac{-b(ax_1 + by_1 + c)}{a^2 + b^2}\right)^2}$$

$$= \sqrt{\frac{(a^2 + b^2)(ax_1 + by_1 + c)^2}{(a^2 + b^2)^2}} = \sqrt{\frac{(ax_1 + by_1 + c)^2}{a^2 + b^2}} = \frac{\sqrt{(ax_1 + by_1 + c)^2}}{\sqrt{a^2 + b^2}} = \frac{|ax_1 + by_1 + c|}{\sqrt{a^2 + b^2}}$$

(2) Case 2. $(a = 0,\ b \neq 0)$ or $(a \neq 0,\ b = 0)$

$a = 0,\ b \neq 0\ \Rightarrow$ The line $ax + by + c = 0$ is parallel to x-axis.

$a \neq 0,\ b = 0\ \Rightarrow$ The line $ax + by + c = 0$ is parallel to y-axis.

Therefore, the distance formula between a point $A(x_1, y_1)$ and a line $ax + by + c = 0$ is

$$d = \frac{|ax_1 + by_1 + c|}{\sqrt{a^2 + b^2}}$$

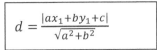

Note: If $l_1 /\!/ l_2$ where $l_1 : ax + by + c = 0$ and $l_2 : ax + by + c' = 0$,

then the distance d between l_1 and l_2 is $d = \frac{|c - c'|}{\sqrt{a^2 + b^2}}$

$\because$ Let a point (x_1, y_1) be on the line l_2. Then $ax_1 + by_1 + c' = 0$

Thus, the didtance between a point (x_1, y_1) and a line l_1 is $d = \frac{|ax_1 + by_1 + c|}{\sqrt{a^2 + b^2}} = \frac{|c - c'|}{\sqrt{a^2 + b^2}}$.

Example Determine the distance between the two parallel lines :

$$l_1 : \ x - 2y - 4 = 0 \ \text{ and } \ l_2 : x - 2y + 6 = 0$$

Since $l_1 /\!/ l_2$, the distance between the two lines is the same as the distance between a point

$(4, 0)$ on a line l_1 and a line l_2.

Thus, the distance d is $d = \frac{|1 \cdot 4 + (-2) \cdot 0 + 6|}{\sqrt{(1)^2 + (-2)^2}} = \frac{|10|}{\sqrt{5}} = \frac{10\sqrt{5}}{5} = 2\sqrt{5}$

OR, $d = \frac{|c - c'|}{\sqrt{a^2 + b^2}} = \frac{|-4 - 6|}{\sqrt{(1)^2 + (-2)^2}} = \frac{|-10|}{\sqrt{5}} = \frac{10}{\sqrt{5}} = \frac{10\sqrt{5}}{5} = 2\sqrt{5}$

4-2 Equations of Circles

1. Equations of Circles

A *circle* is the set of all points in a plane at the same distance from a fixed point, called the *center* of the circle.

A *radius* is a segment connecting the center of a circle and a point on the circle.

A *chord* is a segment connecting two points on a circle.

Note : **Bisecting Chords**

A perpendicular segment from the center of a circle to a chord bisects the chord.

Chord : $\overline{AB}$

($\because$ Since $\overline{PA}$ and $\overline{PB}$ are radii of a circle, $\overline{PA} \cong \overline{PB}$.

Then $\triangle PAB$ is isosceles. So $\angle PAB \cong \angle PBA$.

Since $m \angle PQA = m \angle PQB = 90°$, $\triangle PAQ \cong \triangle PBQ$ by SAA Postulate or HL Theorem.

So $\overline{AQ} \cong \overline{BQ}$ Thus, Q is the midpoint of $\overline{AB}$. Therefore, $\overline{PQ}$ bisects $\overline{AB}$.)

A *tangent* to a circle is a line which intersects the circle at exactly one point.

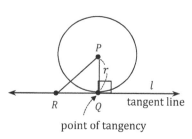

tangent line

point of tangency

Note 1. *If a line is perpendicular to a radius drawn to the point of tangency, then the line is tangent to the circle.*
$\overline{PQ} \perp \overline{RQ}$

Every tangent to a circle is perpendicular to the radius drawn to the point of tangency.

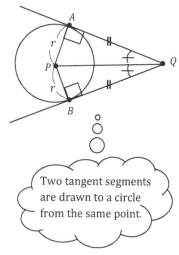

Note 2. *Let $\overline{QA}$ and $\overline{QB}$ be tangents to a circle at points A and B, respectively. Then $\overline{QA} \cong \overline{QB}$ and $\angle AQP \cong \angle BQP$.*

(∵ *Since $\overline{PA}$ and $\overline{PB}$ are radii of a circle, $\overline{PA} \cong \overline{PB}$*

Since every tangent to a circle is perpendicular to the radius drawn to a point of tangency, $\overline{PA} \perp \overline{QA}$ and $\overline{PB} \perp \overline{QB}$

i.e. $m \angle PAQ = m \angle PBQ = 90°$

Since $\overline{PQ}$ is the common hypotenuse of $\triangle APQ$ and $\triangle BPQ$, $\triangle APQ \cong \triangle BPQ$ by HL Theorem.

Therefore, $\overline{QA} \cong \overline{QB}$ and $\angle AQP \cong \angle BQP$.)

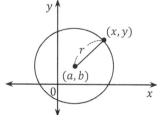

Two tangent segments are drawn to a circle from the same point.

(1) Standard Form of the Equation of a Circle

The distance r between the center $(0, 0)$ and any point (x, y) on the circle can be obtained by the distance formula.

∴ $r = \sqrt{(x-0)^2 + (y-0)^2}$

Squaring both sides, $(x-0)^2 + (y-0)^2 = r^2$

Therefore, $x^2 + y^2 = r^2$

The distance r between the center (a, b) and any point (x, y)

is $r = \sqrt{(x-a)^2 + (y-b)^2}$

∴ $(x-a)^2 + (y-b)^2 = r^2$ ⋯⋯ ①

Conversely, the distance between a point (x, y) satisfying ① and the center (a, b) is r.

∴ The point (x, y) is on the circle.

∴ ① is the equation of a circle with center (a, b) and the length r of radius.

Standard Form of the Equation of a Circle

With center at origin $(0, 0)$ and radius $(r > 0)$: $x^2 + y^2 = r^2$

With center at (a, b) and radius $(r > 0)$: $(x - a)^2 + (y - b)^2 = r^2$

Example Write the standard form of the equation of the circle that passes through the two points $A(-2, 2)$ and $B(4, 4)$ where $\overline{AB}$ is the diameter of the circle.

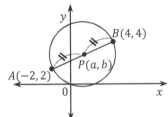

Let $P(a, b)$ be the center of the circle.

Then, $a = \dfrac{-2 + 4}{2} = 1$ and $b = \dfrac{2 + 4}{2} = 3$

$\therefore$ The center of the circle is $P(1, 3)$.

$\therefore$ The length of radius is $d(A, P) = \sqrt{(1 - (-2))^2 + (3 - 2)^2} = \sqrt{9 + 1} = \sqrt{10}$

Therefore, the equation of the circle is $(x - 1)^2 + (y - 3)^2 = 10$

Equations of Circles Which Intersect Axes

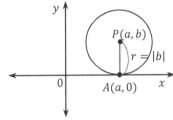

① If a circle with center $P(a, b)$ intersects at only one point on x-axis, then the equation of the circle is

$(x - a)^2 + (y - b)^2 = b^2$

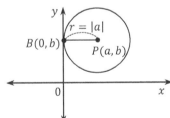

② If a circle with center $P(a, b)$ intersects at only one point on y-axis, then the equation of the circle is

$(x - a)^2 + (y - b)^2 = a^2$

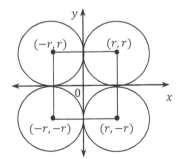

③ If a circle with radius r intersects at only one point on both x-axis and y-axis, then the equation of the circle is

$(x - r)^2 + (y - r)^2 = r^2$ in Quadrant I

$(x + r)^2 + (y - r)^2 = r^2$ in Quadrant II

$(x + r)^2 + (y + r)^2 = r^2$ in Quadrant III

$(x - r)^2 + (y + r)^2 = r^2$ in Quadrant IV

(2) General Form of the Equation of a Circle

The equation of a circle with the center (a, b) and radius r is $(x - a)^2 + (y - b)^2 = r^2$.

$(x - a)^2 + (y - b)^2 = r^2 \Rightarrow x^2 - 2ax + a^2 + y^2 - 2by + b^2 = r^2$

$\Rightarrow x^2 + y^2 - 2ax - 2by + a^2 + b^2 - r^2 = 0$

∴ The equation of a circle is expressed as

$x^2 + y^2 + Ax + By + C = 0$ where A, B, and C are real numbers.

Conversely, $x^2 + y^2 + Ax + By + C = 0 \cdots\cdots$ ①

$$\Rightarrow \left(x + \frac{A}{2}\right)^2 + \left(y + \frac{B}{2}\right)^2 - \frac{A^2}{4} - \frac{B^2}{4} + C = 0$$

$$\Rightarrow \left(x + \frac{A}{2}\right)^2 + \left(y + \frac{B}{2}\right)^2 = \frac{A^2 + B^2 - 4C}{4}$$

If $A^2 + B^2 - 4C > 0$, then ① is the equation of circle with the center $\left(-\frac{A}{2}, -\frac{B}{2}\right)$ and

radius $\sqrt{\frac{A^2 + B^2 - 4C}{4}}$.

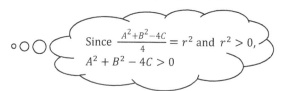

Since $\frac{A^2 + B^2 - 4C}{4} = r^2$ and $r^2 > 0$,

$A^2 + B^2 - 4C > 0$

Note: If $A^2 + B^2 - 4C = 0$, then ① is a point.

If $A^2 + B^2 - 4C < 0$, then there is no point (x, y) so that ① is true.

General Form of the Equation of a Circle

$x^2 + y^2 + Ax + By + C = 0$ where $A^2 + B^2 - 4C > 0$
That is, $\left(x + \frac{A}{2}\right)^2 + \left(y + \frac{B}{2}\right)^2 = \frac{A^2 + B^2 - 4C}{4}$ where $A^2 + B^2 - 4C > 0$

Example Find the center and the length of radius of the equation of a circle:

$$x^2 + y^2 - 4x + 2y - 4 = 0$$

$x^2 + y^2 - 4x + 2y - 4 = 0 \Rightarrow (x - 2)^2 + (y + 1)^2 - 2^2 - 1^2 - 4 = 0$

$$\Rightarrow (x - 2)^2 + (y + 1)^2 = 3^2$$

∴ The center of the circle is $(2, -1)$ and the length of the radius is 3.

Example Determine the equation of a circle which passes through the three points

$(-1, 0)$, $(2, 1)$, and $(7, -4)$.

Consider an equation of a circle $x^2 + y^2 + Ax + By + C = 0$

Substituting the points $(-1, 0)$, $(2, 1)$, and $(7, -4)$, we have

$1 - A + C = 0$; $A - C = 1$ ······①

$4 + 1 + 2A + B + C = 0$; $2A + B + C = -5$ ······②

$49 + 16 + 7A - 4B + C = 0$; $7A - 4B + C = -65$ ······③

From ①: $C = A - 1$

②−③: $-5A + 5B = 60$; $-A + B = 12$; $B = A + 12$

From ③: $7A - 4(A + 12) + (A - 1) = -65$; $4A = -16$; $A = -4$

∴ $A = -4$, $B = 8$, and $C = -5$

Therefore, the equation of the circle is $x^2 + y^2 - 4x + 8y - 5 = 0$.

2. Equation of a Circle Which Pass Through the Intersection Points of Two Circles with a Common Chord

Consider two circles C_1 and C_2.

C_1: $x^2 + y^2 = 1$ ······①

C_2: $(x - 1)^2 + y^2 = 1$ ······②

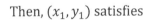

$$k = -\frac{1}{2}$$
$$: (x + 1)^2 + y^2 = 3$$

$k = 3$
$$: (x - \frac{3}{4})^2 + y^2 = \frac{13}{16}$$

$k = -1$; $x = \frac{1}{2}$

Let (x_1, y_1) be the intersection point of ① and ②.

Then, (x_1, y_1) satisfies

C_3: $x^2 + y^2 - 1 + k\{(x - 1)^2 + y^2 - 1\} = 0$ ······③ for any real number k.

Also, ③ passes through the intersection points of C_1 and C_2.

If $k = -1$, then ③ is a line which passes through the intersection points of C_1 and C_2.

If $k \neq -1$, then ③ is a circle which passes through the intersection points of C_1 and C_2.

Therefore, for two circles, $x^2 + y^2 + Ax + By + C = 0$ and $x^2 + y^2 + A'x + B'y + C' = 0$,

which intersect at two points,

---

(1) The equation of a circle which passes through the intersection points is

$\quad x^2 + y^2 + Ax + By + C + k(x^2 + y^2 + A'x + B'y + C') = 0$ where $k \neq -1$

(2) The equation of a line (common chord) which passes through the intersection points is

$\quad (A - A')x + (B - B')y + (C - C') = 0$

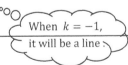

When $k = -1$, it will be a line

Example For a line which passes through the intersection points of two circles:

$\quad x^2 + y^2 = 9$ and $x^2 + y^2 - 3x + ay = 0$ $(a \neq 0)$,

the line is perpendicular to a line $y = x + 2$. Find the value of a.

The equation of the line is $(x^2 + y^2 - 9) - (x^2 + y^2 - 3x + ay) = 0.$ 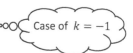 Case of $k = -1$

$\therefore 3x - ay - 9 = 0$ $\quad \therefore y = \dfrac{3x}{a} - \dfrac{9}{a} \cdots\cdots$ ①

Since ① is perpendicular to $y = x + 2$, $\dfrac{3}{a} \times 1 = -1$  $mm' = -1$

Therefore, $a = -3$.

3. Relationships Between a Circle and a Line

(1) Location of a Circle and a Line

For a circle $x^2 + y^2 = r^2$, $r > 0$, and a line $y = mx + n$,

consider a quadratic equation $x^2 + (mx + n)^2 = r^2 \cdots\cdots$ ①

Let D be the discriminant of the equation ① and d be the distance from the center of the

circle to the line. Then, it gives the following results:

Relationship Between a Circle and a Line	Two Different Intersection Points	Only One Intersection Point	No Intersection Point
	$d < r$	$d = r$	$d > r$
The Discriminant D	$D > 0$	$D = 0$	$D < 0$

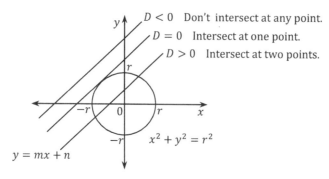

$D < 0$ Don't intersect at any point.
$D = 0$ Intersect at one point.
$D > 0$ Intersect at two points.

$y = mx + n$
$x^2 + y^2 = r^2$

Example For a circle $x^2 + y^2 = 1$ and a line $y = -2x + k$,

determine the relationships of their locations depending on the value of k.

Consider a quadratic equation $x^2 + (-2x + k)^2 = 1$.

Then, $5x^2 - 4kx + k^2 - 1 = 0$.

$\therefore D = (-4k)^2 - 4 \cdot 5 \cdot (k^2 - 1) = 16k^2 - 20k^2 + 20 = -4k^2 + 20 = -4(k^2 - 5)$

i) If $D > 0$, then, $k^2 - 5 < 0$ $\therefore$ $-\sqrt{5} < k < \sqrt{5}$

   $\therefore$ When $-\sqrt{5} < k < \sqrt{5}$, the circle and the line intersect at two different points.

ii) If $D = 0$, then, $k^2 - 5 = 0$ $\therefore$ $k = \pm\sqrt{5}$

   $\therefore$ When $k = \pm\sqrt{5}$, the circle and the line intersect at only one point.

iii) If $D < 0$, then, $k^2 - 5 > 0$ $\therefore$ $k > \sqrt{5}$ or $k < -\sqrt{5}$

   $\therefore$ When $k > \sqrt{5}$ or $k < -\sqrt{5}$, the circle and the line don't intersect at any point.

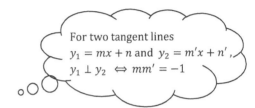

For two tangent lines
$y_1 = mx + n$ and $y_2 = m'x + n'$,
$y_1 \perp y_2 \iff mm' = -1$

(2) Equations of Tangent Lines of Circles

1) Given a Point on a Circle

Consider a circle $x^2 + y^2 = r^2$ $(r > 0)$ and a tangent line $y = mx + n$.

Let $A(x_1, y_1)$ be a point of tangency.

i) $x_1 \neq 0$ and $y_1 \neq 0$ $\Rightarrow$ The slope of a line $\overleftrightarrow{OA}$ is $\frac{y_1}{x_1}$ where $O = (0,0)$.

Since the tangent line passes through the point $A(x_1, y_1)$ is

perpendicular to the line $\overleftrightarrow{OA}$,

the slope of the tangent line is $-\frac{x_1}{y_1}$.

$\therefore$ The equation of the tangent line is $y - y_1 = -\frac{x_1}{y_1}(x - x_1)$.

$\therefore$ $-x_1 x + x_1^2 = y_1 y - y_1^2$

$\therefore$ $x_1^2 + y_1^2 = x_1 x + y_1 y$ $\cdots\cdots$①

Since $A(x_1, y_1)$ is a point on the circle $x^2 + y^2 = r^2$,

$x_1^2 + y_1^2 = r^2$ $\cdots\cdots$②

By ① and ②, $x_1 x + y_1 y = r^2$

ii) $x_1 = 0$ or $y_1 = 0$ $\Rightarrow$ The point $A(x_1, y_1)$ is on axis, $y = \pm r$ or $x = \pm r$

Thus, the equation of the tangent line is $x_1 x + y_1 y = r^2$

Therefore, | the equation of the tangent line

drawn from a point $A(x_1, y_1)$ on a circle $x^2 + y^2 = r^2$ $(r > 0)$ is

$x_1 x + y_1 y = r^2$

Note: *The equation of the tangent line*

   drawn from a point $A(x_1, y_1)$ on a circle $(x - a)^2 + (y - b)^2 = r^2$ $(r > 0)$ is

   $(x_1 - a)(x - a) + (y_1 - b)(y - b) = r^2$

Example Find the equation of a tangent line drawn from a point $(2, -4)$ on a circle

$$x^2 + y^2 = 4.$$

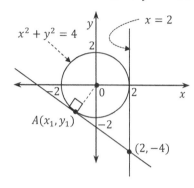

Let $A(x_1, y_1)$ be the point of tangency. Then,

the equation of the tangent line is $x_1 x + y_1 y = 4 \cdots\cdots$ ①

Since ① passes through the point $(2, -4)$,

$2x_1 - 4y_1 = 4 \cdots\cdots$ ②

Since the circle passes through the point $A(x_1, y_1)$,

$x_1^2 + y_1^2 = 4 \cdots\cdots$ ③

From ②, $x_1 = 2 + 2y_1 \cdots\cdots$ ④

Substituting ④ into ③, $(2 + 2y_1)^2 + y_1^2 = 4$

$\therefore\ 5y_1^2 + 8y_1 = 0\ ;\ y_1(5y_1 + 8) = 0 \qquad \therefore\ y_1 = 0$ or $y_1 = -\dfrac{8}{5} \cdots\cdots$ ⑤

Substituting ⑤ into ②, $y_1 = 0 \ \Rightarrow\ x_1 = 2$

$$y_1 = -\frac{8}{5} \ \Rightarrow\ x_1 = -\frac{6}{5}$$

$\therefore$ The equation of the tangent line is $2x = 4$ or $-\dfrac{6}{5}x - \dfrac{8}{5}y = 4$

That is, $x = 2$ or $3x + 4y = -10$.

2) Given the Slope of a Tangent Line

Consider a circle $x^2 + y^2 = r^2$ and a tangent line with slope m.

Since the slope of the tangent line is m, the equation of the line is $y = mx + n$.

Substituting $y = mx + n$ into $x^2 + y^2 = r^2$, we have $x^2 + (mx + n)^2 = r^2$.

$\therefore\ (m^2 + 1)x^2 + 2mnx + n^2 - r^2 = 0 \cdots\cdots$ ①

Since tangent lines intersect a circle at exactly one point,

the discriminant D of ① is $D = 0$.

$D = 4m^2n^2 - 4(m^2 + 1)(n^2 - r^2) = -4n^2 + 4m^2r^2 + 4r^2 = 4r^2(m^2 + 1) - 4n^2 = 0$

$\therefore\ n^2 = r^2(m^2 + 1)$

$\therefore\ n = \pm r\sqrt{m^2 + 1}$

Therefore, $\boxed{\text{the equation of the tangent line is } y = mx \pm r\sqrt{m^2 + 1}}$

For example, the equation of the tangent line with slope 2 of a circle $x^2 + y^2 = 9$ is

$$y = 2x \pm 3\sqrt{(2)^2 + 1} = 2x \pm 3\sqrt{5}$$

Note: The equation of the tangent line with slope m of a circle $(x - a)^2 + (y - b)^2 = r^2$ is

$$\boxed{y - b = m(x - a) \pm r\sqrt{m^2 + 1}}$$

(3) Length of a Common Tangent Line

A *common internal tangent* is a common tangent that intersects the segment that connects the centers of two circles.

A *common external tangent* is a common tangent that does not intersect the segment that connects the centers of two circles.

Let r and r' $(r > r')$ be the lengths of radii and d be the distance between the centers of two circles.

1) Common Internal Tangent Lines

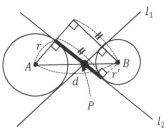

The common internal tangents l_1 and l_2 are tangents to each of the circles.

l_1 and l_2 intersect the segment $\overline{AB}$ of the centers at a point P between the centers of the circles.

Therefore, the length of a common internal tangent of circles is $\boxed{\sqrt{d^2 - (r + r')^2}}$

Common internal tangent

Example Find the length of a common internal tangent of $(x - 1)^2 + (y + 1)^2 = 1$ and $(x - 1)^2 + (y - 5)^2 = 4$

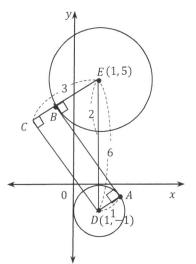

The distance between the centers of the two circles is

$\sqrt{(1 - 1)^2 + (5 - (-1))^2} = 6$

Note that $\overline{AB} = \overline{DC}$ and

$$\overline{EC} = \overline{EB} + \overline{BC} = 2 + \overline{AD} = 1 + 2 = 3$$

By the Pythagorean theorem,

$$\overline{CD} = \sqrt{6^2 - 3^2} = \sqrt{36 - 9} = \sqrt{27} = 3\sqrt{3}$$

Therefore, the length of $\overline{AB}$ is $3\sqrt{3}$.

(OR, by the formula,

$$\sqrt{6^2 - (2 + 1)^2} = \sqrt{6^2 - 3^2} = 3\sqrt{3}\,)$$

2) Common External Tangent Lines

The common external tangents l_1 and l_2 do not intersect the segment $\overline{AB}$ of the centers at a point between the centers of the circles.

Therefore, the length of a common external tangent of circles is $\boxed{\sqrt{d^2 - (r - r')^2}}$

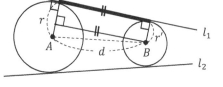 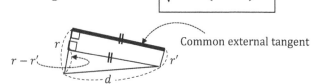

Common external tangent

4-3 Transformations

1. Translations (Slides) in a Coordinate Plane

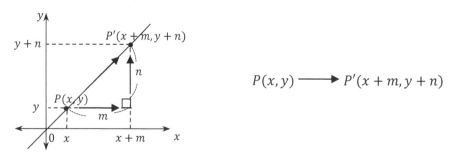

$$P(x, y) \longrightarrow P'(x + m, y + n)$$

The image of a point $P(x, y)$, which is translated m units along the x-axis (in x-axis) and n units along the y-axis (in y-axis), is a point $P'(x + m, y + n)$.

1) The graph of a line $y = ax + b$, $a \neq 0$ is a translation of the graph of a line $y = ax$ with b units in the y-axis.

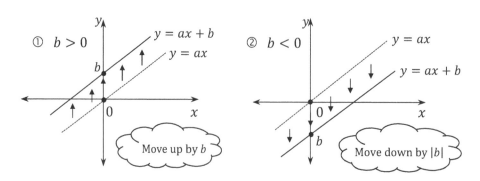

Note: $\boxed{\begin{array}{l} f(x, y) = 0 \longrightarrow f'(x - m, y - n) = 0 \\ : \textit{Translate } m \textit{ units along the } x\textit{-axis and } n \textit{ units along the } y\textit{-axis.} \end{array}}$

2) To find the translation of the graph of $y = ax^2$

with p units along the x-axis and q units along the y-axis,

Substitute $x - p$ into x and substitute $y - q$ into y.

$\Rightarrow \ y - q = a(x - p)^2 \quad \Rightarrow \ y = a(x - p)^2 + q$

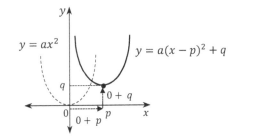

$y = ax^2$

$\Rightarrow \ y - q = a(x - p)^2$

$\Rightarrow \ y = a(x - p)^2 + q$

3) To find the translation of the graph of $y = a(x - p)^2 + q$

with m units along the x-axis and n units along the y-axis,

Substitute $x - m$ into x and $y - n$ into y.

$\Rightarrow y - n = a(x - m - p)^2 + q$

$\Rightarrow y = a(x - (p + m))^2 + q + n$

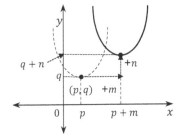

$y = a(x - p)^2 + q$

$\Rightarrow y = a(x - (p + m))^2 + (q + n)$

2. Reflections (Symmetries) in Lines

(1) Reflections of Points in Lines

① The reflection of a point (x, y) along the x-axis is a point $(x, -y)$.

$(x, y) \longrightarrow (x, -y)$

② The reflection of a point (x, y) along the y-axis is a point $(-x, y)$.

$(x, y) \longrightarrow (-x, y)$

③ The reflection of a point (x, y) along the origin $(0, 0)$ is a point $(-x, -y)$.

$(x, y) \longrightarrow (-x, -y)$

④ The reflection of a point (x, y) along a line whose equation is $y = x$ is a point (y, x).

$(x, y) \longrightarrow (y, x)$

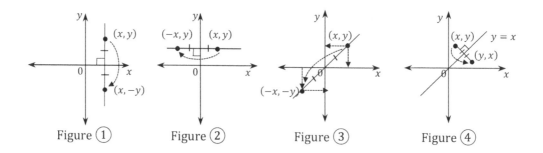

Figure ① Figure ② Figure ③ Figure ④

(2) Reflections of Equations in Lines

① The reflection of an equation $f(x, y) = 0$ along the x-axis is an equation $f(x, -y) = 0$.

$$f(x, y) = 0 \longrightarrow f(x, -y) = 0$$

② The reflection of an equation $f(x, y) = 0$ along the y-axis is an equation $f(-x, y) = 0$

$$f(x, y) = 0 \longrightarrow f(-x, y) = 0$$

③ The reflection of an equation $f(x, y) = 0$ along the origin $(0, 0)$ is an equation

$$f(-x, -y) = 0 \qquad f(x, y) = 0 \longrightarrow f(-x, -y) = 0$$

④ The reflection of an equation $f(x, y) = 0$ along a line whose equation is $y = x$ is

an equation $f(y, x) = 0$ $\qquad f(x, y) = 0 \longrightarrow f(y, x) = 0$

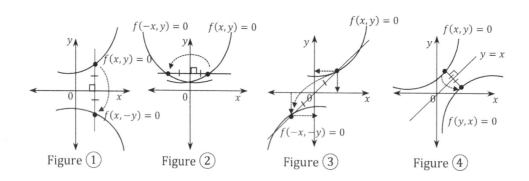

Figure ① Figure ② Figure ③ Figure ④

Note: (*Proof of* ④)

Let $P'(x', y')$ be the reflection point of $P(x, y)$ along a line whose equation is $y = x$.

Since the segment $\overline{PP'}$ is perpendicular to the line $y = x$, $\dfrac{y'-y}{x'-x} = -1$ ……①

Since the midpoint M of $\overline{PP'}$ is on the line $y = x$, $M = (\dfrac{x+x'}{2}, \dfrac{y+y'}{2})$ and $\dfrac{y+y\prime}{2} = \dfrac{x+x\prime}{2}$ ……②

From ①, $x' - x = y - y'$; $x' + y' = x + y$ ······ ③

From ②, $x + x' = y + y'$; $x' - y' = -x + y$ ······ ④

③+④ : $2x' = 2y$ ∴ $x' = y$, $x = y'$

Therefore, $P'(x', y') = P'(y, x)$

Example   Determine the reflections of an equation $x + 2y = 4$ in the x-axis, y-axis, the origin, and a line $y = x$, respectively.

In the x-axis, $(x, y) \longrightarrow (x, -y)$ ∴ $x - 2y = 4$ ∴ $y = \frac{1}{2}x - 2$

In the y-axis, $(x, y) \longrightarrow (-x, y)$ ∴ $-x + 2y = 4$ ∴ $y = \frac{1}{2}x + 2$

In the origin, $(x, y) \longrightarrow (-x, -y)$ ∴ $-x - 2y = 4$ ∴ $y = -\frac{1}{2}x - 2$

In the $y = x$, $(x, y) \longrightarrow (y, x)$ ∴ $y + 2x = 4$ ∴ $y = -2x + 4$

3. Reflections in Points

(1) Reflections of Points in Points

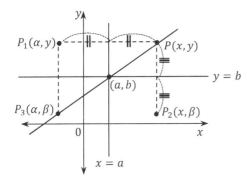

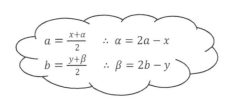

① Let $P_1 = P_1(\alpha, y)$; Reflection of point $P(x, y)$ in the line $x = a$.

Then, the midpoint M_1 of the segment $\overline{PP_1}$ is $M_1 = \left(\frac{x + \alpha}{2}, \frac{y + y}{2} \right) = \left(\frac{x + \alpha}{2}, y \right)$ and

M_1 is on the line $x = a$.

∴ $\frac{x + \alpha}{2} = a$ ∴ $\alpha = 2a - x$

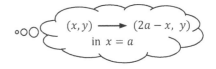

∴ $P_1 = P_1(2a - x, y)$

② Let $P_2 = P_2(x, \beta)$; Reflection of point $P(x, y)$ in the line $y = b$.

Then, the midpoint M_2 of the segment $\overline{PP_2}$ is $M_2 = \left(\frac{x + x}{2}, \frac{y + \beta}{2} \right) = \left(x, \frac{y + \beta}{2} \right)$ and

M_2 is on the line $y = b$.

∴ $\frac{y + \beta}{2} = b$ ∴ $\beta = 2b - y$

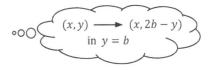

∴ $P_2 = P_2(x, 2b - y)$

③ Since the point $P_3 = P_3(\alpha, \beta)$ is the reflection of point $P(x, y)$

in the lines $x = a$ and $y = b$, $P_3 = P_3(2a - x, \ 2b - y)$

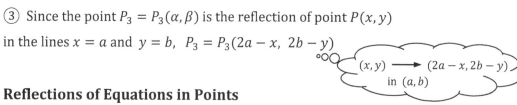

(2) Reflections of Equations in Points

① The reflection of equation $f(x, y) = 0$ in $x = a$ is $f(2a - x, y) = 0$

② The reflection of equation $f(x, y) = 0$ in $y = b$ is $f(x, 2b - y) = 0$

③ The reflection of equation $f(x, y) = 0$ in (a, b) is $f(2a - x, 2b - y) = 0$

4-4 Regions of Inequalities

1. Graphs of Linear Inequalities With Two Variables

(1) Half-Planes

The graph of a linear inequality with two variables, x and y, consists of points in the coordinate plane whose coordinate (x, y) makes the linear inequality true.

For a linear inequality $ax + by \geq c$ or $ax + by \leq c$,

the coordinates on the line $ax + by = c$ are solutions to the linear inequality, so we use a straight line.

For a linear inequality $ax + by > c$ or $ax + by < c$,

the coordinates on the line $ax + by = c$ are not solutions to the linear inequality, so we use a dotted line instead of a straight line.

The straight line or dotted line separates the coordinate plane into two half-planes.

The inequality symbols " $\geq$ " or " $\leq$ " represent *closed half-planes* and the inequality symbols " $>$ " or " $<$ " represent *open half-planes*.

(2) Graphing Linear Inequalities

Each half-plane region is the solution to the inequality.

To find the solution, choose a point in one half-plane and substitute it into the inequality.

If the point makes the inequality true, then the region including the point is the solution to the inequality. Shade the solution region.

Example 1　Graph the linear inequality $2x + y < 4$.

Consider the graph of the line $y = -2x + 4$.

The line $y = -2x + 4$ has slope -2 and y-intercept $(0, 4)$.

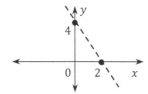

Choose a point $(-1, 0)$ as a testing point in one half-plane and substitute the point into the inequality.

Since $2x + y = 2 \cdot (-1) + 0 = -2$ and $-2 < 4$, the inequality is true.

So the region including the point $(-1, 0)$ is the solution to $-2x + y < 4$.

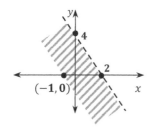

Example 2   Graph the linear inequality $2x - 3y \geq 6$.

$2x - 3y \geq 6 \;\Rightarrow\; 3y \leq 2x - 6 \Rightarrow\; y \leq \frac{2}{3}x - 2$

Graph the line $y = \frac{2}{3}x - 2$.

The line $y = \frac{2}{3}x - 2$ has x-intercept $(3, 0)$ and y-intercept $(0, -2)$.

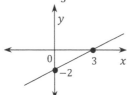

Choose a point $(0, 0)$ as a testing point in one half-plane.

Substituting $x = 0$ and $y = 0$ into the inequality, we get $2x - 3y = 2 \cdot 0 - 3 \cdot 0 = 0$.

Since $0 \geq 6$ is not true, the region including the point $(0, 0)$ is not the solution.

Shade the region which does not include the point $(0, 0)$.

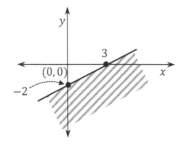

(3) Graphing Systems of Inequalities

To solve a system of inequalities, find the solution region for each inequality. Graph all inequalities on one coordinate plane. The overlapping region of the graph is the solution to the system of inequalities.

Every point in the overlap region makes all the inequalities true.

Example Graph a system of linear inequalities $\begin{cases} 2x - y \geq 1 \\ x - 3y > 4 \end{cases}$.

Since $2x - y \geq 1 \Rightarrow y \leq 2x - 1$, graph the line $y = 2x - 1$.

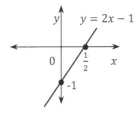

Choose a testing point $(0, 0)$ in one half-plane.

Substituting $x = 0$ and $y = 0$ into the inequality,

$2x - y = 2 \cdot 0 - 0 = 0 \geq 1$: This is not true.

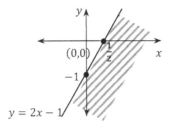

The shaded region is the solution to the linear inequality $2x - y \geq 1$.

Since $x - 3y > 4 \Rightarrow y < \dfrac{1}{3}x - \dfrac{4}{3}$, graph of the line is $y = \dfrac{1}{3}x - \dfrac{4}{3}$.

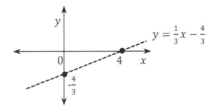

Choose a testing point $(0, 0)$.

Substituting $x = 0$ and $y = 0$ into the inequality,

$x - 3y = 0 - 3 \cdot 0 = 0 > 4$: This is not true.

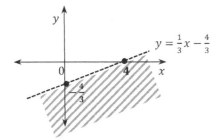

Graph both inequalities on the same coordinate plane.

Shade the overlapping solution region more darkly than the other solution regions.

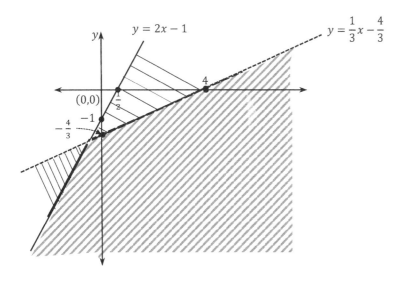

2. Regions of Inequalities

(1) $y > f(x)$ or $y < f(x)$ for a Curved Line f

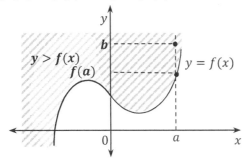

Consider two points $(a, f(a))$ and (a, b).

If (a, b) lies above the graph of $y = f(x)$, then $b > f(a)$

∴ The inequality $y > f(x)$ is always true for a point (a, b).

Conversely, for a point (a, b), if the inequality $y > f(x)$ is true, then $b > f(a)$.

$\therefore$ (a, b) lies above the graph of $y = f(x)$.

Therefore, (a, b) lies above the graph of $y = f(x)$ $\iff$ $y > f(x)$ for point (a, b)

Similarly, a point (a, b) lies below the graph of $y = f(x)$ $\iff$ $y < f(x)$ for a point (a, b)

Hence, ① The region of the inequality $y > f(x)$ is the set of all points in the plane which

lie above the graph of $y = f(x)$

② The Region of the inequality $y < f(x)$ is the set of all points in the plane which

lie below the graph of $y = f(x)$

Example Solve and graph $y - x(x - 5) > 3x$.

$y - x(x - 5) > 3x \Rightarrow y - x^2 + 5x > 3x \Rightarrow y > x^2 - 2x$

The region (solution set) is $\{(x, y) \mid y > x^2 - 2x\}$.

(2) $x^2 + y^2 < r^2$ or $x^2 + y^2 > r^2$

A circle $(x - a)^2 + (y - b)^2 = r^2$ in a coordinate plane divides the plane into 3 parts; interior, exterior, and boundary of the circle.

Let $P(x_1, y_1)$ be an interior point of the circle.

Then, the distance from the center $C(a, b)$ to the point $P(x_1, y_1)$ is less than the length of the radius of the circle.

$\therefore$ $\sqrt{(x_1 - a)^2 + (y_1 - b)^2} < r$ $\therefore$ $(x_1 - a)^2 + (y_1 - b)^2 < r^2$

$\therefore$ For any interior point $P(x_1, y_1)$, $(x - a)^2 + (y - b)^2 < r^2$.

Conversely, if $(x - a)^2 + (y - b)^2 < r^2$ for any interior point $P(x_1, y_1)$,

then $d(C, P) = \sqrt{(x_1 - a)^2 + (y_1 - b)^2} < r$

$\therefore$ The point $P(x_1, y_1)$ is an interior point of the circle.

$\therefore$ The region of the inequalities $(x - a)^2 + (y - b)^2 < r^2$ is the set of all points which lie inside the circle $(x - a)^2 + (y - b)^2 = r^2$.

Similarly, the region of the inequality $(x - a)^2 + (y - b)^2 > r^2$ is the set of all points which lie outside the circle $(x - a)^2 + (y - b)^2 = r^2$.

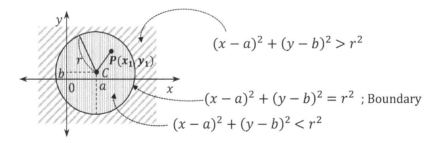

$(x - a)^2 + (y - b)^2 > r^2$

$(x - a)^2 + (y - b)^2 = r^2$; Boundary

$(x - a)^2 + (y - b)^2 < r^2$

Note: For a circle $x^2 + y^2 = r^2$,

    ① $x^2 + y^2 < r^2$ *is the region of the circle's interior (except the boundary of the circle).*

    ② $x^2 + y^2 > r^2$ *is the region of the circle's exterior (except the boundary of the circle).*

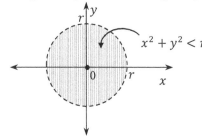 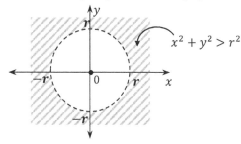

Example   Graph the inequality $x^2 + y^2 - 4x + 4y + 4 < 0$.

$$x^2 + y^2 - 4x + 4y + 4 < 0 \implies (x - 2)^2 + (y + 2)^2 < 4$$

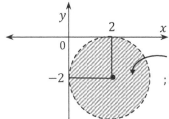

    $(x - 2)^2 + (y + 2)^2 < 4$

    ; Interior region of the circle with center $(2, -2)$ and radius 2

(3) System of Inequalities $\begin{cases} f(x, y) > 0 \\ g(x, y) < 0 \end{cases}$

The region of a system of inequalities $\begin{cases} f(x, y) > 0 \\ g(x, y) < 0 \end{cases}$ is the common region of inequalities

$f(x, y) > 0$ and $g(x, y) < 0$.

Example   Graph the system $\begin{cases} x + 2y \geq 2 \\ x^2 + y^2 \leq 4 \end{cases}$

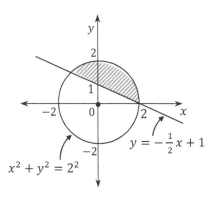

For a system $\begin{cases} x + 2y \geq 2 & \cdots\cdots ① \\ x^2 + y^2 \leq 4 & \cdots\cdots ② \end{cases}$,

the region of ① is the above part of $y = -\dfrac{1}{2}x + 1$

and the region ② is the interior part of the circle $x^2 + y^2 = 2^2$.

∴ The common region is the shaded part including the boundary.

(4) The Region of Product of Polynomials:

$f(x, y)\, g(x, y) > 0$ or $f(x, y)\, g(x, y) < 0$

The region of the product of polynomials $f(x, y)\, g(x, y) > 0$ is the region of a system:

$$\begin{cases} f(x, y) > 0 \\ g(x, y) > 0 \end{cases} \text{ or } \begin{cases} f(x, y) < 0 \\ g(x, y) < 0 \end{cases}.$$

The region of the product of polynomials $f(x, y)\, g(x, y) < 0$ is the region of a system:

$$\begin{cases} f(x, y) > 0 \\ g(x, y) < 0 \end{cases} \text{ or } \begin{cases} f(x, y) < 0 \\ g(x, y) > 0 \end{cases}.$$

Example   Graph the inequality $(x + y)(x - y) < 0$.

$$\Rightarrow \begin{cases} x + y > 0 \\ x - y < 0 \end{cases} \text{ or } \begin{cases} x + y < 0 \\ x - y > 0 \end{cases}.$$

$\Rightarrow$
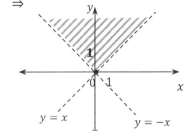   OR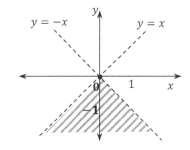

$\Rightarrow$
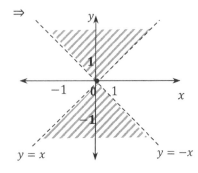

The union regions of two systems of inequalities is the region of the product polynomial $(x + y)(x - y) < 0$.

3. Maxima and Minima in Regions of Inequalities

To find maximum value or minimum value of $f(x, y)$ for a point (x, y) in the region of given inequality, consider the following steps:

Step 1. Graph the region of the given inequality in a coordinate plane.

Step 2. Let $f(x, y) = k$ for a constant k.

Step 3. Move the graph of f for any k in the region.

Step 4. Find maximum or minimum value of k.

Example For any real numbers x and y such that $x \le 3$, $y \le 3$ and $x + y \ge 3$,

find maximum value of $y - x$.

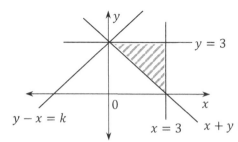

Let $y - x = k$. Then $y = x + k$

When the line $y = x + k$ passes through a point $(0, 3)$, k has maximum value.

Therefore, maximum value of $y - x$ is 3.

$y - x = k$

$x = 3$　$x + y = 3$; $y = -x + 3$

Example For any real numbers x and y such that $x^2 + y^2 - 4x - 4y \le 0$,

find maximum value of $3x + y$.

$x^2 + y^2 - 4x - 4y \le 0 \Rightarrow (x - 2)^2 + (y - 2)^2 \le 8$

Let $3x + y = k$. Then $y = -3x + k$

The distance between the center $(2, 2)$ of the circle

and the line $3x + y - k = 0$ is $2\sqrt{2}$ which is the

length of the radius of the circle.

$y = -3x + k$

$2\sqrt{2}$

$(x - 2)^2 + (y - 2)^2 = 8$

$\therefore \dfrac{|3\cdot2 + 1\cdot2 - k|}{\sqrt{3^2 + 1^2}} = 2\sqrt{2}$

$\therefore \dfrac{|8 - k|}{\sqrt{10}} = 2\sqrt{2}$; $|8 - k| = 2\sqrt{20}$; $8 - k = \pm 2\sqrt{20}$

$\therefore k = 8 \mp 2\sqrt{20}$ 　 $\therefore k = 8 - 4\sqrt{5}$ or $k = 8 + 4\sqrt{5}$

Since $k > 0$, $k = 8 + 4\sqrt{5}$

Therefore, maximum value of $3x + y$ is $8 + 4\sqrt{5}$.

Exercises

#1 Find the distance between two points A and B on a number line or a coordinate plane.

(1) $A\left(-2 + \sqrt{3}\right)$, $B(-6)$

(2) $A(2,3)$, $B(-5,-2)$

(3) $A(-4,-1)$, $B(-2,-7)$

(4) $A(0,0)$, $B(3,4)$

#2 Find the coordinates of the point P.

(1) For two points $A(2,2)$ and $B(-2,4)$, $\overline{AP} = \overline{BP}$ and P is on the x-axis.

(2) For two points $A(1,1)$ and $B(2,3)$, $\overline{AP} = \overline{BP}$ and P is on the line $y = x + 2$.

(3) For a triangle $\triangle ABC$, the midpoints of $\overline{AB}, \overline{BC}$, and $\overline{CA}$ are $D\left(2,\frac{7}{2}\right)$, $E\left(\frac{3}{2},4\right)$, and $F\left(\frac{5}{2},\frac{9}{2}\right)$, respectively. Find the centroid P.

(4) For two points $A(-2,a)$ and $B(b,3)$, the midpoint of the segment $\overline{AB}$ is $P = (-2,-1)$. Find the point which divides $\overline{AB}$ externally in the ratio $\overline{AP} : \overline{PB} = 2 : 1$

(5) The line $(1 + k)x - (1 + 3k)y - 2 - 3k = 0$ passes through a point P for any k. Find the coordinates of P.

#3 For two points $A(-3,4)$ and $B(2,-1)$ on a coordinate plane, find the coordinates of the point P.

(1) P is the point which divides the segment $\overline{AB}$ internally in the ratio $\overline{AP} : \overline{PB} = 2 : 3$

(2) P is the point which divides the segment $\overline{AB}$ externally in the ratio $\overline{AP} : \overline{PB} = 2 : 1$

(3) P is the midpoint of segment $\overline{AB}$

#4 Find the value of real number a.

(1) The distance between two points $A(-1,3)$ and $B(a,-3)$ is $4\sqrt{3}$. Find the value of a.

(2) For three points $A(-1,2)$, $B(7,1)$, and $P(x,0)$, find the value of $a = \alpha + \beta$ where α is minimum value of $\overline{AP^2} + \overline{BP^2}$ and β is minimum value of $\overline{AP} + \overline{BP}$.

(3) For two points $A(-6,4)$ and $B(-3,-5)$, a line $y = 2x + a$ passes through a point which divides the segment $\overline{AB}$ internally in the ratio $\overline{AP} : \overline{PB} = 2 : 1$ Find the value of a.

(4) For three points $A(-2,1)$, $B(-3,-2)$, and $C(2,7)$, let M_1, M_2, and M_3 be the midpoints of $\overline{AB}, \overline{BC}$, and $\overline{AC}$, respectively. When $P(\alpha,\beta)$ is the centroid of the triangle $\triangle M_1 M_2 M_3$, find the value of $a = \alpha + \beta$.

(5)

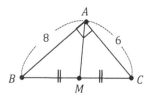

For a triangle with $m(\angle A) = 90^\circ$,

$\overline{AB} = 8$, $\overline{AC} = 6$, and M is the midpoint of the

segment $\overline{BC}$. Find the length a of $\overline{AM}$.

(6)

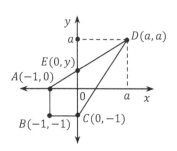

For a quadrilateral $\square ABCD$ with vertexes

$A(-1,0), B(-1,-1), C(0,-1)$, and $D(a,a)$, y-axis

divides the area of $\square ABCD$ into 2 halves.

Find the value of a.

(7) For four points $A(2,5), B(-1,1), C(\alpha,-2)$, and $D(\beta,2)$, the quadrilateral $\square ABCD$ is a

rhombus. (where $\alpha > 2$) Find the value of the real number $a = \alpha\beta$.

(8) The area of a triangle $\triangle ABC$ which is surrounded by a line $ax + 3y = 3a$, x-axis, and y-axis

is 24. Find the value of positive number a.

#5 For a regular triangle $\triangle ABC$, the coordinate of the point A is $A(-4,6)$.

When the centroid of $\triangle ABC$ is $P(-1,2)$, find the area of the $\triangle ABC$.

#6 Find an equation in the standard form for each line.

(1) with y-intercept -3 and slope 2

(2) with y-intercept 5 and slope 0

(3) with x-intercept 5 and slope $-\dfrac{2}{3}$

(4) with x-intercept -3 and slope -2

(5) through $(1,2)$ with slope 3

(6) through $(3,-4)$ with slope -2

(7) through $(2,3)$ with undefined slope

(8) through $(-2,3)$ with y-intercept -1

(9) through $(2,4)$ with x-intercept -5

(10) through $(3,1)$ and $(-2,4)$

(11) through $(-2,-3)$ and $(-1,5)$

(12) with x-intercept -3 and y-intercept 3

(13) with x-intercept $\frac{3}{2}$ and y-intercept -4

(14) Vertical line through $(-1, 2)$

(15) Horizontal line through $(3, -4)$

#7 Find an equation for the line through $(2, 3)$ which is:

(1) parallel to the line $y = 2x - 5$

(2) parallel to the line $y = -3x + 1$

(3) parallel to the line $x = 4$

(4) parallel to the line $y = -2$

(5) parallel to the line $3x + 4y = 5$

(6) perpendicular to the line $y = \frac{2}{3}x - 1$

(7) perpendicular to the line $x + 3y = -3$

(8) perpendicular to the line $x = 5$

(9) perpendicular to the line $y = -2$

#8 Find the value of a for each line:

(1) through $(2, 3)$ and $(1, -a)$ with slope 2

(2) through $(2a - 1, -2)$ and $(-1, 1)$ with slope -2

(3) through $(1, -2)$, $(-3, 2)$, and $(-a + 1, -5)$

(4) through $(2a + 1, -4)$, $(2, 5)$, and $(2, -3)$

(5) through $(-3, 3)$, $(3, a - 1)$, and $(0, 3)$

(6) through $(a, 2a - 3)$ and $(-a - 1, 3 + 4a)$ and parallel to the x-axis

(7) through $(-3a + 1, -5)$ and $(2a - 1, a + 3)$ and perpendicular to the x-axis

(8) through $(3, -2a)$ and $(2a - 1, -3a + 2)$ and parallel to the y-axis

(9) through $(-1, 5)$ and $(2, -4)$ and parallel to the line $ax + 3y + 5 = 0$

#9 Find the value of a such that the line $ax + 2y = 5$:

(1) is parallel to the line $2x + 3y = -2$.

(2) is perpendicular to the line $y = -2x + 3$.

(3) coincides with the line $6y = -4x + 15$.

#10 Find the value of ab for which :

(1) the system $\begin{cases} x - 3y = a \\ 2x + by = 3 \end{cases}$ has the intersection point $(2, 3)$.

(2) the system $\begin{cases} -ax + by = 4 \\ 2ax + 3by = 2 \end{cases}$ has the intersection point $(-1, 2)$.

(3) the system $\begin{cases} px + y = 3 \\ 2x - 3y = q \end{cases}$ has no intersection when $p = a, \ q \neq b$.

(4) a system $\begin{cases} 2ax + 4y = -3 \\ 3x + 6y = 2b \end{cases}$ has unlimited numbers of intersections.

#11 Find the value of a such that :

(1) the system $\begin{cases} ax + y = -2 \\ -3x + 2y = 4 \end{cases}$ has no solution.

(2) the system $\begin{cases} 2x - ay + 3 = 0 \\ x + 3y - 2 = 0 \\ 2x + y + 1 = 0 \end{cases}$ has one solution.

(3) the system $\begin{cases} x - 3y = 2 \\ 2x + y = -3 \end{cases}$ has a solution $(2a, -1)$.

(4) the line $2ax + 3y - 1 = 0$ passes through the intersection of the system $\begin{cases} x - 2y = 3 \\ 2x + 2y = 1 \end{cases}$.

#12 Find the equation of each line such that :

(1) the line passes through the intersection of the system $\begin{cases} x + 2y = 3 \\ 3x + y = -2 \end{cases}$

and runs parallel to the y-axis.

(2) the line passes through the intersection of the system $\begin{cases} -x + y + 2 = 0 \\ 2x + y - 3 = 0 \end{cases}$

and runs perpendicular to the x-axis.

(3) the line passes through the intersection of the system $\begin{cases} 2x - y + 3 = 0 \\ x + 2y + 4 = 0 \end{cases}$

and runs parallel to the line $3x + 2y = 5$.

#13 For a line $\dfrac{x}{a} + \dfrac{y}{b} = 1 \ (a > 0, \ b > 0)$ which passes through a point $A(4, 3)$,

let B be the intersection point of the line and y-axis, and C be the intersection point of the line and x-axis. Find minimum value of the area of triangle $\triangle\, OBC$ (where O is the origin$(0,0)$).

#14 Find the value of the real number a.

(1) For the two points $A(1, 3)$ and $B(4, -3)$, a line $y = 2x + a$ passes through the intersection
point P such that $\overline{AP} : \overline{PB} = 2 : 3$

(2) The three points $A(-2, 6), B(4, -3)$, and $C(a, 3)$ are on the same line.

#15 For two lines l_1: $ax + 2y - 2 = 0$ and l_2: $x + (a - 1)y + 2 = 0$,

find the value of the constant a when $l_1 /\!/ l_2$ (parallel) and $l_1 \perp l_2$ (perpendicular).

#16 For a line l_1 which passes through the intersection point of two lines l_2: $2x + y + 8 = 0$ and

l_3: $x - 2y - 6 = 0$, l_1 is parallel to the line $l_4 : x - 3y = 0$ and passes through a point $(a, -2)$.

Find the real number a.

#17 Determine the distance between the two parallel lines.

(1) $2x + 3y - 4 = 0$, $2x + 3y + 6 = 0$

(2) $x + 3y - 4 = 0$, $-x - 3y + 2 = 0$

#18 For two parallel lines $12x - 5y - 2 = 0$ and $12x - 5y - 3k = 0$,

find the real number k so that the distance between the lines is 3.

#19 Find an equation of a line so that the line is perpendicular to the line $2x + 3y = 0$ and

the distance from the origin is 2.

#20 For a triangle $\triangle OAB$ with vertices $O(0, 0)$, $A(3, 6)$, and $B(5, -4)$, find the area of the triangle.

#21 Find maximum value of the distance between the origin and the line

$2x - 3y - 2 - k(x - y) = 0$ (where k is a real number).

#22 For any real numbers a and b, a quadratic equation $x^2 - 2(a - b)x - 2(a - b) - 1 = 0$ has a

double root. Find minimum value of the distance between the two points $A(a, b)$ and $B(1, -1)$.

#23

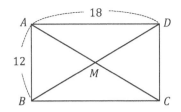

For a rectangle $\Box ABCD$ with $\overline{AB} = 12$ and $\overline{AD} = 18$, let the intersection point of the two diagonals be M. When the centroid of the triangle $\triangle ABC$ is P_1 and the centroid of the triangle $\triangle CDM$ is P_2, find the distance between P_1 and P_2.

#24 Determine the equation of a circle.

(1) The diameter is the segment joining the points $A(-2, 2)$ and $B(4, 4)$.

(2) The equation is $x^2 + y^2 - 2x + 6y - 6 = 0$.

(3) The circle passes through the three points $(-1, 4), (1, 2),$ and $(-1, 0)$.

(4) The center of the circle is on the y-axis and the circle passes through two points $(2, 2)$, $(1, -1)$.

#25 The equation of a circle is $x^2 + y^2 - 4ax + 2ay + 20a - 28 = 0$. Find the coordinates of the center of the circle when the area of the circle has minimum value.

#26 When two circles pass through a point $A(-3, 5)$ and intersect at only one point on both the x-axis and the y-axis, find the distance between their centers.

#27 For two real numbers x and y such that $(x - 4)^2 + (y - 3)^2 = 4$, find maximum and minimum values of $\sqrt{x^2 + y^2}$.

#28 For a line which passes through the intersection points of two circles $x^2 + y^2 - 8 = 0$ and $x^2 + y^2 + 6x + ay = 0$, the line is perpendicular to $y = 2x + 1$.
Find the value of the constant a.

#29 State how the relations between a circle $x^2 + y^2 = 1$ and a line $y = -2x + k$ are different depending on the value of k.

#30 For a circle with center $(2, 2)$ and radius 2, a line $y = ax + 1$ intersects the circle at 2 different points. Find the range of a.

#31 For real numbers a and b, a line $\dfrac{x}{a} + \dfrac{y}{b} = 1$ intersects a circle $x^2 + y^2 = 3$ at only one point in the first quadrant. Find minimum value of ab.

#32 Find x and y intercepts of a tangent line at $(a, 2)$ of a circle $x^2 + y^2 = 9$ $(a > 0)$.

#33 For a point P such that $\overline{AP} : \overline{PB} = 1 : 2$ where $A = A(3, 0)$ and $B = B(0, 3)$, find maximum value of the area of $\triangle ABP$.

#34 When a circle $x^2 + y^2 = 9$ and a line $ax - y + a + 1 = 0$ intersect at two points A and B, find maximum and minimum values of $\overline{AB}$.

#35 For a point $P(x, y)$ on a circle $x^2 + y^2 - 8x + 6y + 21 = 0$, find maximum distance between the point P and a line $3x + 4y - 24 = 0$.

#36 When two circles $x^2 + (y - 1)^2 = 20$ and $(x - a)^2 + (y - 1)^2 = 4$ intersect each other at two different points, find the positive number a so that the common chord has maximum value.

#37 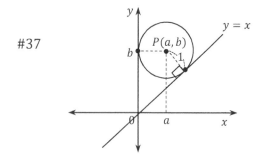 For a circle with radius 1, the center P of the circle is in quadrant I and the circle intersects y-axis at one point and has tangent line $y = x$. Find the coordinates of P.

#38 When a circle $x^2 + y^2 + 4ax - 2ay + 8a - 4 = 0$ passes through two points A and B, not depending on the value of real number a, find the length of the segment $\overline{AB}$.

#39 Find a tangent line at $(2, 1)$ of a circle $(x - 1)^2 + (y + 2)^2 = 10$.

#40 For points of tangency A and B of a circle, find the length of the segment $\overline{AB}$.

(1) When two tangent lines of a circle $x^2 + y^2 - 6x - 2y + 6 = 0$ are drawn from a point $P(3, 5)$.

(2) When two tangent lines of a circle $x^2 + y^2 = 5$ are drawn from a point $P(4, 2)$.

#41 Two tangent lines of a circle $(x-3)^2 + y^2 = 6$ are drawn from a point $P(a,b)$ which lies on the x-axis. When the lines are perpendicular, find the coordinates of $P(a,b)$. (where $a > 0$)

#42 Find the value of $a + b$ for which:

    (1) The graph of $y = ax + 2$ is translated by b along the y-axis from a graph of $y = 3x - 5$.

    (2) The graph is translated by a along the y-axis from a graph of $y = 2x + 4$ and passes through both points $(a+1, -2)$ and $\left(-\frac{1}{3}, b\right)$.

    (3) A point $(-1,1)$ is on the graph of $y = -2x + a$. If the graph is translated by b along the y-axis, then the translated line will pass through the point $(3, -4)$.

#43 Determine the equation whose graph is translated from the graph of $x^2 + y^2 - 2x + 4y = 0$ by -1 unit in x-axis and 2 units in y-axis.

#44 Find the value of the real number a.

    (1) When a line $2x + 3y = 5$ is translated by -3 units in the x-axis and $+2$ units in the y-axis, the translated line will divide a circle $(x+a)^2 + (y-3)^2 = 6$ in half.

    (2) When a line $2x - y + 1 = 0$ is translated by a units in the x-axis, the translated line will be a tangent line of a circle $(x-2)^2 + (y-3)^2 = 12$.

    (3) If a circle $x^2 + y^2 = 4$ is translated by a $(a > 0)$ units in the x-axis, then it will meet a line $2x - 4y - 6 = 0$ at one point.

    (4) If a circle $x^2 + y^2 + 2ax - 4y + a^2 = 0$ is translated by 1 unit in the x-axis and -2 units in the y-axis, then it will be divided in half by a line $2x - y + 4 = 0$.

#45 If a line $y = ax + b$ is translated by -2 units in the x-axis and $+1$ unit in the y-axis, then the translated line and a line $x - 2y - 4 = 0$ will be perpendicular to each other on y-axis. Find the values of the real numbers a and b.

#46 Find the value of the real number a.

    (1) The reflection of a circle $(x-3)^2 + (y+2)^2 = 4$ on line $y = x$ has the center which lies on $y = ax + 1$.

    (2) The reflection of a line $2x - 3y + a = 0$ in line $y = x$ is the tangent line of a circle $x^2 + (y-2)^2 = 4$.

    (3) The reflection of a line $y = ax + 1$ in the y-axis divides the area of a circle $(x+2)^2 + (y-3)^2 = 4$ in half.

#47 For three points $A(-2, 2)$, $B(4, 3)$, and $C(a, 0)$, find minimum value of $\overline{AC} + \overline{CB}$.

#48 Find a reflection point of a point $A(2, 3)$ in line $y = 2x + 3$.

#49 For two points $A(2, 2)$ and $B(3, 2)$, a point $P(x, y)$ lies on a line $y = -2x + 1$.
 Find the coordinates of $P(x, y)$ so that $\overline{PA} + \overline{PB}$ has minimum value.

#50 For any positive real numbers a and b, two lines $y = ax$ and $y = bx$ are translations of each
 other in line $y = x$. For a point $P(a, b)$, find minimum value of the distance between the origin
 O and the point P.

#51 Find an equation of a circle C which is translated by a circle $x^2 + y^2 = 4$ in line $y = 2x + 1$.

#52 Graph the region of the following inequalities on a coordinate plane.
 (1) $\begin{cases} x^2 + y^2 - 9 < 0 \\ x + y - 2 > 0 \end{cases}$
 (2) $|x - y| \leq 2$
 (3) $(x - y + 2)(x^2 + y^2 - 4) < 0$
 (4) $xy(x - y + 2) \geq 0$

#53 When a point $(1, a)$ is in the region of a circle $(x - 2)^2 + y^2 < 13$ and a point $(-1, a)$ is not in
 the region of the circle, determine the range of the real number a.

#54 When a positive number a satisfies the inequality $x^2 + y^2 \leq a$,
 the inequality $3x + 4y - 20 \leq 0$ is always true. Find maximum value of a.

#55 Find the range of a constant a.
 (1) By a line $ax + y - 2 = 0$, two points $A(1, 1)$ and $B(3, -4)$ lie in different regions.
 (2) A line $x + ay + 2 + a = 0$ runs through between two points $A(1, 1)$ and $B(-1, 1)$.

#56 Find maximum value.
 (1) When a point $P(x, y)$ satisfies all inequalities: $x \geq 0$, $y \geq 0$, $2x + y \leq 5$, and $3x + 4y \leq 10$,
  find maximum value of $x + y$.
 (2) When a point $P(x, y)$ satisfies all inequalities: $x - y \geq 0$, $2x - 3y \leq 0$, and $x + 3y \leq 4$,
  find maximum value of $x + y$.

(3) When a point $P(x, y)$ satisfies the inequality: $x^2 + y^2 - 2x - 4y \leq 0$,

find maximum value of $2x + y$.

(4) When a point $P(x, y)$ satisfies the system of inequalities: $\begin{cases} y \geq x^2 \\ y \leq -x + 2 \end{cases}$,

find maximum value of $y - 2x$.

#57 When a point $P(x, y)$ satisfies the inequality $x^2 - 4y^2 - 2x + 1 \leq 0$,

find minimum value of $x^2 + y^2$.

#58 Two points $A(a, b)$ and $B(c, d)$ satisfy a system of inequalities $\begin{cases} y \geq x \\ x + y \leq 1 \\ y - 2x \leq 0 \end{cases}$

Find maximum and minimum values of $\dfrac{b+d}{a+c}$, $(a + c \neq 0)$.

Chapter 5. Functions

5-1 Relations and Functions

1. Identifying Functions

(1) Relations

Consider a set $X \times Y$: a set of all ordered pairs of real numbers (x, y).

A *relation* is a subset of $X \times Y$.

The *domain*, which is a subset of all values (inputs) of X, and the *range*, which is a subset of all values (outputs) of Y, are defined by a relation.

A *function* f is a relation in which each element of the domain is paired with exactly one element of the range.

That is, for a given x in the domain of a relation f, only one value y in Y is associated with the given x. A function f is represented by

$$f : X \to Y \quad \text{or} \quad X \xrightarrow{f} Y \quad \text{or} \quad y = f(x)$$

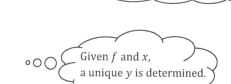

$y = f(x)$ means
y is a function of x or y is f of x.

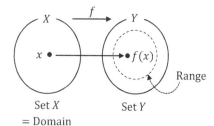

Given f and x,
a unique y is determined.

For example, given functions $y = x + 2$, $y = x^2$, and $y = \dfrac{1}{x}$, let R be the set of all real numbers. Then, the domains of the functions are R, R, and $R - \{0\}$, respectively, and the ranges of the functions are R, $\{y|y \geq 0\}$, and $R - \{0\}$, respectively.

The range is a subset of Y
; i.e., $f(x) = \{y|y = f(x), \ x \in X\}$
: A set of all function values.

Note : (***Dependent and Independent variables***)

Since each value of y of the range on the selected value of x,
we call y the dependent variable.

When f is a function of x, x is the independent variable of f.

(2) Vertical Line Test for Functions

A set of points in a coordinate plane is the graph of y as a function of x.

$\xLeftrightarrow{\text{if and only if}}$ No vertical line intersects the graph at more than one point.

Example

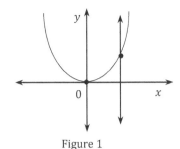

Figure 1

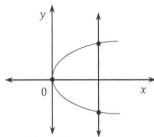

Figure 2

From the graphs of Figure 1 and Figure 2, we can determine which of the relation is a function.

In Figure 1, there are no two different pairs which have the same element in the domain.

That is, every vertical line intersects the graph at most once. Therefore, it is a function.

However, in Figure 2, there are two points which have the same element in the domain.

That is, a vertical line intersects the graph in more than one point. Therefore, it is not a function.

Note : (***Bounded and Unbounded Intervals***)

For any real numbers a and b such that a < b, bounded and unbounded intervals on the real number line

are as follows (where a and b are the end points of each interval).

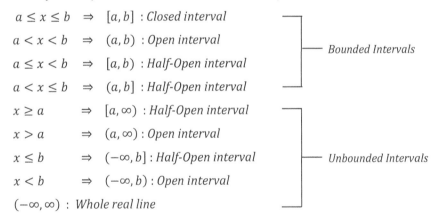

$a \leq x \leq b \;\; \Rightarrow \;\; [a, b] : Closed\ interval$

$a < x < b \;\; \Rightarrow \;\; (a, b) : Open\ interval$

$a \leq x < b \;\; \Rightarrow \;\; [a, b) : Half\text{-}Open\ interval$

$a < x \leq b \;\; \Rightarrow \;\; (a, b] : Half\text{-}Open\ interval$

$x \geq a \qquad\;\; \Rightarrow \;\; [a, \infty) : Half\text{-}Open\ interval$

$x > a \qquad\;\; \Rightarrow \;\; (a, \infty) : Open\ interval$

$x \leq b \qquad\;\; \Rightarrow \;\; (-\infty, b] : Half\text{-}Open\ interval$

$x < b \qquad\;\; \Rightarrow \;\; (-\infty, b) : Open\ interval$

$(-\infty, \infty) \; : \; Whole\ real\ line$

Bounded Intervals

Unbounded Intervals

Example   Find the domain and the range from the graph of a function.

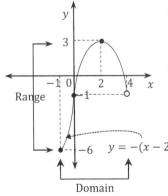

$y = -(x - 2)^2 + 3$

① Since the closed dot indicates that $x = -1$ is in the domain of f,
whereas the open dot indicates $x = 4$ is not in the domain of f,
the domain of f is all x in the interval $[-1, 4)$.

② Since a point $(-1, -6)$ is on the graph of f, $f(-1) = -6$
and a point $(2, 3)$ is on the graph of f, $f(2) = 3$.
Since the graph does not extend below $f(-1) = -6$
or above $f(2) = 3$, the range of f is the interval $[-6, 3]$.

(3) Equality of Two Functions

For any two functions f and g,

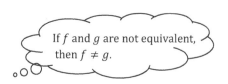

If f and g are not equivalent, then $f \neq g$.

if ① their domains and ranges are the same and

② $f(x) = g(x)$ for given x of the domain,

then the functions f and g are called equivalent and denoted by $f = g$

When both the domain and range of a function consist of real numbers, we can picture the function by drawing its graph on a coordinate plane.

2. Graphing Functions

The graph of a relation whose ordered pairs are (x, y) in

the XY-plane, is the set of points whose coordinates are in the relation.

Note: (***Intercepts***)

The x-intercepts are the x-coordinates of the points where the graph crosses (intersects) the x-axis.

The x-intercepts correspond to x's for which the value of the function is zero.

Example   Find the intercepts of the graph of $y = x^2 - 4x + 3$

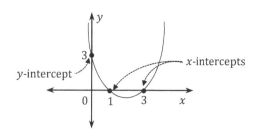

y-intercept

x-intercepts

To find the x-intercepts, let $y = 0$ and solve for x.

Then, $y = x^2 - 4x + 3 = (x - 1)(x - 3) = 0$

∴ $x = 1$ or $x = 3$

Thus, the x-intercepts are 1 and 3.

Letting $x = 0$, we find $y = 3$, which is the y-intercept.

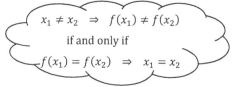

$x_1 \neq x_2 \ \Rightarrow \ f(x_1) \neq f(x_2)$

if and only if

$f(x_1) = f(x_2) \ \Rightarrow \ x_1 = x_2$

(1) One-to-One Correspondence

For a function $f: X \to Y$, if, for any x_1 and x_2 in X, $x_1 \neq x_2$ implies $f(x_1) \neq f(x_2)$, then

the function f is called a *one-to-one*.

When a function $f: X \to Y$ is ① one-to-one and ② the range = the set Y, the function f is called

a *one-to-one correspondence*.

Example

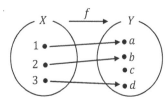

One-to-one function

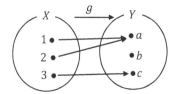

$1 \neq 2$, but $g(1) = g(2)$ ∴ g is not a one-to-one.

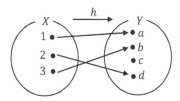

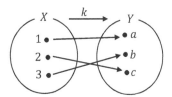

One-to-one function
but, not one-to-one correspondence

One-to-one function
and also, one-to-one correspondence

Example Which functions are one-to-one ?

$$(1)\ f(x) = x^3 - 1 \qquad (2)\ g(x) = x^2 - x \qquad (3)\ h(x) = \sqrt{x}$$

∵ (1) Let a and b be real numbers such that $f(a) = f(b)$.

Then, $a^3 - 1 = b^3 - 1$

∴ $a^3 = b^3$ ∴ $a = b$

Therefore, $f(a) = f(b) \Rightarrow a = b$

Hence, f is a one-to one.

(2) Since $g(-1) = (-1)^2 - (-1) = 1 + 1 = 2$ and $g(2) = (2)^2 - 2 = 4 - 2 = 2$,

$-1 \neq 2$ but $g(-1) = g(2)$

Therefore, g is not a one-to one.

(3) Let a and b be non-negative real numbers such that $h(a) = h(b)$.

Then, $\sqrt{a} = \sqrt{b}$

Squaring both sides gives $a = b$

Therefore, $h(a) = h(b) \Rightarrow a = b$

Hence, h is a one-to one.

Horizontal Line Test for One-to-One Function

A function is a one-to-one if no horizontal line intersects its graph in more than one point.

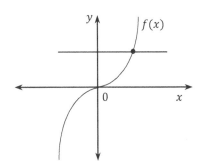

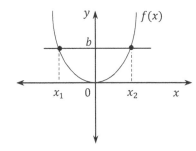

One-to-one function

Not a one-to-one function
($\because x_1 \neq x_2$ but $f(x_1) = f(x_2) = b$)

(2) Increasing and Decreasing Functions

> The graphs of one-to-one correspondence are increasing or decreasing functions.

For a function $f: X \to Y$,

① the function f is *increasing* on an interval if, for any x_1 and x_2 in the interval,

$x_1 < x_2 \Rightarrow f(x_1) < f(x_2)$

② the function f is *decreasing* on an interval if, for any x_1 and x_2 in the interval,

$x_1 < x_2 \Rightarrow f(x_1) > f(x_2)$

③ the function f is *constant* on an interval if, for any x_1 and x_2 in the interval,

$f(x_1) = f(x_2) = c$ (c is a constant.)

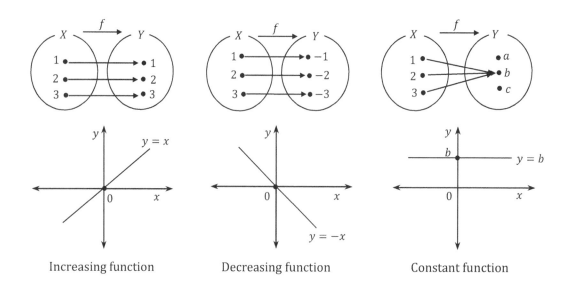

Increasing function　　　Decreasing function　　　Constant function

Example　Determine the intervals on which each function is increasing, decreasing, or constant.

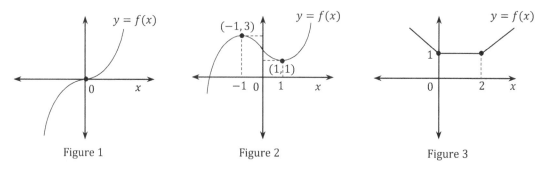

Figure 1　　　　Figure 2　　　　Figure 3

In Figure 1, if $x_1 < x_2$, then $f(x_1) < f(x_2)$.

Therefore, the function is increasing over the entire real number line.

In Figure 2, the function is increasing on the interval $(-\infty, -1]$, decreasing on the interval $[-1, 1]$, and increasing on the interval $[1, \infty)$.

In Figure 3, the function is decreasing on the interval $(-\infty, 0]$, constant on the interval $[0, 2]$, and increasing on the interval $[2, \infty)$.

(3) Identity Functions

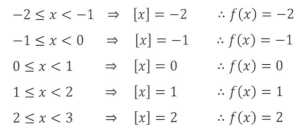

For a function $f: X \to Y$, if $f(x) = x$ for any x in X,

then the function f is called *identity* function and denoted by I.

For example, the graph of $y = x$ shows an identity function.

(4) Step Functions

The *greatest integer function* is denoted by $[x]$ and is defined by $f(x) = [x]$: the greatest integer less than or equal to x. That is, for any real number x and integer n, $n \le x < n+1 \Leftrightarrow [x] = n$

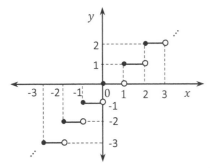

Greatest integer function $f(x) = [x]$

$$-2 \le x < -1 \Rightarrow [x] = -2 \quad \therefore f(x) = -2$$
$$-1 \le x < 0 \Rightarrow [x] = -1 \quad \therefore f(x) = -1$$
$$0 \le x < 1 \Rightarrow [x] = 0 \quad \therefore f(x) = 0$$
$$1 \le x < 2 \Rightarrow [x] = 1 \quad \therefore f(x) = 1$$
$$2 \le x < 3 \Rightarrow [x] = 2 \quad \therefore f(x) = 2$$

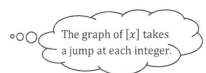
The graph of $[x]$ takes a jump at each integer.

If $g(x) = [x - 1]$, then

the graph of the function g is translated by 1

along the x-axis from a graph of $f(x) = [x]$.

That is, the graph of g is translated by -1

along the y-axis from a graph of f.

Therefore, the graph of $g(x) = [x - 1]$ is equal to

the graph of $h(x) = [x] - 1$. Hence, $[x - 1] = [x] - 1$.

$g(x) = [x - 1] = [x] - 1$

(5) Even and Odd Functions

1) Even Functions

For any point (x, y) on the graph, the symmetry with respect to the y-axis is the point $(-x, y)$.

When the graph of a function is symmetric with respect to the y-axis, the function is an *even* function. That is, for each x in the domain of f, if $f(-x) = f(x)$, then the function is even.

2) Odd Functions

For any point (x, y) on the graph, the symmetry with respect to the origin $(0, 0)$ is the point $(-x, -y)$. When the graph of a function is symmetric with respect to the origin, the function is an *odd* function.

That is, for each x in the domain of f, if $f(-x) = -f(x)$, then the function is odd.

Note : For any point (x, y) on the graph, the symmetry with respect to the x-axis is the point $(x, -y)$.
Thus, the graph contains two points that have the same x-value but two different y-values.
Therefore, the graph of a non-zero function cannot be symmetric with respect to the x-axis.

Example Determine whether the following functions are even, odd, or neither.

(1) $f(x) = x^2 + 1$ (2) $g(x) = x^3 + 1$ (3) $h(x) = x^3 - x$ (4) $k(x) = |x|$

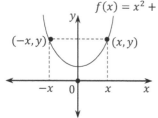

(1) Substituting $-x$ for x gives

$$f(-x) = (-x)^2 + 1 = x^2 + 1 = f(x)$$

$\therefore$ f is an even function.

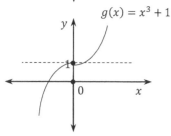

(2) Substituting $-x$ for x gives

$$g(-x) = (-x)^3 + 1 = -x^3 + 1.$$

Since $-g(x) = -(x^3 + 1) = -x^3 - 1$, $g(-x) \neq -g(x)$

Also, $g(-x) \neq g(x)$

$\therefore$ g is neither odd nor even function.

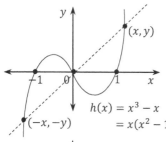

(3) Substituting $-x$ for x gives

$$h(-x) = (-x)^3 - (-x) = -x^3 + x = -(x^3 - x) = -h(x)$$

$\therefore$ h is an odd function.

$$h(x) = x^3 - x$$
$$= x(x^2 - 1) = x(x + 1)(x - 1)$$

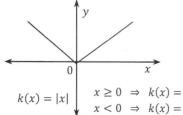

$$k(x) = |x| \quad \begin{array}{l} x \geq 0 \implies k(x) = x \\ x < 0 \implies k(x) = -x \end{array}$$

(4) Substituting $-x$ for x gives

$$k(-x) = |-x| = |x| = k(x)$$

$\therefore$ k is an even function.

3. Function Operations

(1) Operations on Functions

For any two functions f and g with the same domain (overlapping domains),

a new function h is defined by addition, subtraction, multiplication, and division on f and g.

For all x, $h(x) = (f + g)(x) = f(x) + g(x)$ \hspace{2cm} Addition

$h(x) = (f - g)(x) = f(x) - g(x)$ \hspace{2cm} Subtraction

$h(x) = (f \times g)(x) = f(x) \times g(x)$ \hspace{2cm} Multiplication

$h(x) = (f \div g)(x) = \left(\dfrac{f}{g}\right)(x) = \dfrac{f(x)}{g(x)}$, $g(x) \neq 0$ \hspace{1cm} Division

Example Let $f(x) = 2\sqrt{x}$ and $g(x) = -4\sqrt{x}$.

Find $(f + g)(x)$, $(f - g)(x)$, $(fg)(x)$, and $\left(\dfrac{f}{g}\right)(x)$. State their domains.

$(f + g)(x) = f(x) + g(x) = 2\sqrt{x} + (-4\sqrt{x}) = 2\sqrt{x} - 4\sqrt{x} = (2 - 4)\sqrt{x} = -2\sqrt{x}$

$(f - g)(x) = f(x) - g(x) = 2\sqrt{x} - (-4\sqrt{x}) = 2\sqrt{x} + 4\sqrt{x} = (2 + 4)\sqrt{x} = 6\sqrt{x}$

$(fg)(x) = f(x) \times g(x) = (2\sqrt{x}) \times (-4\sqrt{x}) = (2 \times -4)(\sqrt{x} \times \sqrt{x}) = -8(\sqrt{x})^2 = -8x$

$(f \div g)(x) = \left(\dfrac{f}{g}\right)(x) = \dfrac{f(x)}{g(x)} = \dfrac{2\sqrt{x}}{-4\sqrt{x}} = -\dfrac{1}{2}$

Since f and g have the same domain which consists of all non-negative real numbers,

the domains of $f + g$, $f - g$, and fg also consist of all non-negative real numbers.

Since $g(0) = 0$, the domain of $\dfrac{f}{g}$ is restricted to the set of all positive real numbers to avoid

division by zero.

(2) Compositions of Functions

1) Composition of Two Functions

For any two functions $f: X \to Y$ and $g: Y \to Z$, a new function, called the *composition* of the

function g with the function f, is denoted by $g \circ f: X \to Z$, $(g \circ f)(x) = g(f(x))$

Note : The domain of the composition function $g \circ f$ is the set of all x-values such that x is in the domain

of f and $f(x)$ is in the domain of g.

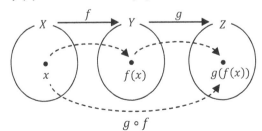

$g \circ f$

The composition of a function g followed by a function f.

$$g \circ f \neq f \circ g$$

The composition of a function f followed by a function g.

For example,

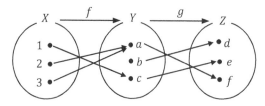

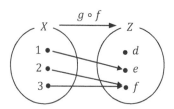

$(g \circ f)(1) = g\big(f(1)\big) = g(c) = e$

$(g \circ f)(2) = g\big(f(2)\big) = g(a) = f$

$(g \circ f)(3) = g\big(f(3)\big) = g(a) = f$

Note : $(f \circ g)(x) = (g \circ f)(x) = x$

⇒ The composite functions $f \circ g$ and $g \circ f$ are equal, and represent the identity function.

2) Properties of Composite Functions

① For any two functions f and g, $f \circ g \neq g \circ f$ in general.

(Commutative law is not satisfied.)

② For any three functions f, g, and h, $h \circ (g \circ f) = (h \circ g) \circ f$

(Associative law is satisfied.)

③ $I \circ f = f \circ I = f$ where I is the identity function

Example Let $f(x) = \dfrac{2}{x}$ and $g(x) = 3x + 1$. Find composite functions $f \circ g$ and $g \circ f$.

$(f \circ g)(x) = f\big(g(x)\big) = f(3x + 1) = \dfrac{2}{3x+1}$

$(g \circ f)(x) = g\big(f(x)\big) = g\left(\dfrac{2}{x}\right) = 3\left(\dfrac{2}{x}\right) + 1 = \dfrac{6}{x} + 1$

The domain of $f \circ g$ consists of all real numbers except $x = -\dfrac{1}{3}$.

because $g\left(-\dfrac{1}{3}\right) = 0$ is not in the domain of f.

The domain of $g \circ f$ consists of all real numbers except $x = 0$.

because 0 is not in the domain of f.

Note that $(f \circ g)(x)$ and $(g \circ f)(x)$ are not equal.

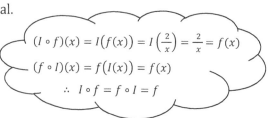

$(I \circ f)(x) = I\big(f(x)\big) = I\left(\dfrac{2}{x}\right) = \dfrac{2}{x} = f(x)$

$(f \circ I)(x) = f\big(I(x)\big) = f(x)$

∴ $I \circ f = f \circ I = f$

Example Let $f(x) = x + 2$, $g(x) = x^2 - 1$, and $h(x) = 2x + 1$

Find composite functions $h \circ (g \circ f)$ and $(h \circ g) \circ f$.

$(g \circ f)(x) = g(f(x)) = g(x + 2) = (x + 2)^2 - 1 = x^2 + 4x + 3$

$\therefore$ $(h \circ (g \circ f))(x) = h((g \circ f)(x)) = h(x^2 + 4x + 3) = 2(x^2 + 4x + 3) + 1 = 2x^2 + 8x + 7$

$(h \circ g)(x) = h(g(x)) = h(x^2 - 1) = 2(x^2 - 1) + 1 = 2x^2 - 1$

$\therefore$ $((h \circ g) \circ f)(x) = (h \circ g)(f(x)) = 2(f(x))^2 - 1 = 2(x + 2)^2 - 1 = 2(x^2 + 4x + 4) - 1$

$$= 2x^2 + 8x + 7$$

Note that $h \circ (g \circ f)$ and $(h \circ g) \circ f$ are equal.

4. Inverse Functions

(1) The Inverse of a Function

For a function $f : X \to Y$, if the function is one-to-one correspondence, then there is only one x in X such that $y = f(x)$ for any y in Y.

Therefore, we can define a new function called the *inverse function* of f, denoted by f^{-1},

by interchanging the first and second coordinates of each of these ordered pairs.

$f^{-1} : Y \to X$, $f^{-1}(y) = x$ (: Inverse function of $f : X \to Y$, $f(x) = y$)

The domain of the inverse function f^{-1} is the range of the original function f and the range of the inverse function is the domain of the original function.

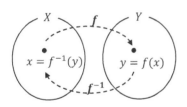

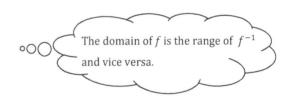

The domain of f is the range of f^{-1} and vice versa.

For a function $y = f(x)$, if the inverse function f^{-1} of f exists, then

$[y = f(x) \iff x = f^{-1}(y)]$.

For all y in the domain of f^{-1}, $(f \circ f^{-1})(y) = f(f^{-1}(y)) = f(x) = y$

Thus, $f \circ f^{-1}$ is an identity function defined in Y.

Similarly, for all x in the domain of f, $(f^{-1} \circ f)(x) = f^{-1}(f(x)) = f^{-1}(y) = x$

Thus, $f^{-1} \circ f$ is an identity function defined in X.

Therefore, the inverse function of f^{-1} is $(f^{-1})^{-1} = f$.

Note: If f and g are functions such that

① $(f \circ g)(x) = f\big(g(x)\big) = x$ *for all x in the domain of g and*

② $(g \circ f)(x) = g\big(f(x)\big) = x$ *for all x in the domain of f,*

then g is the inverse function of f and $g = f^{-1}$ (f inverse).

Example   Find an equation for the inverse of the relation $y = 2x - 1$.

$\quad y = 2x - 1$ Original relation

$\Rightarrow \quad x = 2y - 1$ Swich x and y

$\Rightarrow \quad 2y = x + 1$ Solve for y

$\Rightarrow \quad y = \dfrac{1}{2}x + \dfrac{1}{2}$ Simplify

$\therefore$ The inverse relation is $y = \dfrac{1}{2}x + \dfrac{1}{2}$.

Example   Verify that $f(x) = 2x - 1$ and $f^{-1}(x) = \dfrac{1}{2}x + \dfrac{1}{2}$ are inverses.

$f\big(f^{-1}(x)\big) = f\left(\dfrac{1}{2}x + \dfrac{1}{2}\right) = 2\left(\dfrac{1}{2}x + \dfrac{1}{2}\right) - 1 = x + 1 - 1 = x$

$f^{-1}\big(f(x)\big) = f^{-1}(2x - 1) = \dfrac{1}{2}(2x - 1) + \dfrac{1}{2} = x - \dfrac{1}{2} + \dfrac{1}{2} = x$

Therefore, $f\big(f^{-1}(x)\big) = f^{-1}\big(f(x)\big) = x$

$$f^{-1}(x) \neq \dfrac{1}{f(x)}$$
$$[f(x)]^{-1} = \dfrac{1}{f(x)}$$

(2) Properties of Inverse Functions

Suppose two functions $f: X \to Y$ and $g: Y \to Z$ are one-to-one correspondences. Then,

1) There exists an inverse function $f^{-1}: Y \to X$

2) $y = f(x) \xleftrightarrow[\text{if and only if}]{} x = f^{-1}(y)$

3) $(f \circ f^{-1})(y) = y, \ y \in Y \qquad (f^{-1} \circ f)(x) = x, \ x \in X$

4) $(f^{-1})^{-1} = f(x), \ x \in X$

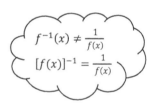

$$I_X : X \to X \qquad I_Y : Y \to Y$$

5) $f^{-1} \circ f = I_X$ (Identity function in X) $\qquad f \circ f^{-1} = I_Y$ (Identity function in Y)

6) $(g \circ f)^{-1} = f^{-1} \circ g^{-1}$

Note: For all x in the domain $(-\infty, \infty)$ of f, the function $f(x) = x^2$ does not have an inverse.

To have an inverse, a function must be one-to-one. So we have to find the domain for x such that

the restricted function does have an inverse.

A function f has an inverse function f^{-1}. $\xleftrightarrow[\text{if and only if}]{}$ f is one-to-one.

A function f is one-to-one :
$$f(a) = f(b) \ \Rightarrow \ a = b$$
for a and b in its domain.

(3) Finding the Inverse of a Function

Step 1. Rewrite $f(x)$ with y.

Step 2. Switch x and y.

Step 3. If the new equation represents y as a function of x, then solve it for y.

Step 4. Replace y by $f^{-1}(x)$.

Step 5. Check $f(f^{-1}(x)) = f^{-1}(f(x)) = x$ so that f and f^{-1} are inverses of each other.

Example   For two functions $f(x) = 2x + 1$ and $g(x) = 3x - 2$,

verify that $(g \circ f)^{-1} = f^{-1} \circ g^{-1}$

$(g \circ f)(x) = g(f(x)) = g(2x + 1) = 3(2x + 1) - 2 = 6x + 1$

Since $g \circ f$ is one-to-one correspondence, $(g \circ f)^{-1}$ exists.

Let $y = 6x + 1$.

Switching x and y gives $x = 6y + 1$; $6y = x - 1$

$\therefore \ y = \dfrac{1}{6}x - \dfrac{1}{6}$

$\therefore \ (g \circ f)^{-1}(x) = \dfrac{1}{6}x - \dfrac{1}{6}$

Since f and g are one-to-one correspondence, f^{-1} and g^{-1} exist.

$f(x) = 2x + 1 \ \Rightarrow \ y = 2x + 1 \qquad$ Rewrite $f(x)$ with y

$\Rightarrow \ x = 2y + 1 \qquad$ Switch x and y

$\Rightarrow \ y = \dfrac{1}{2}x - \dfrac{1}{2} \qquad$ Solve for y

$\therefore \ f^{-1}(x) = \dfrac{1}{2}x - \dfrac{1}{2}$

$g(x) = 3x - 2 \ \Rightarrow \ y = 3x - 2 \qquad$ Rewrite $g(x)$ with y

$\Rightarrow \ x = 3y - 2 \qquad$ Switch x and y

$\Rightarrow \ y = \dfrac{1}{3}x + \dfrac{2}{3} \qquad$ Solve for y

$\therefore \ g^{-1}(x) = \dfrac{1}{3}x + \dfrac{2}{3}$

Thus, $(f^{-1} \circ g^{-1})(x) = f^{-1}(g^{-1}(x)) = f^{-1}(\dfrac{1}{3}x + \dfrac{2}{3}) = \dfrac{1}{2}(\dfrac{1}{3}x + \dfrac{2}{3}) - \dfrac{1}{2}$

$= \dfrac{1}{6}x + \dfrac{1}{3} - \dfrac{1}{2} = \dfrac{1}{6}x - \dfrac{1}{6}$

Therefore, $(g \circ f)^{-1} = f^{-1} \circ g^{-1}$

(4) The Graph of the Inverse Functions

f^{-1} is the set of all ordered pairs (b, a) such that (a, b) is an ordered pair of f.

Thus, f and f^{-1} are symmetric with respect to the line $y = x$.

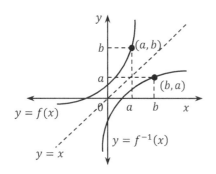

∵ The graph of a function $y = f(x)$ is

the graph of an equation $y - f(x) = 0 \cdots\cdots$ ①

For the inverse function $y = f^{-1}(x)$,

the relation is given by $f(y) = x$.

Thus, the graph of the inverse function $y = f^{-1}(x)$

is the graph of an equation $x - f(y) = 0 \cdots\cdots$ ②

Swiching x and y in ① gives ②.

Therefore, the graph of the function $y = f(x)$ and the graph of the inverse function $y = f^{-1}(x)$ are symmetric with respect to the line $y = x$.

Example Find the inverse of the function $f(x) = x^2$, $x \geq 0$

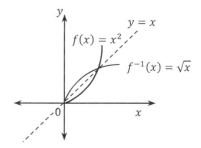

$f(x) = x^2$

$\Rightarrow\ y = x^2$ Rewrite $f(x)$ with y

$\Rightarrow\ x = y^2$ Switch x and y

$\Rightarrow\ \pm\sqrt{x} = y$ Take square roots of each side

Since the domain of f is restricted to non-negative numbers,

the inverse function is $f^{-1}(x) = \sqrt{x}$.

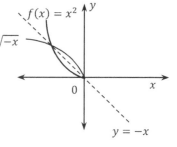

Note that the inverse of the function $f(x) = x^2$, $x < 0$,

is $f^{-1}(x) = \sqrt{-x}$

Note: The graphs of f^{-1} are reflections (symmetries) of the graphs of f in the line $y = x$.

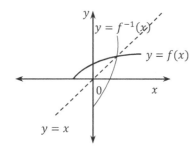
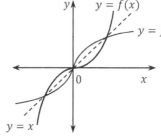
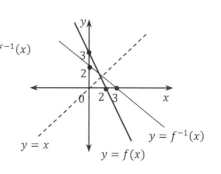

5-2 Polynomial Functions

Let n ($n \geq 0$) be an integer and let a_n, a_{n-1}, $\cdots\cdots$, a_2, a_1, a_0 be real numbers with $a_n \neq 0$.

A polynomial function of x with degree n is defined by

$$f(x) = a_n x^n + a_{n-1}x^{n-1} + \cdots\cdots + a_2 x^2 + a_1 x + a_0$$

where a_n is the *leading coefficient*, a_0 is the *constant term*, and n is the *degree*.

If the terms of a polynomial function are written in descending order of exponents from left to right, then the polynomial function is in *standard form*.

For $a \neq 0$,

(1) If the function $f(x) = a$ has degree 0, then f is called a *constant function*.

(2) If the function $f(x) = ax + b$ has degree 1, then f is called a *linear function*.

(3) If the function $f(x) = ax^2 + bx + c$ has degree 2, then f is called a *quadratic function*.

(4) If the function $f(x) = ax^3 + bx^2 + cx + d$ has degree 3, then f is called a *cubic function*.

(5) If the function $f(x) = ax^4 + bx^3 + cx^2 + dx + e$ has degree 4, then f is called a *quartic function*.

1. Linear Functions

(1) Graphs of Linear Functions

A linear function is a function in the form $f(x) = ax + b$ where a and b are real numbers, $a \neq 0$.

The graph of a linear function $f(x) = ax + b$ is a straight line with slope a and y-intercept b.

If $a > 0$, then f is an increasing function.

If $a < 0$, then f is a decreasing function.

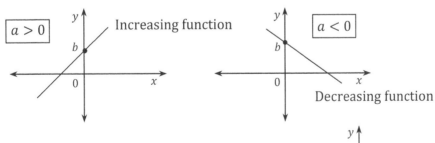

Note : For a function $y = ax + b$, if $a = 0$, then $y = b$.

This is not a linear function.

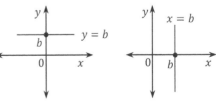

(2) Graphs of Linear Functions Involving Absolute Values

1) Graphing $y = |x|$

Case 1. $x \geq 0 \Rightarrow y = x$ Case 2. $x < 0 \Rightarrow y = -x$

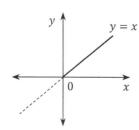

Therefore, the graph of $y = |x|$ is:

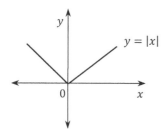

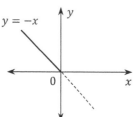

To find the range, $y = |x| \geq 0$

Domain = All real numbers

Range = $\{ y | y \geq 0 \}$: All non-negative real numbers.

2) Graphing $y = |x + 2| - 1$

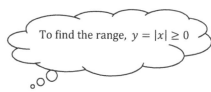

To get the range,
Since $|x + 2| \geq 0$,
$y = |x + 2| - 1 \geq -1$

Case 1. $x + 2 \geq 0$ $(x \geq -2)$ Case 2. $x + 2 < 0$ $(x < -2)$
$\Rightarrow y = x + 1$ $\Rightarrow y = -x - 3$

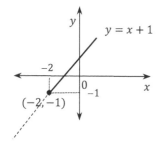

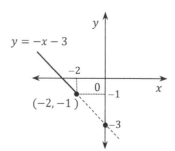

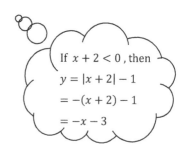

If $x + 2 < 0$, then
$y = |x + 2| - 1$
$= -(x + 2) - 1$
$= -x - 3$

Therefore, the graph of $y = |x + 2| - 1$ is:

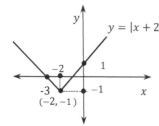

Domain = All real numbers

Range = $\{ y | y \geq -1 \}$: All real numbers greater than or equal to -1.

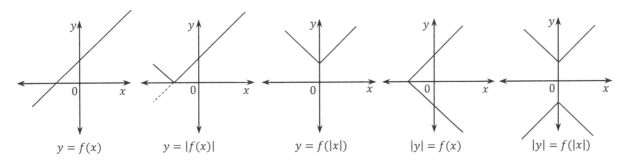

$y = f(x)$ $y = |f(x)|$ $y = f(|x|)$ $|y| = f(x)$ $|y| = f(|x|)$

2. Quadratic Functions

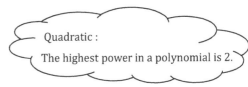

Quadratic :
The highest power in a polynomial is 2.

For any real numbers a, b, and c with $a \neq 0$, the function $f(x) = ax^2 + bx + c$ is a quadratic

function. For example, $y = 3x^2$, $y = -x^2 + 1$, $y = 2x^2 + 3x$, $y = \frac{1}{2}x^2 - x + 2$

A graph of a quadratic function on a coordinate plane is called a *parabola*.

The symmetrical axis (central line) of a parabola is called the *axis* of the parabola or *the axis of

symmetry*. A parabola is symmetrical with respect to the axis of symmetry.

The intersection point of the parabola and the axis of symmetry is called the *vertex*.

There are two types of parabola

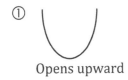

① Opens upward

The vertex is a minimum point.

(Lowest point)

② Opens downward

The vertex is a maximum point.

(Highest point)

*Note : On the graph of the simplest quadratic equation $y = x^2$, the equation of the axis of symmetry is the

line $x = 0$ (y- axis) and the coordinates of the vertex are (0,0).*

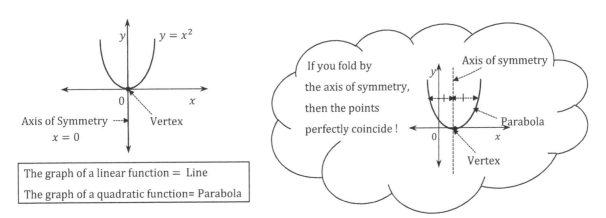

$y = x^2$

Axis of Symmetry ---> Vertex
$x = 0$

If you fold by
the axis of symmetry,
then the points
perfectly coincide !

Axis of symmetry

Parabola

Vertex

The graph of a linear function = Line
The graph of a quadratic function= Parabola

(1) Graphs of Quadratic Functions

1) Graphs of $y = ax^2$, $a \neq 0$

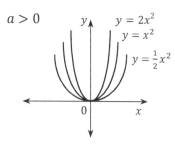

$a > 0$

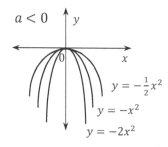

$a < 0$

① Vertex : $(0, 0)$

② Axis of symmetry : $x = 0$ (y-axis)

③ $a > 0 \Rightarrow$ Open upward and $y = f(x) \geq 0$

  $a < 0 \Rightarrow$ Open downward and $y = f(x) \leq 0$

The point (a, b) is on the graph of $y = f(x)$.
$\Rightarrow f(a) = b$

④ The larger the magnitude of the absolute value of a, the narrower the width of the parabola.

Note : The graphs of $y = ax^2$ and $y = -ax^2$ are symmetric along the x-axis.

When $a > 0$,	When $a < 0$,
$x \uparrow \Rightarrow f(x) \uparrow$	$x \uparrow \Rightarrow f(x) \downarrow$
(If the value of x increases, then the value of $f(x)$ increases.)	(If the value of x increases, then the value of $f(x)$ decreases.)

2) Graphs of $y = ax^2 + q$, $a \neq 0$

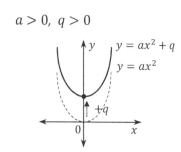

$a > 0, q > 0$

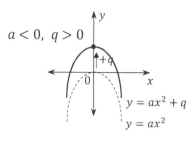

$a < 0, q > 0$

The graph of $y = ax^2 + q$ is a translation of the graph of $y = ax^2$, with the q units along the y-axis.

① Vertex : $(0, q)$

② Axis of symmetry : $x = 0$ (y-axis)

The graphs of $y = ax^2$ and $y = ax^2 + q$ are equivalent, except that $y = ax^2 + q$ is q units above $y = ax^2$.

3) Graphs of $y = a(x - p)^2$, $a \neq 0$

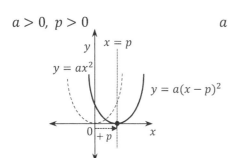

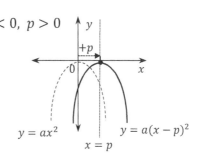

The graph of $y = a(x - p)^2$ is a translation of the graph of $y = ax^2$,

with the p units along the x- axis.

① Vertex : $(p, 0)$

② Axis of symmetry : $x = p$

Shift the graph of $y = ax^2$ p units to the right and q units up.

4) Graphs of $y = a(x - p)^2 + q$, $a \neq 0$

$a > 0$, $p > 0$, $q > 0$

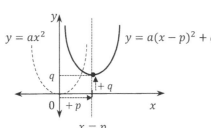

$a < 0$, $p > 0$, $q > 0$

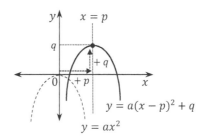

The graph of $y = a(x - p)^2 + q$ is a translation of the graph of $y = ax^2$,

with the p units along the x-axis and the q units along the y-axis.

① Vertex : (p, q)

② Axis of symmetry : $x = p$

A parabola with vertex (m, n)
$\Leftrightarrow y = a(x - m)^2 + n$

Note : Graph of $y = a(x + p)^2 - q$, $a \neq 0$

$a < 0$, $p > 0$, $q > 0$

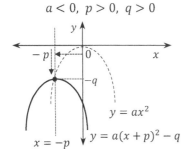

The graph of $y = a(x + p)^2 - q$ is a translation

of the graph of $y = ax^2$, with the $-p$ units along the x-axis

and the $-q$ units along the y-axis.

① Vertex : $(-p, -q)$

② Axis of symmetry : $x = -p$

(2) The Standard Form of a Quadratic Function

The quadratic function $f(x) = a(x - p)^2 + q$, $a \neq 0$ is said to be in *standard form*.

To find the vertex and the axis of symmetry for a graph, transform the graph of the general quadratic function form $y = ax^2 + bx + c$, $a \neq 0$ to the graph of a standard form.

Example   Describe the graph of $f(x) = 3x^2 + 6x - 2$.

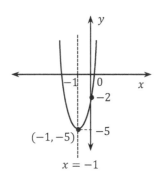

$$f(x) = 3x^2 + 6x - 2$$
$$= 3(x^2 + 2x) - 2 \qquad \text{Factor 3 out of } x \text{ terms}$$
$$= 3((x + 1)^2 - 1) - 2 \quad \text{Complete the square}$$
$$= 3(x + 1)^2 - 5 \qquad \text{Standard form}$$

From the standard form, we see the graph of f is a parabola that opens upwards with vertex $(-1, -5)$ and the axis of symmetry $x = -1$.

$$\textit{Note}: \quad y = ax^2 + bx + c = a\left(x^2 + \frac{b}{a}x\right) + c = a\left(\left(x + \frac{b}{2a}\right)^2 - \left(\frac{b}{2a}\right)^2\right) + c$$
$$= a\left(x + \frac{b}{2a}\right)^2 - \frac{b^2}{4a} + c = a\left(x + \frac{b}{2a}\right)^2 - \frac{b^2 - 4ac}{4a}$$

① $y = ax^2 + bx + c \xrightarrow[\text{Transform}]{} y = a\left(x + \frac{b}{2a}\right)^2 - \frac{b^2 - 4ac}{4a}$

② Vertex : $\left(-\dfrac{b}{2a}, -\dfrac{b^2 - 4ac}{4a}\right)$

③ Axis of symmetry : $x = -\dfrac{b}{2a}$

Note : (**_Intercepts_**)

(1) x-intercept:

   The $x\text{-}intercept$ is the x-coordinate of the point where the graph intersects the x-axis.

   $x\text{-}intercept$ is the value of x when $y = 0$.

(2) y-intercept:

   The $y\text{-}intercept$ is the y-coordinate of the point where the graph intersects the y-axis.

   $y\text{-}intercept$ is the value of y when $x = 0$.

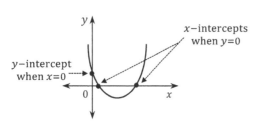

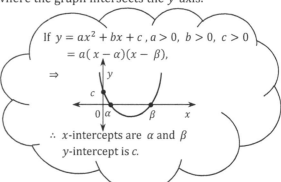

If $y = ax^2 + bx + c = a(x - \alpha)(x - \beta)$,

the equation $ax^2 + bx + c = 0$ has two different solutions.

  (using $D = b^2 - 4ac > 0$)

$\Rightarrow$ The parabola has 2 different x-intercepts.

 or

If $y = ax^2 + bx + c = a(x - \alpha)^2$,

the equation $ax^2 + bx + c = 0$ has one solution (a double root).

  (using $D = b^2 - 4ac = 0$)

$\Rightarrow$ The parabola has only 1 x-intercept.

 or

If $y = ax^2 + bx + c$,

the equation $ax^2 + bx + c = 0$ has no solution.

  (using $D = b^2 - 4ac < 0$)

$\Rightarrow$ The parabola has no x-intercept.

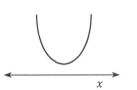 or

Example   Find an equation for the parabola that has its vertex at $(2, 4)$ and that passes

through the origin $(0, 0)$.

The standard form of the parabola with vertex $(2, 4)$ is $f(x) = a(x - 2)^2 + 4$

Since the parabola passes through the point $(0, 0)$, $f(0) = 0$.

$\therefore\ \ 0 = a(0 - 2)^2 + 4 = 4a + 4$ $\therefore\ \ a = -1$

Thus, the equation is $f(x) = -(x - 2)^2 + 4 = -x^2 + 4x$

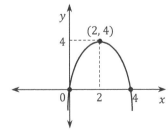

To find the x-intercepts of the graph of $f(x) = -x^2 + 4x$,

  solve the equation $-x^2 + 4x = 0$

  $\therefore\ \ -x^2 + 4x = -x(x - 4) = 0$

  $\therefore\ \ $ x-intercepts are $x = 0$ and $x = 4$.

(3) Properties of Quadratic Functions

$$y = ax^2 + bx + c = a(x - p)^2 + q, \ a \neq 0 \quad : \text{Standard form}$$

1) Conditions for the y-value

① $a > 0$

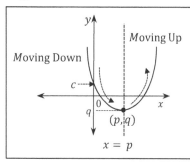

i) When $x < p$,

If the x−values are increasing, $\Rightarrow$ the y−values are decreasing.
(Moving right)   (Moving down)

ii) When $x > p$,

If the x−values are increasing, $\Rightarrow$ the y−values are increasing.
(Moving right)   (Moving up)

② $a < 0$

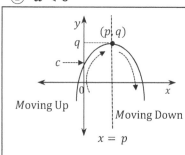

i) When $x < p$,

If the x−values are increasing, $\Rightarrow$ the y−values are increasing.
(Moving right)   (Moving up)

ii) When $x > p$,

If the x−values are increasing, $\Rightarrow$ the y−values are decreasing.
(Moving right)   (Moving down)

2) The Signs of the x^2-Coefficient

① $a > 0$

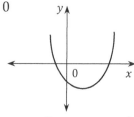

$\Rightarrow$ Opens upward

② $a < 0$

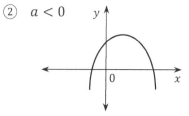

$\Rightarrow$ Opens downward

$y = ax^2 + bx + c$

① Sign of a:

$a > 0$ $a < 0$

② Sign of b:

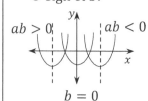

$ab > 0$ $ab < 0$

$b = 0$

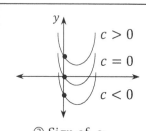

$c > 0$

$c = 0$

$c < 0$

③ Sign of c:

3) The Sides of the Axis of Symmetry

$$y = a(x-p)^2 + q, \ a > 0 \ \Rightarrow \ \text{Axis of symmetry}: \ x = p$$

① $p > 0$

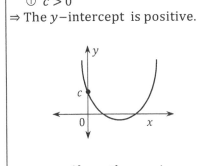

On the right side of the y-axis

② $p < 0$

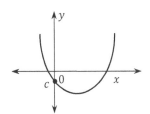

On the left side of the y-axis

③ $p = 0$

$\Rightarrow$ The y-axis is the axis of symmetry.

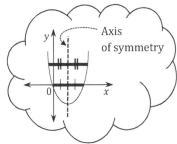

Note: $y = ax^2 + bx + c \ \Rightarrow \ y = a\left(x + \dfrac{b}{2a}\right)^2 - \dfrac{b^2 - 4ac}{4a}$

$\Rightarrow$ *Axis of symmetry is* $x = -\dfrac{b}{2a}$

i) *If* $-\dfrac{b}{2a} > 0$ ($\dfrac{b}{2a} < 0$; $\dfrac{b}{a} < 0$; a *and* b *have different signs*),

then the axis of symmetry is on the right side of the y-axis

ii) *If* $-\dfrac{b}{2a} < 0$ ($\dfrac{b}{2a} > 0$; $\dfrac{b}{a} > 0$; a *and* b *have the same sign*),

then the axis of symmetry is on the left side of the y-axis

iii) *If* $-\dfrac{b}{2a} = 0, \ then \ b = 0$

4) The Signs of the y-intercept

If $y = ax^2 + bx + c = a(x-p)^2 + q, \ a > 0$, then the y-intercept is c. ($c = ap^2 + q$)

① $c > 0$	② $c < 0$	③ $c = 0$
$\Rightarrow$ The y-intercept is positive.	$\Rightarrow$ The y-intercept is negative.	$\Rightarrow$ The y-intercept is 0.
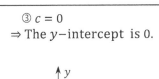	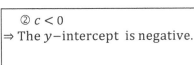	
Above the x-axis	Below the x-axis	Pass through the origin

(4) Solving Quadratic Functions

1) If the vertex (p, q) and one point are given,

$\Rightarrow \quad y = a(x - p)^2 + q$

Find a by substituting the point in the equation.

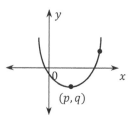

2) If the axis of symmetry $x = p$ and two different points are given,

$\Rightarrow \quad y = a(x - p)^2 + q$

Find a and q

by substituting the two points in the equation.

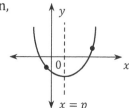

3) If three different points are given,

$\Rightarrow \quad y = ax^2 + bx + c$

Find a, b and c

by substituting the points in the equation.

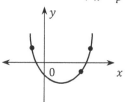

4) If the x-intercepts α, β and one point are given,

$\Rightarrow \quad y = a(x - \alpha)(x - \beta)$

Find a by substituting the point in the equation.

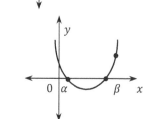

(5) Maximum and Minimum Values of Quadratic Functions

To get the maximum value or the minimum value of a quadratic function, transform the general form of the quadratic function $y = ax^2 + bx + c$ to the standard form of a quadratic function $y = a(x - p)^2 + q$.

If $a > 0 \Rightarrow$ minimum at the vertex, there is no maximum

If $a < 0 \Rightarrow$ maximum at the vertex, there is no minimum

1) When $y = a(x - p)^2 + q$

① $a > 0$

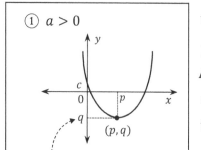

Minimum value (The lowest value)

When $x = p$,

$y = q$ is the minimum value of this quadratic function.

All y-values of this function are greater than or equal to q ($f(x) \geq q$). Since the graph opens upward, there is no maximum value.

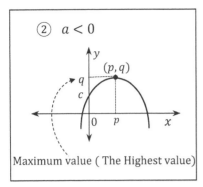

Maximum value (The Highest value)

When $x = p$,

$y = q$ is the maximum value of this quadratic function.

All y-values of this function are less than or

equal to q ($f(x) \leq q$). Since the graph opens downward,

there is no minimum value.

Note : $y = a(x - \alpha)(x - \beta)$

$\Rightarrow$ *This parabola has its maximum or minimum value at* $x = \frac{\alpha+\beta}{2}$

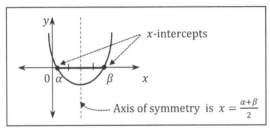

2) When $y = ax^2 + bx + c$

At first, transform to the standard form !

$$y = ax^2 + bx + c = a\left(x + \frac{b}{2a}\right)^2 - \frac{b^2-4ac}{4a}$$

① $a > 0$ $\Rightarrow$ Minimum value : $-\dfrac{b^2-4ac}{4a}$ when $x = -\dfrac{b}{2a}$

② $a < 0$ $\Rightarrow$ Maximum value : $-\dfrac{b^2-4ac}{4a}$ when $x = -\dfrac{b}{2a}$

3) When Maximum Value or Minimum Value is given:

① The minimum value q at $x = p$

$\Rightarrow$ $y = a(x - p)^2 + q$, $a > 0$

② The maximum value q at $x = p$

$\Rightarrow$ $y = a(x - p)^2 + q$, $a < 0$

Note: *Maximum and minimum values of a quadratic function* $f(x) = a(x - p)^2 + q$

in the limited interval $\alpha \leq x \leq \beta$

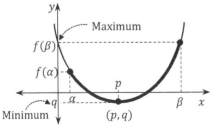

i) *If the vertex* (p, q) *is in the interval* $[\alpha, \beta]$,

then the y-coordinate $f(p) = q$ *of the vertex or*

the end points , $f(\alpha)$ and $f(\beta)$,

determine maximum or minimum values.

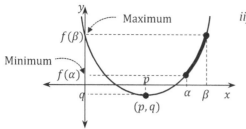

ii) *If the vertex is not in the interval* $[\alpha, \beta]$

; i.e., $p < \alpha$ or $p > \beta$, then

the end points, $f(\alpha)$ and $f(\beta)$,

determine maximum or minimum values.

(6) The Graph of a Quadratic Function and a Line

The relationships between a graph of a quadratic function $y = ax^2 + bx + c$ and a line $y = mx + n$ are determined by the discriminant D of the quadratic equation

$ax^2 + bx + c = mx + n$; i.e., $ax^2 + (b - m)x + c - n = 0$

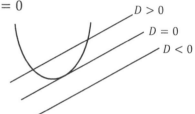

1) $D > 0 \iff$ Intersect at two different points

2) $D = 0 \iff$ Intersect at exactly one point

3) $D < 0 \iff$ Do not intersect at any point

(7) Quadratic Functions and Equations

For a quadratic function $y = ax^2 + bx + c$, let $D = b^2 - 4ac$ be the discriminant of a quadratic equation $ax^2 + bx + c = 0$. Then, we have the results as follows.

The sign of D	$D > 0$		$D = 0$		$D < 0$	
Graph of $y = ax^2 + bx + c$	$a > 0$	$a < 0$	$a > 0$	$a < 0$	$a > 0$	$a < 0$
Relationship	Intersect at two different points		Intersect at one point		Don't intersect at any point	
Solution of $ax^2 + bx + c = 0$	$x = \alpha$ or $x = \beta$ (Real number roots)		$x = \alpha$ (A double root) (Real, rational, root)		Not real number root (Imaginary roots)	

When $D = b^2 - 4ac > 0$,

if D is a perfect square, then the solution is rational.

if D is not a perfect square, then the solution is irrational.

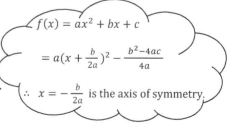

$f(x) = ax^2 + bx + c$

$= a(x + \dfrac{b}{2a})^2 - \dfrac{b^2 - 4ac}{4a}$

$\therefore \ x = -\dfrac{b}{2a}$ is the axis of symmetry.

<u>Properties of Solutions of a Quadratic Equation</u>

For a quadratic equation $ax^2 + bx + c = 0$, $a > 0$, let $D = b^2 - 4ac$.

Then, the quadratic function $f(x) = ax^2 + bx + c$ has the following results:

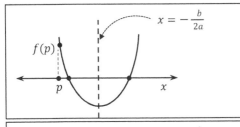

① Two solutions are greater than a point p.

$\iff D \geq 0$, $f(p) > 0$, and $-\dfrac{b}{2a} > p$

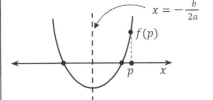

② Two solutions are less than a point p.

$\iff D \geq 0$, $f(p) > 0$, and $-\dfrac{b}{2a} < p$

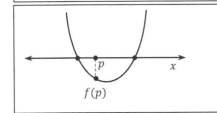

③ A point p is between the two solutions

(One solution is greater than p and

the other solution is less then p).

$\iff f(p) < 0$

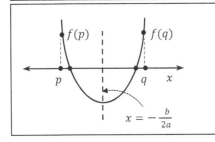

④ Two solutions are between p and q $(p < q)$.

$\iff D \geq 0$, $f(p) > 0$, $f(q) > 0$, and $p < -\dfrac{b}{2a} < q$

For any real number x, the quadratic inequality $ax^2 + bx + c > 0$, $a \neq 0$, is always true.

$\iff$ A quadratic function $y = ax^2 + bx + c$ takes on every value above the x-axis.

$\iff a > 0$, $D = b^2 - 4ac < 0$

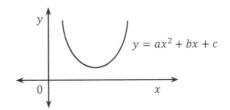

(8) Quadratic Functions and Inequalities

For a quadratic function $y = ax^2 + bx + c,\ a > 0,$

let $D = b^2 - 4ac$ be the discriminant of a quadratic equation $ax^2 + bx + c = 0.$

Then, we have the results as follows.

The sign of D	$D > 0$	$D = 0$	$D < 0$
Graph of $y = ax^2 + bx + c$			
Solution of $ax^2 + bx + c > 0$	$x < \alpha$ or $x > \beta$	All real numbers except $x = \alpha$ ($x < \alpha$, $\alpha < x$)	All real numbers
Solution of $ax^2 + bx + c \geq 0$	$x \leq \alpha$ or $x \geq \beta$	All real numbers	All real numbers
Solution of $ax^2 + bx + c < 0$	$\alpha < x < \beta$	No solution	No solution
Solution of $ax^2 + bx + c \leq 0$	$\alpha \leq x \leq \beta$	$x = \alpha$	No solution

3. Polynomial Functions of Higher Degree

(1) Graphs of Polynomial Functions

Consider a polynomial function $f(x) = x^n$, n is an integer greater than zero.

When n is even, the graph of $f(x) = x^n$ is similar to the graph of $f(x) = x^2$.

When n is odd, the graph of $f(x) = x^n$ is similar to the graph of $f(x) = x^3$.

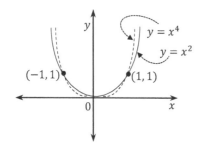
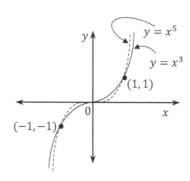

(2) The Leading Coefficient Test

The graph of a polynomial function $f(x) = a_n x^n + a_{n-1} x^{n-1} + a_{n-2} x^{n-2} + \cdots\cdots + a_1 x + a_0$

rises (increasing) or falls (decreasing) without bound as x moves to the right or left.

By the function's degree (even or odd) and its leading coefficient (positive or negative), the

graph of a polynomial function is determined.

Leading coefficient / Function's degree	$a_n > 0$ (Positive)	$a_n < 0$ (Negative)
n is even	Start here!	Start here!
n is odd	Start here!	Start here!

When the leading coefficient a_n is positive ($a_n > 0$), start graphing from the above the x-axis on

the right side and travel down and to the left in a wave shape.

When the leading coefficient a_n is negative ($a_n < 0$), start graphing from the below the x-axis

on the right side and travel up and to the left in a wave shape.

If the function's degree is 3, the graph may have 2 turning points.

Example   Graph the polynomial function $f(x) = -x^3 + 4x$.

$-x^3 + 4x = 0$

$\Rightarrow \quad -x(x^2 - 4) = -x(x + 2)(x - 2) = 0 \quad \therefore \ x = 0, \ x = -2, \text{ or } x = 2$

$\therefore \ x$-intercepts are $(0,0), \ (-2,0), \text{ and } (2,0)$.

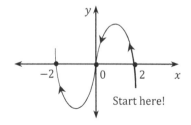

Since the leading coefficient of the function $f(x)$ is

negative, we start graphing below the x-axis from

right to left passing through the x-intercepts.

Example   Use x-intercepts to graph a polynomial function $f(x) = \frac{1}{2}(x + 2)(x - 1)^2$.

Since $x + 2$ and $x - 1$ are factors of $f(x)$, $x = -2$ and $x = 1$ are solutions of the function.

$\therefore$ $(-2, 0)$ and $(1, 0)$ are x-intercepts.

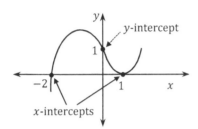

Since the leading coefficient of $f(x)$ is positive, we start graphing above the x-axis from right to left passing through the x-intercepts.

Since $x = 1$ is a double root, the point $(1, 0)$ will be a turning point.

(3) Real Zeros of Polynomial Functions

Let $f(x) = a_n x^n + a_{n-1} x^{n-1} + a_{n-2} x^{n-2} + \cdots\cdots + a_1 x + a_0$ be a polynomial.

Then, the following statements are equivalent.

① $x = a$ is a zero of the polynomial function $f(x)$.

② $x = a$ is a solution of the polynomial equation $f(x) = 0$.

③ $(x - a)$ is a factor of the polynomial equation $f(x) = 0$.

④ $(a, 0)$ is a x-intercept of the graph of the polynomial function $f(x)$ where a is a real number.

Turning Points

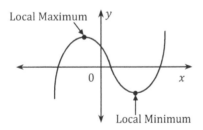

The y-coordinate of a turning point is a *local maximum* of the function if the point is higher than all nearby points.

The y-coordinate of a turning point is a *local minimum* of the function if the point is lower than all nearby points.

5-3 Rational Functions (Fractional Functions)

1. Rational Functions

(1) Direct and Inverse Variation

1) Direct Variation

If $y = ax$ for some fixed real number a $(a \neq 0)$, then

y varies directly as x or y is directly proportional to x.

(a is called the *constant of variation* or the *constant of proportionality*.)

2) Inverse Variation

If $y = \dfrac{a}{x}$ for some fixed real number a ($a \neq 0$), then

y varies inversely as x or y is inversely proportional to x.

(Two variables x and y show inverse variation.)

For example, $\dfrac{y}{2} = x \;\Rightarrow\; y = 2x$; y varies directly with x.

$\qquad\qquad y = 2x + 1 \;\Rightarrow\;$ Neither direct nor inverse.

$\qquad\qquad xy = 3 \;\Rightarrow\; y = \dfrac{3}{x}$; y varies inversely with x.

$\qquad\qquad z = 2xy \;\Rightarrow\; z$ varies jointly with x and y.

$\qquad\qquad y = \dfrac{4}{x^2} \;\Rightarrow\; y$ varies inversely with the square of x.

$\qquad\qquad z = \dfrac{5y}{x} \;\Rightarrow\; z$ varies directly with y and inversely with x.

Example The variables x and y vary inversely, and $y = 5$ when $x = 2$.

$\qquad\qquad$ Find an equation that relates x and y.

The general equation for inverse variation is $y = \dfrac{a}{x}$, $a \neq 0$.

Substituting 5 for y and 2 for x gives $5 = \dfrac{a}{2}$.

Solving for a, $a = 10$.

Therefore, the inverse variation equation is $y = \dfrac{10}{x}$.

(2) Graphing $y = \dfrac{a}{x}$, $(a \neq 0)$

The graph of this function is known as an inverse variation : A pair of smooth curves which are symmetric along the origin (0,0).

1) $a > 0$	2) $a < 0$
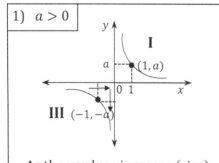	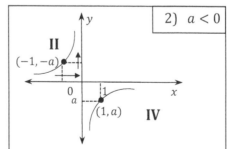
As the x values increase (rise) from left to right, the y values decrease in the coordinate plane.	As the x values increase (rise) from left to right, the y values increase as well in the coordinate plane.

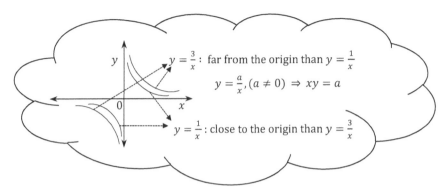

Note : From the graph of $y = \frac{a}{x}$, $(a \neq 0)$,

① *As the value of x moves away from 0, the graph moves closer to the x-axis, but it never crosses the x-axis.*

② *As the value of x approaches 0, it gets closer to the y-axis, but it never crosses the y-axis.*

(3) Rational Functions

A function f denoted by $y = f(x) = \frac{P(x)}{q(x)}$,

where $P(x)$ and $q(x)$ are polynomials and $q(x) \neq 0$ (not the zero polynomial),

is called a *rational function*.

Note that the rational function f is not defined at points where $q(x) = 0$. That is, the domain of a rational function of x includes all real numbers except x-value that makes the denominator zero.

For example, $f(x) = \frac{1}{x-2}$, $g(x) = \frac{1}{(x-1)(x+2)}$, and $h(x) = \frac{1}{2x^2+1}$

⇒ The domain of f excludes $x = 2$, and the domain of g excludes $x = 1$ and $x = -2$.

The domain of h is all real numbers.

(∵ There are no real numbers of x which make the denominator $2x^2 + 1$ equal to zero.)

Note:

① *$y = \frac{x}{x}$ is not defined at $x = 0$. However, $\frac{x}{x} = 1$ for other values of x.*

Therefore, the two functions $\frac{x}{x}$ and 1 are not identical.

② *Similarly, for two rational functions $y = \frac{x(x-1)}{x-1}$ and $y = x$,*

when $x \neq 1$, $\frac{x(x-1)}{x-1} = x$. That is, the two functions have the same values.

However, when $x = 1$, the function $y = \frac{x(x-1)}{x-1}$ is not defined, whereas the other function $y = x$ has

the value 1. Therefore, the two functions are not identical.

2. Graphing Rational Functions

(1) The Hyperbola

Consider a simple rational function $y = \dfrac{1}{x}$.

The graph of the function is called a *hyperbola*. Form the graph of the function, we can see that:

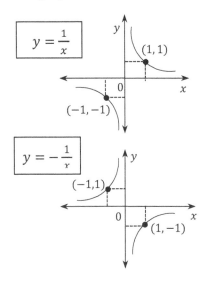

① As the value of x gets away from 0, the graph gets closer to the x-axis. But, it never crosses the x-axis.

② As the value of x approaches 0, the graph gets closer to the y-axis. But, it never crosses the y-axis.

When $x = 0$, the denominator is zero. Thus, the domain of f is all real numbers except $x = 0$.

(2) The Asymptotes

1) Horizontal and Vertical Asymptotes

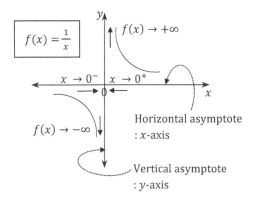

From the graph of $f(x) = \dfrac{1}{x}$,

we see $f(x)$ decreases without bound

as x approaches 0 from the left

$(f(x) \to -\infty$ as $x \to 0^{-})$ and

$f(x)$ increases without bound

as x approaches 0 from the right

$(f(x) \to \infty$ as $x \to 0^{+})$.

The line which a graph approaches indefinitely is called an *asymptote* of the graph.

The line $x = 0$ (the y-axis) is a *vertical asymptote* of the graph of f and

the line $y = 0$ (the x-axis) is a *horizontal asymptote* of the graph of f.

Note that the values of $f(x) = \dfrac{1}{x}$ approach zero as x increases or decreases without bound,

that is, $f(x) \to 0$ as $x \to -\infty$; $f(x) \to 0$ as $x \to \infty$.

2) Graphing $y = \dfrac{k}{x-p} + q$, $k \neq 0$

For all rational functions of the form $y = \dfrac{k}{x-p} + q$, $k \neq 0$,

the functions' graphs are hyperbolas with asymptotes at $x = p$ and $y = q$.

That is, the equations of the asymptotes of the graph of $y = \dfrac{k}{x-p} + q$, $k \neq 0$, are $x = p$ and

$y = q$. The graph of $y = \dfrac{k}{x-p} + q$ has been translated from the graph of $y = \dfrac{k}{x}$, p units along

the x-axis and q units along the y-axis. The domain is all real numbers except $x = p$ and the

range is all real numbers except $y = q$.

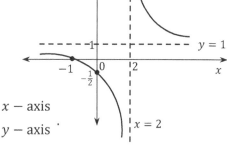

If $c = 0$, then
y is not a rational function.
If $ad - bc = 0$, then
y is a constant function.

Note: $y = \dfrac{ax+b}{cx+d}$, $c \neq 0$, $ad - bc \neq 0$

$\Rightarrow$ *Transform the equation to the form* $y = \dfrac{k}{x-p} + q$, $k \neq 0$.

Therefore, all rational functions of the form $y = \dfrac{ax+b}{cx+d}$ *also have graphs that are hyperbolas.*

The vertical asymptote occurs at the x-value that makes the dominator zero.

The horizontal asymptote is the line $y = \dfrac{a}{c}$

Note: $y = \dfrac{ax+b}{cx+d} = \dfrac{b - \frac{ad}{c}}{cx+d} + \dfrac{a}{c} = \dfrac{\frac{b}{c} - \frac{ad}{c^2}}{x + \frac{d}{c}} + \dfrac{a}{c}$

$\therefore$ $x = -\dfrac{d}{c}$ *and* $y = \dfrac{a}{c}$ *are asymptotes.*

Example Find the equations of the asymptotes of the rational function $y = \dfrac{x+1}{x-2}$

and graph the function.

$y = \dfrac{x+1}{x-2} = \dfrac{x-2+3}{x-2} = 1 + \dfrac{3}{x-2}$

$\therefore$ The equations of the asymptotes are $x = 2$, $y = 1$.

Since $x = 0 \Rightarrow y = -\dfrac{1}{2}$ and $y = 0 \Rightarrow x = -1$,

the points of intersections with the axes are $\begin{cases} (-1, 0) : x - \text{axis} \\ \left(0, -\dfrac{1}{2}\right) : y - \text{axis} \end{cases}$.

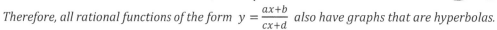

The domain is all real numbers except $x = 2$ and the range is all real numbers except $y = 1$.

Example Identify the horizontal and vertical asymptotes of the graph of the function

$$y = \frac{-2x+3}{2x+1}.$$ Then state the domain and range.

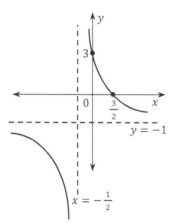

$$y = \frac{-2x+3}{2x+1} = \frac{-(2x+1)+4}{2x+1} = -1 + \frac{4}{2x+1}$$

∴ The equations of the asymptotes are

$$x = -\frac{1}{2} \text{ (vertical asymptote)}, \quad y = -1 \text{ (horizontal asymptote)}.$$

Since $x = 0 \Rightarrow y = 3$ and $y = 0 \Rightarrow x = \frac{3}{2}$,

the points of intersections with the axes are $\begin{cases} \left(\frac{3}{2}, 0\right) : x - \text{axis} \\ (0, 3) : y - \text{axis} \end{cases}$.

The domain is all real numbers except $-\frac{1}{2}$ and the range is all real numbers except -1.

Example State how the graph of $y = \frac{2x-3}{2x-4}$ has been translated from the graph of $y = \frac{1}{2x}$.

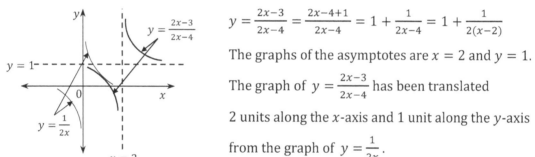

$$y = \frac{2x-3}{2x-4} = \frac{2x-4+1}{2x-4} = 1 + \frac{1}{2x-4} = 1 + \frac{1}{2(x-2)}$$

The graphs of the asymptotes are $x = 2$ and $y = 1$.

The graph of $y = \frac{2x-3}{2x-4}$ has been translated

2 units along the x-axis and 1 unit along the y-axis

from the graph of $y = \frac{1}{2x}$.

3) Graphing General Rational Functions

Let $f(x) = \frac{P(x)}{q(x)} = \frac{a_n x^n + a_{n-1}x^{n-1} + a_{n-2}x^{n-2} + \cdots\cdots + a_1 x + a_0}{b_m x^m + b_{m-1}x^{m-1} + b_{m-2}x^{m-2} + \cdots\cdots + b_1 x + b_0}$ be the rational function

Where $p(x)$ and $q(x)$ are polynomials with no common factors other than 1. Then,

① the x-intercepts of the graph of f are the real zeros of $p(x)$.

② the graph of f has vertical asymptotes at the zeros of $q(x)$.

③ the graph of f has at most one horizontal asymptote, as follows.

    i) If $n < m$, then the x-axis ($y = 0$) is a horizontal asymptote.

    ii) If $n = m$, then the line $y = \frac{a_n}{b_m}$ is a horizontal asymptote.

    iii) If $n > m$, then the graph of f has no horizontal asymptote.

Example   Graph each function:

$$① \ y = \frac{2}{x^2+1} \quad ② \ y = \frac{5x^2}{x^2-1} \quad ③ \ y = \frac{x^2-3x+2}{x+2}$$

① $y = \frac{2}{x^2+1}$

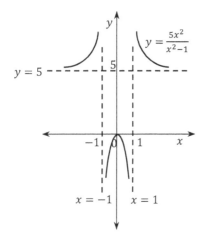

Since the numerator has no zeros, there is no x-intercept. Since the denominator has no real number zeros, there is no vertical asymptote. Since the numerator has degree of 0 and the denominator has degree of 2, the horizontal asymptote is $y = 0$ (the x-axis).

The domain is all real numbers and the range is $\{y \mid 0 < y \le 2\}$.

② $y = \frac{5x^2}{x^2-1}$

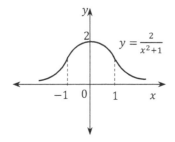

Since the numerator has 0 as its zero, x-intercept is $(0, 0)$.

Since the denominator $x^2 - 1$ is factored as $(x + 1)(x - 1)$, there are two zeros $x = 1$ and $x = -1$.

That means vertical asymptotes are $x = 1$ and $x = -1$.

Since the degrees of the numerator and denominator are the same, the horizontal asymptote is the ratio of the leading coefficients; i.e., $y = \frac{5}{1} = 5$.

Plotting some points between and beyond the vertical asymptotes, we have the graph of y.

③ $y = \frac{x^2-3x+2}{x+2}$

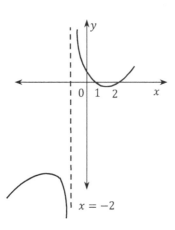

Since the numerator is factored as $(x - 1)(x - 2)$, x-intercepts are $(1, 0)$ and $(2, 0)$.

Since the denominator $x + 2$ has zero $x = -2$, the vertical asymptotes is $x = -2$.

Since the numerator has degree of 2 and the denominator has degree of 1; i.e., the degree of the numerator is greater than the degree of the denominator, there is no horizontal asymptote.

(3) Slant Asymptotes

Consider a graph of a function $y = \dfrac{x^2+1}{x}$.

The degree of the numerator of the rational function is exactly one more than the degree of its denominator. In this case, the graph of the function has a slant asymptote. Using long division,

we have $y = \dfrac{x^2+1}{x} = \boxed{x} + \dfrac{1}{x}$

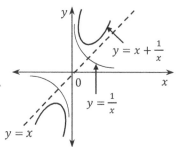

Slant asymptote ; $y = x$

For the graph of $y = ax + \dfrac{b}{x}$, $y = ax$ is called a *slant asymptote*.

∴ The graph has asymptotes $x = 0$ and $y = ax$.

Note: The value of the y-coordinate of $y = x + \dfrac{1}{x}$ is

   (the value of the y-coordinate of $y = \dfrac{1}{x}$) + (the value of the y-coordinate of $y = x$).

$y = x + \dfrac{1}{x}$	$y = -x - \dfrac{1}{x}$	$y = -x + \dfrac{1}{x}$	$y = x - \dfrac{1}{x}$
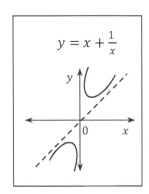	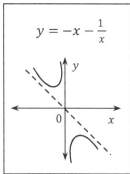	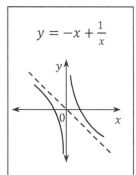	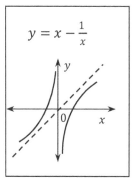

Example   Graph the function $f(x) = \dfrac{x^2-x-2}{x-1}$.

$f(x) = \dfrac{x^2-x-2}{x-1} = \dfrac{(x-2)(x+1)}{x-1}$

∴ x-intercepts are $(2, 0)$ and $(-1, 0)$.

Using long division, $f(x) = x - \dfrac{2}{x-1}$.   ∴ $y = x$ is a slant asymptote.

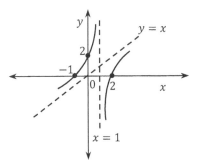

Therefore, x-intercepts are $(2, 0)$, $(-1, 0)$,

   y-intercept (when $x = 0$) is $(0, 2)$,

   vertical asymptote is $x = 1$, and

   slant asymptote is $y = x$.

Example   Find the number of common points between the graphs of the functions:

$$y = \frac{x}{x-1} \ , \ \ y = -x + k$$

Let $\frac{x}{x-1} = -x + k$

Then, $x = (-x + k)(x - 1), \ x \neq 1$ Cross product

$$= -x^2 + kx + x - k$$

$$\therefore \ \ x^2 - kx + k = 0 \cdots\cdots \ ①$$

The discriminant D of $①$ is $D = (-k)^2 - 4 \cdot 1 \cdot k = k^2 - 4k = k(k - 4)$.

If $D > 0$; i.e., $k(k - 4) > 0$; $k > 4$ or $k < 0$, then there are 2 different common points.

If $D = 0$; i.e., $k(k - 4) = 0$; $k = 4$ or $k = 0$, then there is only one common point.

If $D < 0$; i.e., $k(k - 4) < 0$; $0 < k < 4$, then there is no common point.

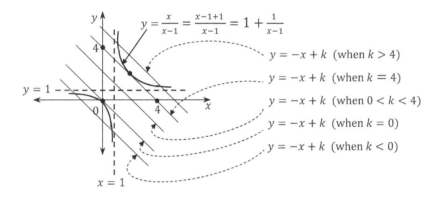

5-4 Graphing Fractional and Radical Equations

1. Fractional Equations

Note: ___Operations with Rational Expressions___

For any polynomials $A, B, C,$ and $D,$

① $\frac{A}{C} \pm \frac{B}{C} = \frac{A \pm B}{C}$, $(C \neq 0)$

② $\frac{A}{B} \times \frac{C}{D} = \frac{AC}{BD}$, $(B \neq 0, \ D \neq 0)$

③ $\frac{A}{B} \div \frac{C}{D} = \frac{A}{B} \times \frac{D}{C} = \frac{AD}{BC}$, $(B \neq 0, \ C \neq 0, \ D \neq 0)$

(1) Fractional Equations

A fractional equation is an equation which contains a variable in the denominator.

For example, $\frac{2}{x} + \frac{1}{3} = \frac{5}{4x}$, $\frac{x}{x+1} - \frac{1}{x-1} = \frac{2}{x^2-1}$, $\cdots$

Rational equations consist of polynomial eqautions and fractional equations.

(2) Steps for Solving Fractional Equations

Step 1. Multiply each member of both sides of the fractional equation

by the least common multiple (LCM) of the denominators of its fractions.

Step 2. Solve the polynomial equation obtained from Step 1.

Step 3. If the solution of the polynomial equation includes extra roots, called *extraneous roots*,

which are not roots of the original fractional equation, then exclude them from the

solution.

Step 4. Check for solution (except the extraneous roots) that satisfies the original fractional

equation.

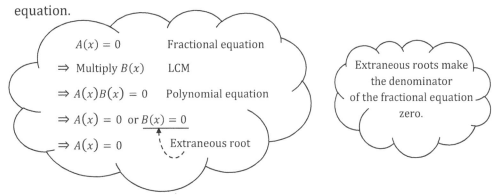

Example   Solve the fractional equation $\dfrac{x}{x+3} - \dfrac{1}{x-3} = -\dfrac{6}{x^2-9}$.

$\dfrac{x}{x+3} - \dfrac{1}{x-3} = -\dfrac{6}{x^2-9}$ Original equation

$\Rightarrow \dfrac{x}{x+3} - \dfrac{1}{x-3} = -\dfrac{6}{(x+3)(x-3)}$ Rewrite the equation with its denominator factored

$\Rightarrow$ Multiplying both sides by the LCM of the denominator, $(x+3)(x-3)$,

$\dfrac{x}{x+3} \cdot (x+3)(x-3) - \dfrac{1}{x-3} \cdot (x+3)(x-3) = -\dfrac{6}{(x+3)(x-3)} \cdot (x+3)(x-3)$

$\Rightarrow x(x-3) - (x+3) = -6$

$\Rightarrow x^2 - 4x + 3 = 0$

$\Rightarrow (x-1)(x-3) = 0$

$\Rightarrow x-1 = 0 \text{ or } x-3 = 0$

$\Rightarrow x = 1 \text{ or } x = 3$

If $x = 3$, then the fractions $\dfrac{1}{x-3}$ and $\dfrac{6}{x^2-9}$ in the original equation are not defined.

($\because$ The denomiantors $x - 3$ and $x^2 - 9$ will be zero.)

Thus, $x = 3$ is an extraneous root. Therefore, $x = 3$ is rejected.

Hence, $x = 1$ is the only root of the fractional equation .

2. Radical Equations

Radical equations are equations that contain radicals or rational exponents whose radicands include variables. For example, $\sqrt{x} + 2 = x$, $\sqrt{x+1} + \sqrt{2x} = 1$, $\cdots$

Radical equations are considered in real number system.

Thus, in the radical equation $\sqrt{f(x)}$, $f(x) \geq 0$.

(1) Graphing Radical Equations (Square Root and Cube Root Functions)

Consider a radical function $y = \sqrt{x}$ (*Square root function*).

The function is one-to-one from $X = \{x \mid x \geq 0\}$ to $Y = \{y \mid y \geq 0\}$.

Thus, the inverse function exists and it is $y = x^2$, $x \geq 0$.

Therefore, the graph of rational function $y = \sqrt{x}$ and the graph of $y = x^2$, $x \geq 0$ are

symmetric with respect to the line $y = x$.

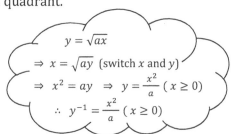

1) $y = \sqrt{ax}$, $a \neq 0$

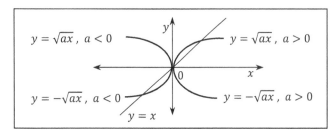

① If $a > 0$, then the graph of $y = \sqrt{ax}$ is in the first quadrant.

   The domain is $\{x \mid x \geq 0\}$ and the range is $\{y \mid y \geq 0\}$.

② If $a < 0$, then the graph of $y = \sqrt{ax}$ is in the second quadrant.

The domain is $\{x \mid x \leq 0\}$ and the range is $\{y \mid y \geq 0\}$.

The inverse function of $y = \sqrt{ax}$ is $y = \dfrac{x^2}{a}$, $x \geq 0$.

Therefore, the graph of $y = \sqrt{ax}$ and the graph of

$y = \dfrac{x^2}{a}$ ($x \geq 0$) are symmetric along the line $y = x$.

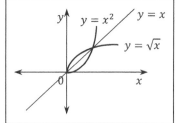

2) $y = \sqrt{a(x-p)} + q$, $a \neq 0$

The graph of $y = \sqrt{a(x-p)} + q$, $a \neq 0$ is a translation of the graph of $y = \sqrt{ax}$, $a \neq 0$

with p units along the x-axis and q units along the y-axis.

① If $a > 0$, then the domain is $\{x \mid x \geq p\}$ and the range is $\{y \mid y \geq q\}$.

② If $a < 0$, then the domain is $\{x \mid x \leq p\}$ and the range is $\{y \mid y \geq q\}$.

3) $y = \sqrt{ax + b} + c$, $a \neq 0$

Transform the equation $y = \sqrt{ax + b} + c$, $a \neq 0$ to the equation $y = \sqrt{a(x + \frac{b}{a})} + c$, $a \neq 0$.

The graph of $y = \sqrt{a(x + \frac{b}{a})} + c$ is a translation of the graph of $y = \sqrt{ax}$

with $-\frac{b}{a}$ units along the x-axis and c units along the y-axis.

① If $a > 0$, then the domain is $\{x \mid x \geq -\frac{b}{a}\}$ and the range is $\{y \mid y \geq c\}$.

② If $a < 0$, then the domain is $\{x \mid x \leq -\frac{b}{a}\}$ and the range is $\{y \mid y \geq c\}$.

Consider a radical function $y = \sqrt[3]{x}$ (*Cube root function*).

The function is one-to-one from $X = \{x \mid x \text{ is all real numbers}\}$ to $Y = \{y \mid y \text{ is all real numbers}\}$.

Thus, the inverse function exists and it is $y = x^3$.

Therefore, the graph of the radical function $y = \sqrt[3]{x}$ and

the graph of $y = x^3$ are symmetric along the line $y = x$.

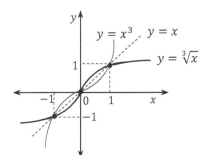

To graph $y = a\sqrt[3]{x - p} + q$,

Step 1. Sketch the graph of $y = a\sqrt[3]{x}$.

Step 2. Shift the graph p units horizontally and q units vertically.

Example   Graph the rational function $y = \sqrt{2x + 4} + 1$ and give its domain and range.

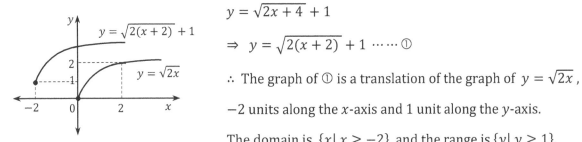

$y = \sqrt{2x + 4} + 1$

$\Rightarrow\ y = \sqrt{2(x + 2)} + 1 \ \cdots\cdots ①$

∴ The graph of ① is a translation of the graph of $y = \sqrt{2x}$,

-2 units along the x-axis and 1 unit along the y-axis.

The domain is $\{x \mid x \geq -2\}$ and the range is $\{y \mid y \geq 1\}$.

(2) Steps for Solving Radical Equations

Raise each side of the equation to the same power.

> Raising both sides of an eqaution to the same power may include extraneous roots. So, you have to check each solution in the original equation.

Note: ***Powers Property of Equality***

If $a = b$, then $a^n = b^n$.

Step 1. Isolate the radical expression on one side of the equation.

Step 2. Raise (square or cube, or $\cdots$) both sides of the equation to the power that eliminates the radical.

Step 3. Solve the polynomial equation obtained from Step 2.

Step 4. If the solution of the polynomial equation includes extraneous roots, which do not satisfy the original radical equation, then exclude them.

Step 5. Check for solution that satisfies the original radical equation.

> If all of the solutions of a radical equation are extraneous, then the equation has no solution.

Example   Solve $\sqrt[3]{x} - 2 = 0$ and $2x^{\frac{3}{2}} = 16$.

$\sqrt[3]{x} - 2 = 0 \quad \Rightarrow \quad \sqrt[3]{x} = 2$ Isolate radical

$\qquad\qquad \Rightarrow \quad (\sqrt[3]{x})^3 = 2^3$ Cube each side

$\qquad\qquad \Rightarrow \quad x = 8$ Simplify

Therefore, the solution is 8.

$2x^{\frac{3}{2}} = 16 \quad \Rightarrow \quad x^{\frac{3}{2}} = 8$ Isolate power

$\qquad\qquad \Rightarrow \quad (x^{\frac{3}{2}})^{\frac{2}{3}} = 8^{\frac{2}{3}}$ Raise each side to $\dfrac{2}{3}$ power

$\qquad\qquad \Rightarrow \quad x = (2^3)^{\frac{2}{3}} = 2^2 = 4$ Simplify

Therefore, the solution is 4.

(3) Solving Radical Equations by using Graphs

The solution of the radical equation $\sqrt{f(x)} = g(x)$ is the x-coordinates of the intersection points of the graphs of two functions $y = \sqrt{f(x)}$ and $y = g(x)$.

① Type of $\sqrt{A} = B$: Eliminate the radical by squaring both sides of the equation.

② Type of $\sqrt{A} + \sqrt{B} = C$

: Transform in form of $\sqrt{A} = C - \sqrt{B}$ and then square both sides of the equation to have the type of ①

③ Type of $\sqrt{A} + \sqrt{B} = \sqrt{C} + \sqrt{D}$

: Square both sides of the equation to have the type of ① or ②.

④ If the equation has the same term, replace it with one variable.

Example   Solve a radical equation $\sqrt{x} + 2 = x$.

$\sqrt{x} + 2 = x \Rightarrow \sqrt{x} = x - 2$ Isolate radicals

$\Rightarrow (\sqrt{x})^2 = (x-2)^2$ Square both sides

$\Rightarrow x = x^2 - 4x + 4$ Simplify left side and expand right side

$\Rightarrow x^2 - 5x + 4 = 0$ Combine like terms and write in standard form

$\Rightarrow (x-1)(x-4) = 0$ Factor

$\Rightarrow x - 1 = 0$ or $x - 4 = 0$ Zero product property

$\Rightarrow x = 1$ or $x = 4$ Simplify

> Solution check:
> Substitute 1 for x.
> Then, $\sqrt{1} + 2 = 1$;
> $3 = 1$ (It's false.)
> Substitute 4 for x.
> Then, $\sqrt{4} + 2 = 4$;
> $2 + 2 = 4$ (It's true.)

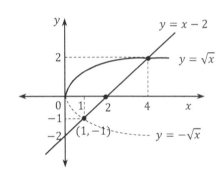

Since $\sqrt{x} \geq 0$, $\sqrt{x} = x - 2 \geq 0$ $\therefore x \geq 2$

Thus, $x = 1$ is an extraneous root, and is rejected.

Therefore, the solution of $\sqrt{x} + 2 = x$ is $x = 4$

$x = 4$ is the x-coordinate of the intersection point of

$y = \sqrt{x}$ and $y = x - 2$.

But, the extraneous root $x = 1$ is the x-coordinate

of the intersection point of $y = -\sqrt{x}$ and $y = x - 2$.

Example   Solve a radical equation $\sqrt{2x + 1} + \sqrt{x} = 1$.

$\sqrt{2x+1} + \sqrt{x} = 1 \Rightarrow \sqrt{2x+1} = 1 - \sqrt{x}$ Subtract $\sqrt{x}$ to each side

$\Rightarrow (\sqrt{2x+1})^2 = (1 - \sqrt{x})^2$ Square both sides

$\Rightarrow 2x + 1 = 1 - 2\sqrt{x} + x$ Simplify left side and expand right side

$\Rightarrow 2\sqrt{x} = -x$ Type of ①

$\Rightarrow (2\sqrt{x})^2 = (-x)^2$ Square both sides

$\Rightarrow 4x = x^2$ Simplify

$\Rightarrow x(x - 4) = 0$ Factor

$\Rightarrow x = 0$ or $x = 4$ Solve for x

Check : $x = 0 \Rightarrow \sqrt{2 \cdot 0 + 1} + \sqrt{0} = 1$ Substitute 0 for x

$\Rightarrow \sqrt{1} = 1$ It's true. ($\because$ $x = 0$ is a solution.)

$x = 4 \Rightarrow \sqrt{2 \cdot 4 + 1} + \sqrt{4} = 1$ Substitute 4 for x

$\Rightarrow 3 + 2 = 1$ It's false. ($\because$ $x = 4$ (extraneous root) is rejected.)

Therefore, $x = 0$ is the solution of the equation $\sqrt{2x + 1} + \sqrt{x} = 1$.

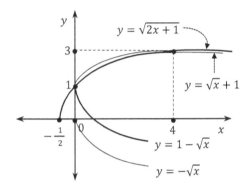

$x = 0$ is the x-coordinate of the intersection point

of $y = \sqrt{2x + 1}$ and $y = 1 - \sqrt{x}$.

But, the extraneous root $x = 4$ is

the x-coordinate of the intersection point of

$y = \sqrt{2x + 1}$ and $y = 1 + \sqrt{x}$.

Example   Solve a radical equation $x^2 + x - \sqrt{x^2 + x - 2} = 4$.

Let $\sqrt{x^2 + x - 2} = t$, $t \geq 0$

Then, $x^2 + x - 2 = t^2$ $\therefore$ $x^2 + x = t^2 + 2$

$\therefore$ The given equation is $(t^2 + 2) - t = 4$; i.e., $t^2 - t - 2 = 0$

$\therefore$ $(t - 2)(t + 1) = 0$

$\therefore$ $t = 2$ or $t = -1$

Since $t \geq 0$, $t = 2$

Thus, $\sqrt{x^2 + x - 2} = 2$ $\cdots\cdots$ ①

Squaring both sides of ①, $x^2 + x - 2 = 4$

$\therefore$ $x^2 + x - 6 = 0$

$\therefore$ $(x + 3)(x - 2) = 0$

$\therefore$ $x = -3$ or $x = 2$

Check : If $x = -3$, then $(-3)^2 + (-3) - \sqrt{(-3)^2 + (-3) - 2} = 4$

$9 - 3 - \sqrt{4} = 4$; $4 = 4$ (It's true).

If $x = 2$, then $2^2 + 2 - \sqrt{2^2 + 2 - 2} = 4$

$4 + 2 - \sqrt{4} = 4$; $4 = 4$ (It's true).

Therefore, $x = -3$ and $x = 2$ are the roots of the equation.

(4) Relationship Between a Radical Function $y = \sqrt{x - p}$ and a Line $y = x + k$

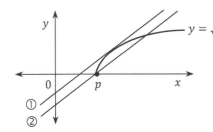

① is the line that is intersecting the graph of $y = \sqrt{x - p}$ at one point.

② is the line that is passing through a point $(p, 0)$.

i)   The rational function $y = \sqrt{x - p}$ and the line $y = x + k$ intersect at two different points.

⟺ The line $y = x + k$ is between the line ① and the line ②, or the line ②.

ii) The rational function $y = \sqrt{x - p}$ and the line $y = x + k$ intersect at one point.

⟺ The line ① or the right side of the line ②.

iii) The rational function $y = \sqrt{x - p}$ and the line $y = x + k$ don't intersect at any point.

⟺ The left side of the line ①.

Example   Determine the range of the real number k so that the radical equation $\sqrt{x - 1} = x + k$ has two different real number solutions.

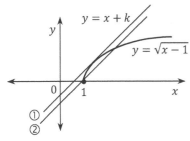

When the line $y = x + k$ is between the lines ① and ②, the equation $\sqrt{x - 1} = x + k$ has two different real number solutions.

Case 1. Find the equation of the line ①.

Let $\sqrt{x - 1} = x + k$

Squaring both sides, $x - 1 = x^2 + 2kx + k^2$ ∴ $x^2 + (2k - 1)x + k^2 + 1 = 0$

Since the discriminant D of the equation is $D = 0$, $D = (2k - 1)^2 - 4(k^2 + 1) = 0$

∴ $4k^2 - 4k + 1 - 4k^2 - 4 = 0$ ∴ $k = -\dfrac{3}{4}$

Case 2. Find the equation of the line ②.

Since the line $y = x + k$ passes through the point $(1, 0)$, $0 = 1 + k$ ∴ $k = -1$

Therefore, by Case 1 and Case 2, $-1 \leq k < -\dfrac{3}{4}$

5-5 Graphing Inequalities

1. Inequalities with Higher Degree

Consider the polynomial expression $f(x)$ of degree n, $n \geq 3$

(1) Steps for Solving Inequalities with Higher Degree

Step 1. Rearrange all terms to the left side of the algebraic inequality symbol such that
$$f(x) > 0, \ f(x) \geq 0, \ f(x) < 0, \ \text{or} \ f(x) \leq 0$$

Step 2. In the real number system, solve the equation $f(x) = 0$ by using factorization formula, factor theorem, or synthetic division.

Step 3. Divide intervals by the solution of $f(x) = 0$.

If $f(x) > 0$, then find the range of x such that $f(x) > 0$.

If $f(x) < 0$, then find the range of x such that $f(x) < 0$.

Example   Solve the inequality $f(x) = (x - 1)(x + 1)(x + 2) > 0$.

$(x - 1)(x + 1)(x + 2) = 0 \Rightarrow x = 1, \ x = -1, \ \text{or} \ x = -2$

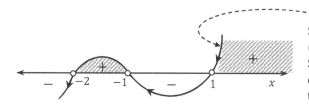

Start here, above the x-axis.
($\because$ The leading coefficient is positive.)
Since the degree of the function is 3, draw 3 turns from right to left passing the x-intercepts.

Since $f(x) > 0$, the shaded parts (above the x-axis) are the solution.

That is, $-2 < x < -1$ or $x > 1$

Example   Solve the inequality $f(x) = x^4 + 2x^3 - 9x^2 - 2x + 8 \leq 0$

Let $f(x) = x^4 + 2x^3 - 9x^2 - 2x + 8$.

Since $f(1) = 0$ and $f(-1) = 0$, by the factor theorem and synthetic division,

$$
\begin{array}{r|rrrrr}
1 & 1 & 2 & -9 & -2 & 8 \\
  &   & 1 & 3 & -6 & -8 \\
\hline
-1 & 1 & 3 & -6 & -8 & \boxed{0} \\
   &   & -1 & -2 & 8 & \\
\hline
   & 1 & 2 & -8 & \boxed{0} &
\end{array}
$$

$\therefore \ f(x) = (x - 1)(x + 1)(x^2 + 2x - 8)$

$\therefore\ f(x) = (x - 1)(x + 1)(x + 4)(x - 2)$

$(x - 1)(x + 1)(x + 4)(x - 2) = 0\ \Rightarrow\ x = 1,\ x = -1,\ x = -4,\ \text{or}\ x = 2$

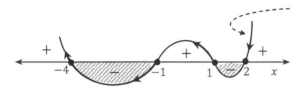

Start here, above the x-axis.
($\because$ The leading coefficient is positive.)
Since the degree of the function is 4,
draw 4 turns from right to left passing
the x-intercepts.

Since $f(x) \leq 0$, the shaded parts (below the x-axis) are the solution.

That is, $-4 \leq x \leq -1$ or $1 \leq x \leq 2$

(2) Solving Inequalities that contain Special Factors

1) For any two polynomials $f(x)$ and $g(x)$,

① When $f(x) > 0$ is always true for all real number x, $f(x) > g(x) > 0\ \Leftrightarrow\ g(x) > 0$

② $(x - a)^2 f(x) > 0\ \Leftrightarrow\ f(x) > 0$ and $x \neq a$

③ $(x - a)^2 f(x) \geq 0\ \Leftrightarrow\ f(x) \geq 0$ or $x = a$

④ $(x - a)^3 f(x) > 0\ \Leftrightarrow\ (x - a)f(x) > 0$

⑤ $(x - a)^3 f(x) \geq 0\ \Leftrightarrow\ (x - a)f(x) \geq 0$

When n is even, $(x - a)^n f(x) > 0\ \Leftrightarrow\ f(x) > 0$ and $x \neq a$
$(x - a)^n f(x) \geq 0\ \Leftrightarrow\ f(x) \geq 0$ or $x = a$
When n is odd, $(x - a)^n f(x) > 0\ \Leftrightarrow\ (x - a)f(x) > 0$
$(x - a)^n f(x) \geq 0\ \Leftrightarrow\ (x - a)f(x) \geq 0$

2) When $a < b < c$,

Inequalities of degree 3:

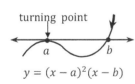

$y = (x - a)^2(x - b)$

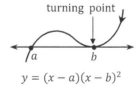

$y = (x - a)(x - b)^2$

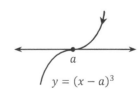

$y = (x - a)^3$

Inequalities of degree 4:

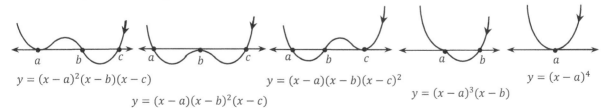

$y = (x - a)^2(x - b)(x - c)$

$y = (x - a)(x - b)^2(x - c)$

$y = (x - a)(x - b)(x - c)^2$

$y = (x - a)^3(x - b)$

$y = (x - a)^4$

2. Fractional Inequalities

For a fractional inequality $\dfrac{f(x)}{g(x)} \le 0$ or $\dfrac{f(x)}{g(x)} < 0$,

if $g(x) \neq 0$, then $\{g(x)\}^2 > 0$.

Thus, we can transform the fractional inequality to the polynomial inequality,

by multiplying $\{g(x)\}^2$, in form of $f(x)g(x) \le 0$ or $f(x)g(x) < 0$.

(1) Steps for Solving Fractional Inequalities

Step 1.   Rewrite all terms to the left side of the algebraic inequality symbol such that

$$\frac{f(x)}{g(x)} > 0\,,\frac{f(x)}{g(x)} \ge 0\,,\frac{f(x)}{g(x)} < 0\,, \text{ or } \frac{f(x)}{g(x)} \le 0$$

Step 2.   Multiply both sides by the square of the denominator, $\{g(x)\}^2$,

(if $g(x) > 0$, then multiply both sides by the denominator $g(x)$)

and find the solution of $f(x)g(x) > 0$, $f(x)g(x) \ge 0$, $f(x)g(x) < 0$, or $f(x)g(x) \le 0$

Step 3.   From the obtained solution, exclude the extraneous roots which make the denominator

$g(x)$ zero.

Example   Solve the fractional inequality $\dfrac{x^2+3x-8}{x-1} \ge 5$.

$$\frac{x^2+3x-8}{x-1} \ge 5$$

$\Rightarrow \dfrac{x^2+3x-8-5(x-1)}{x-1} \ge 0$ Subtract 5 to each side

$\Rightarrow \dfrac{x^2-2x-3}{x-1} \ge 0$ Simplify

$\Rightarrow \dfrac{(x-3)(x+1)}{x-1} \ge 0$ Factor

$\Rightarrow (x-3)(x+1)(x-1) \ge 0$, $x-1 \neq 0$ Multiply both sides by $(x-1)^2$

If $x = 1$, then the denominator of the given fractional inequality will be zero.

Thus, $x = 1$ (extraneous root) is rejected.

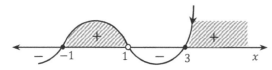

Therefore, the solution is $-1 \le x < 1$ or $x \ge 3$.

(2) Solving Fractional Inequalities that contain Special Factors

① $\dfrac{f(x)}{(x-a)^2} \geq 0 \iff f(x) \geq 0$ and $x \neq a$

② $\dfrac{(x-a)^2}{f(x)} \geq 0 \iff f(x) > 0$ or $x = a$

③ $\dfrac{|f(x)|}{g(x)} \geq 0 \iff g(x) > 0$ or $f(x) = 0$

④ $\dfrac{|f(x)|}{g(x)} \leq 0 \iff g(x) < 0$ or $f(x) = 0$

⑤ $\dfrac{f(x)}{|g(x)|} < 0 \iff f(x) < 0$ and $g(x) \neq 0$

Exercises

#1 For a function $f(x) = -\dfrac{1}{2}x + 4$, find each value.

(1) $f(0)$

(2) $f(2)$

(3) $f(-4)$

(4) $f\left(\dfrac{1}{2}\right)$

#2 The domain of a function $f(x) = -2x + 3$ is $\{0, 1, 2, 3\}$. Find the range of $f(x)$.

#3 Find the domain.

(1) The range of a function $f(x) = 2x$ is $\{-8, 0, 4, 8\}$. Find the domain of $f(x)$.

(2) The range of a function $g(x) = ax$ is $\{-2, 0, 2\}$ when $g(2) = -1$.

Find the domain of the function.

(3) Find the domain of $f(x) = \dfrac{1}{x-3}$.

(4) Find the domain of $g(x) = \sqrt{9 - x^2}$.

#4 Find the value of $f(4)$.

(1) For any real number x, a function $f(x)$ satisfies $f\left(\dfrac{3x-1}{2}\right) = 3x - 4$.

(2) For any real number x, a function $f(x)$ satisfies $f(x) + 3f\left(\dfrac{1}{x}\right) = 2$.

#5 A function $f(x)$ satisfies $f(x) + 2f\left(\dfrac{1}{x}\right) = 6x$ for any non-zero real number x.

Find the value of real number x such that $f(x) = f(-x)$.

#6 Is $f(x) = \dfrac{x^3 + 3x}{x^4 - 2x^2 + 3}$ even, odd, or neither?

#7 Specify whether the given function is even, odd, or neither, and then sketch its graph.

(1) $f(x) = -2$

(2) $f(x) = 2x + 1$

(3) $f(x) = 3x^2 + 2x - 1$

(4) $f(x) = \sqrt{x - 1}$

(5) $f(x) = |2x|$

(6) $f(x) = \left[\dfrac{x}{2}\right]$

(7) $f(x) = x^3 - 2x$

#8 For a function $f(x)$, $f(x+3) = f(x)$ for any real number x

and $f(x) = 3 - |x|$ when $-\dfrac{3}{2} \le x \le \dfrac{3}{2}$. Find the value of $f(11)$.

#9 Find the range of a.

(1) For a function $f: X = \{x|1 \le x \le 2\} \rightarrow Y = \{y|0 \le y \le 5\}$, $f(x) = ax - 1$

when a is a constant.

(2) For a function $f(x) = \begin{cases} -x & , \ x \le 0 \\ (a^2 - a)x, & x > 0 \end{cases}$, $f(x)$ is one-to-one correspondence.

#10 Find the value.

(1) For a function $f(x) = ax$, $f(3) = -4$. Find the value of $f(9)$.

(2) Find the value of $f(3) - f(2) + f(4)$ for the function $f(x) = \dfrac{3}{x}$.

(3) For the two functions $f(x) = ax + 2$ and $g(x) = \dfrac{b}{x} - 2$, $f(1) = g(-1) = 3$.

Find the value of $a + b$.

(4) For the two functions $f(x) = \dfrac{a}{x} + 2$ and $g(x) = -\dfrac{3}{x} + 5$, $3f(-2) = 2g(-3)$.

Find the value of b which satisfies $f(b) = g(b)$.

(5) For the function $f(3x - 2) = 2x - a$, $f(4) = 3$. Find the value of $f(1)$.

(6) For the two functions $f(x) = 2ax$ and $g(x) = \dfrac{2}{x} - 1$, $g\big(f(2)\big) = 3$. Find the value of a.

(7) Two points $P(a + 2, 4 - 2a)$ and $Q(2 - 2b, 3b + 1)$ are on the x-axis and y-axis,

respectively. Find the value of $a + b$.

(8) The function $f(x) = -\dfrac{3}{2}x$ passes through a point $(a + 1, 2a - 3)$. Find the value of a.

(9) The function $y = ax$ passes through a point $(3, -15)$ and $(b, 10)$. Find the value of $a - b$.

(10) For any constants a and b, the function $f(x) = \dfrac{2a}{x}$ passes through the points

$(-2, 8)$ and $(4, b)$. Find the value of $a + b$.

(11) For any constants a, b, and c, the function $f(x) = \dfrac{a}{x}$ passes through the points $(b, 1)$,

$(1, c)$, and $(3, -1)$. Find the value of $a + b + c$.

(12) Two functions $f(x) = ax$ and $g(x) = \dfrac{b}{x}$ intersect at the points $(3, 9)$ and $(-3, c)$.

Find the value of $a + b + c$.

(13) Two functions $y = -ax$ and $y = -\dfrac{2}{x}$ intersect at Point $A(b, 8)$. Find the value of ab.

#11 Find the following values for the given linear functions with a condition:

(1) $f(1)$ for $f(x) = 2ax + 1$ with $f(-1) = 3$

(2) $\frac{a}{2}$ for $f(x) = \frac{1}{2}x + 5$ with $f\left(\frac{a}{2}\right) = -a$

(3) $a + b$ for $f(x) = 3ax - 2$ with $f(-1) = 4$ and $f(b) = 1$

(4) $a + \frac{1}{a}$ when $f(x) = 3x - 1$ passes through the point $(a, a + 3)$.

(5) $a - b$ when $f(x) = ax + 2$ passes through both point $(1, 3)$ and point $(2, b)$.

#12 Find the value of $a + b$ for which :

(1) The graph of $y = ax + 2$ is translated by b along the y-axis from a graph of $y = 3x - 5$.

(2) The graph is translated by a along the y-axis from a graph of $y = 2x + 4$ and passes through both point $(a + 1, -2)$ and point $\left(-\frac{1}{3}, b\right)$.

(3) A point $(-1, 1)$ is on the graph of $y = -2x + a$. If the graph is translated by b along the y-axis, then it will pass through the point $(3, -4)$.

#13 Find the x-intercept and y-intercept.

(1) The linear function $y = ax + b$ passes through both point $(1, 2)$ and point $(-1, 4)$.

(2) The graph of $y = ax + b$ intersects the graph of $y = 2x + 3$ on the x-axis.
It also intersects the graph of $y = -5x - 6$ on the y-axis.

(3) The area surrounded by the graph of $y = \frac{1}{2}x + a$ $(a > 0)$, the x-axis, and the y-axis is 36.

#14 Find an equation in the standard form for each line.

(1) with y-intercept -3 and slope 2

(2) with y-intercept 5 and slope 0

(3) with x-intercept 5 and slope $-\frac{2}{3}$

(4) with x-intercept -3 and slope -2

(5) through $(1, 2)$ with slope 3

(6) through $(3, -4)$ with slope -2

(7) through $(2, 3)$ with undefined slope

(8) through $(-2, 3)$ with y-intercept -1

(9) through $(2, 4)$ with x-intercept -5

(10) through $(3, 1)$ and $(-2, 4)$

(11) through $(-2, -3)$ and $(-1, 5)$

(12) with x-intercept -3 and y-intercept 3

(13) with x-intercept $\frac{3}{2}$ and y-intercept -4

(14) Vertical line through $(-1, 2)$

(15) Horizontal line through $(3, -4)$

#15 Find an equation for the line through $(2, 3)$ which is:

(1) parallel to the line $y = 2x - 5$

(2) parallel to the line $y = -3x + 1$

(3) parallel to the line $x = 4$

(4) parallel to the line $y = -2$

(5) parallel to the line $3x + 4y = 5$

(6) perpendicular to the line $y = \frac{2}{3}x - 1$

(7) perpendicular to the line $x + 3y = -3$

(8) perpendicular to the line $x = 5$

(9) perpendicular to the line $y = -2$

#16 Find the value of a for the following lines:

(1) through $(2, 3)$ and $(1, -a)$ with slope 2

(2) through $(2a - 1, -2)$ and $(-1, 1)$ with slope -2

(3) through $(1, -2)$, $(-3, 2)$, and $(-a + 1, -5)$

(4) through $(2a + 1, -4)$, $(2, 5)$, and $(2, -3)$

(5) through $(-3, 3)$, $(3, a - 1)$, and $(0, 3)$

(6) through $(a, 2a - 3)$ and $(-a - 1, 3 + 4a)$, and parallel to the x-axis

(7) through $(-3a + 1, -5)$ and $(2a - 1, a + 3)$, and perpendicular to the x-axis

(8) through $(3, -2a)$ and $(2a - 1, -3a + 2)$, and parallel to the y-axis

(9) through $(-1, 5)$ and $(2, -4)$, and parallel to the line $ax + 3y + 5 = 0$

#17 Find the value of a such that the line $ax + 2y = 5$:

(1) is parallel to the line $2x + 3y = -2$.

(2) is perpendicular to the line $y = -2x + 3$.

(3) coincides with the line $6y = -4x + 15$.

#18 Find the value of ab for which:

(1) the system $\begin{cases} x - 3y = a \\ 2x + by = 3 \end{cases}$ has the intersection point $(2, 3)$.

(2) the system $\begin{cases} -ax + by = 4 \\ 2ax + 3by = 2 \end{cases}$ has the intersection point $(-1, 2)$.

(3) the system $\begin{cases} px + y = 3 \\ 2x - 3y = q \end{cases}$ has no intersection when $p = a$, $q \neq b$.

(4) the system $\begin{cases} 2ax + 4y = -3 \\ 3x + 6y = 2b \end{cases}$ has unlimited numbers of intersections.

#19 Find the value of a such that:

(1) the system $\begin{cases} ax + y = -2 \\ -3x + 2y = 4 \end{cases}$ has no solution.

(2) the system $\begin{cases} 2x - ay + 3 = 0 \\ x + 3y - 2 = 0 \\ 2x + y + 1 = 0 \end{cases}$ has one solution.

(3) the system $\begin{cases} x - 3y = 2 \\ 2x + y = -3 \end{cases}$ has a solution $(2a, -1)$.

(4) the line $2ax + 3y - 1 = 0$ passes through the intersection of the system $\begin{cases} x - 2y = 3 \\ 2x + 2y = 1 \end{cases}$.

#20 Find the equation of each line such that:

(1) the line passes through the intersection of the system $\begin{cases} x + 2y = 3 \\ 3x + y = -2 \end{cases}$

and runs parallel to the y-axis.

(2) the line passes through the intersection of the system $\begin{cases} -x + y + 2 = 0 \\ 2x + y - 3 = 0 \end{cases}$

and runs perpendicular to the x-axis.

(3) the line passes through the intersection of the system $\begin{cases} 2x - y + 3 = 0 \\ x + 2y + 4 = 0 \end{cases}$

and runs parallel to the line $3x + 2y = 5$.

#21 Find the value.

(1) Two functions $f(x) = x^2 - 1$ and $g(x) = ax + b$ have the same domain $\{-2, 1\}$.

Find the value of ab such that $f(x) = g(x)$ for any real number x.

(2) For a function $f : X = \{x \mid -1 \le x \le 2\} \to Y = \{y \mid -2 \le y \le 4\}$,

$f(x) = ax + b$ is one-to-one correspondence. Find the value of ab (where $a < 0$).

(3) For a function $f(x) = ax + b$, $-1 \le f(0) \le 1$, $0 \le f(1) \le 3$, and $m \le f(2) \le n$.

Find the value of $m + n$.

#22 Find the number of a function f.

(1) For a function $f : X = \{-1, 0, 1\} \to Y = \{-2, -1, 0, 1, 2, 3\}$, $xf(x) = 0$ for any x in X.

(2) For a function $f : X = \{-1, 0, 1\} \to Y = \{-1, 0, 1\}$, $\{f(-1) + 1\} \cdot \{f(1) - 1\} \ne 0$

(3) For a function $f : X = \{0, 1, 2\} \to Y = \{1, 2, 3, 4, 5, 6\}$, $f(1) = 3$,

and if $x_1 < x_2$, then $f(x_1) < f(x_2)$

(4) For a function $f : X = \{1, 2, 3, 4\} \to Y = \{1, 2, 3, 4\}$, $f(x) \ge x$ for any x in X.

#23 A function such that $f(x) = \begin{cases} x, & x \text{ is rational} \\ 1-x, & x \text{ is irrational} \end{cases}$ is defined in the domain

$X = \{x \mid 0 \leq x \leq 1\}$. Find the value of $f(x) + f(1-x)$.

#24 For two functions $f(x) = 2x+1$ and $g(x) = x-3$, find each composite function.

(1) $(f \circ f)(x)$ (3) $(f \circ f \circ f)(x)$

(2) $(f \circ g)(x)$ (4) $(g \circ f \circ g)(x)$

#25 For two functions $f(x) = 2x-1$ and $g(x) = x+2$, find the function $h(x)$ such that:

(1) $(f \circ h)(x) = g(x)$ (2) $(g \circ h)(x) = f(x)$

#26 Find the value.

(1) For a function f, $f_1 = f$, $f_2 = f \circ f_1$, $f_3 = f \circ f_2$, $\cdots\cdots$, $f_n = f \circ f_{n-1}$

When $f(x) = \dfrac{x-1}{x}$, find the value of $f_{10}(2)$.

(2) For two functions $f(x) = \begin{cases} x^2, & x \geq 0 \\ \dfrac{x}{2}, & x < 0 \end{cases}$ and $g(x) = x+2$, a function h satisfies $h \circ g = f$.

Find the value of $h(-2)$.

(3) For two functions $f(x) = x - a$ and $g(x) = x^2 + 1$,

find the value of a constant a such that $(f \circ g)(x) = (g \circ f)(x)$.

(4) For any real number x, $f(f(x)) = x$ and $f(0) = 1$. Find the value of $f(-2)$.

#27 Sketch all possible graphs of a function $y = f(x)$ such that $(f \circ f)(x) = x$.

#28 Tell whether x and y show direct variation, inverse variation, or neither.

(1) $xy = 5$ (4) $\dfrac{y}{2} = x$

(2) $x = \dfrac{3}{y}$ (5) $2x = y$

(3) $y = x + 1$ (6) $x = 3y$

#29 Determine which of the functions have inverse functions.

(1) $f(x) = x^2$, for all x (3) $h(x) = x^2$, $x \leq 0$

(2) $g(x) = x^2$, $x \geq 0$

#30 Find the inverse of each function and sketch the graphs of f and f^{-1}.

(1) $f(x) = \dfrac{3-5x}{2}$

(2) $f(x) = \sqrt{2x-1}$

(3) $f(x) = x^3$

(4) $f(x) = \dfrac{1}{x-1}$

#31 Find the value.

(1) A function f with the real number domain is one-to-one correspondence.

   When $f\left(\dfrac{3x-1}{3}\right) = -4x + 5$, find the value of $f^{-1}(-3)$.

(2) For an inverse function $f^{-1}(x) = ax + b$ of $f(x)$, $f(-2) = 1$ and $f^{-1}(3) = -4$.

   Find the value of $a + b$ $(a, b$ are real numbers).

(3) For two functions $f(x) = x + a$ and $g(x) = bx + c$, $(f^{-1} \circ g)(x) = 2x + 3$ and $g^{-1}(3) = 2$.

   Find the value of $a + b + c$.

(4) For two functions $f(x) = \dfrac{1}{2}x + a$ and $g(x) = bx - 2$ with the real number domains,

   $(f \circ g)(x) = x + 3$. Find the value of $f^{-1}(g(-1))$.

(5) For a function $f(x) = x|x| + a$, a is a real number, $f^{-1}(3) = -1$.

   Find the value of $(f^{-1} \circ f^{-1})(3)$.

(6) For a function $f(x) = \begin{cases} x + a & (x \geq 3) \\ 2x + 1 & (x < 3) \end{cases}$ with the real number domain, the inverse function

   of f exists. Find the value of $(f^{-1} \circ f^{-1})(10)$. (Where a is a constant.)

(7) When two functions $f(x)$ and $g(x)$ are one-to-one correspondences,

   $g(x) = f(3x - 1)$ and $f^{-1}(1) = 5$. Find the value of $g^{-1}(1)$.

(8) For a function $f(x)$, $f^{-1}(x)$ is the inverse function of $f(x)$ such that $f^{-1}(0) = 3$.

   For a function $h(x)$ such that $h(x) = f(2x - 1)$, $h^{-1}(x)$ is the inverse function of $h(x)$.

   Find the value of $h^{-1}(0)$.

(9) For an inverse function $g(x)$ of $f(x)$, $f\left(2g(x) - \dfrac{2x-3}{x+2}\right) = x$. Find the value of $f(1)$.

#32 Find the range.

(1) For a function $f(x) = |2x - 1| + ax + 3$ which is defined in the real number system, find the range of a so that f has an inverse function.

(2) When $g(x)$ is an inverse function of $f(x) = \dfrac{x^2}{2} + a$ $(x \geq 0)$, find the range of a so that the equation $f(x) = g(x)$ has two different non-negative real number solutions.

(3) For two functions $f(x) = x^2 - x - 12$ and $g(x) = x^2 + ax + 5$, find the range of a real number a such that $(f \circ g)(x) \geq 0$ for any real number x.

#33 Find the distance.

(1) When a function $f(x) = \dfrac{1}{2}(x^2 + x - 2)$, $x \geq -1$, and its inverse function intersect at two points A and B, find the length of the segment $\overline{AB}$.

(2) When a function $g(x)$ is an inverse of $f(x) = x^2 - 3x + 3$ $(x \geq 1)$, find the distance between the intersection points of the graph of $y = f(x)$ and a line $y = x$.

#34 When $g(x)$ is an inverse function of $f(x)$, find the inverse function of $f(2x)$.

#35 For a function $f(x) = \begin{cases} x, & x \geq 0 \\ 2x, & x \leq 0 \end{cases}$, answer the question.

(1) Find the value of $f(f(10))$.

(2) Find the value of $f^{-1}(-4)$.

(3) Find the number of intersection points of the graphs of the function $y = f(x)$ and its inverse function $y = f^{-1}(x)$.

#36 Let $\mathbb{R} = \{x| x$ is a real number$\}$. For a function $f: \mathbb{R} \to \mathbb{R}$, the inverse function f^{-1} exists. When $f(a + b) = f^{-1}(a) + f^{-1}(b)$ for any real numbers a and b, determine whether the following statements are true or false.

(1) $f(f(a) + f(b)) = a + b$

(2) If $f(-1) = 2$, then $f(-4) = -1$

(3) $f^{-1}(a + b) = f(a) + f(b)$

#37 Minimum and Maximum values.

(1) When the domain of a function $y = (x - 1)^2 - 2(x - 1) - 2$ is $\{x| 0 \leq x \leq 2\}$, find minimum and maximum values for the function.

(2) When the domain of a function $y = -3x + a$ is $\{x|-1 \leq x \leq 3\}$, minimum value of the function is 1. Find maximum value of the function.

(3) For a function $y = |x - 1| + |2x - 3|$, find minimum value of the function.

(4) For a function $f(x) = x^2 - 2x + 4$ with the domain $\{x|0 \leq x \leq 3\}$,

   find maximum value of $f(f(x))$.

(5) The graph of a function $ax + by + 1 = 0$ is as shown in Figure.

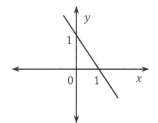

Find maximum and minimum values of

$y = ax^2 + 2bx$, $0 \leq x \leq 2$

where a and b are real numbers.

(6) For a function $y = |x^2 - 4x + 3|$ with the domain $\{x|1 \leq x \leq 3\}$, find maximum and minimum values of the function.

(7) For a quadratic function $y = f(x)$, the leading coefficient of the function is positive.

   When $f(4 - x) - f(x) = 0$ for any real number x, find minimum value of $f(x)$.

(8) When $x > 0$, find maximum value of $y = -(x + \frac{1}{x})^2 + 4\left(x + \frac{1}{x}\right) + 5$.

(9) For any real numbers x and y, $x^2 + y = 1$.

   Find maximum and minimum values of $2x - y$, $y \geq 0$.

(10) For any real number x, $f(x) + 2f(1 - x) = x^2$. Find minimum value of $f(x)$, $-1 \leq x \leq 1$.

#38 Find the range of a.

(1) For a function $y = 2ax - a + 1$, find the range of a such that $y > 0$ for $1 < x < 3$.

(2) For any real number x, $(a + 2)x^2 + 2(a + 2)x - 1 < 0$ is always true.

   Find the range of the real number a.

(3) When the graph of a function $y = x^2 + ax + b$ and the x-axis intersect at one point,

   find the range of the real number a so that the graph of the function and a line $y = 4x$ don't

   intersect at any point.

(4) For a function $f(x) = x^2 - (2a + 1)x + 4$, find the range of a such that $f(x) > x$ for any

   real number x.

(5) When an equation $x^2 + ax - 3a + 3 = 0$ has two different roots, one is less than 2 and the

   other one is greater than 2. Find the range of the real number a.

(6) Find the range of the real number a so that one root of an equation $x^2 + 2x + a = 0$ is in-

   between the two roots of an equation $x^2 - 4x + 3 = 0$.

(7) Find the range of the real number a so that an equation $|x^2 - 1| = x + a$ has four different real number solutions.

(8) Find the range of the real number a so that the solution of an inequality

$x^2 + ax + a^2 - 4 \leq 0$ includes the interval $[0, 2] = \{x | 0 \leq x \leq 2\}$.

(9) Find the range of the real number a so that the vertex of the graph of a function $y = x^2 + 4ax - 2a + 1$ lies on the second quadrant.

(10) Find the range of the real number a so that an equation $|x^2 - 1| = ax - 3a + 2$ has four different real number solutions.

#39 Let $[x]$ be the greatest integer less than or equal to x. Find the range of f.

(1) f is a function such that $f(x) = [2x] - 2[x]$ with the domain $\{x | x$ is a real number$\}$.

(2) f is a function such that $f(x) = \left[\frac{[x]}{x} \right]$ for any positive real number x.

(3) f is a function such that $f(x) = [x] + [-x]$ for any positive real number x.

#40 For a function $f(x) = |x - 1| + |x - a|$, find all possible numbers for an integer k so that $f(x)$ has minimum value 3 when $x = k$.

#41 Solve the following inequalities by using graphs.

(1) $x^2 + 2x + 1 > 0$

(2) $x^2 + 4x + 4 \geq 0$

(3) $x^2 - 10x + 25 \leq 0$

(4) $4x^2 < 4x - 1$

#42 Determine the values of the coefficients a and b so that the following inequalities have the given solutions.

(1) $x^2 + ax + b > 0$, the solution is $x < -1$, $x > 2$

(2) $x^2 + ax + b \geq 0$, the solution is $x \geq -2$, $x \leq -4$

(3) $ax^2 + bx - 24 \leq 0$, $(a > 0)$, the solution is $-2 \leq x \leq 3$

(4) $ax^2 - bx + 6 > 0$, $(a < 0)$, the solution is $-1 < x < 3$

#43 Determine the range of values of the constant a so that the solution of each inequality is all real numbers.

(1) $x^2 - (a + 1)x + a + 2 > 0$

(2) $-x^2 + (a + 1)x - a^2 < 0$

(3) $ax^2 - 2ax + 3 \geq 0$, $a > 0$

#44 Find the range of a constant a.

(1) The graph of $y = x^2 + ax + 4$ and a line $y = x - 5$ do not intersect at any points.

(2) The graph of $y = x^2 - ax + 3$ and the x-axis do not intersect at any points.

(3) The two graphs of $f(x) = 2x^2 + ax + 3$ and $g(x) = x^2 - 2x - a$ intersect at two different points.

(4) The two roots of the equation $x^2 - 2ax + a + 12 = 0$ are less than 1.

(5) One root of the equation $2x^2 - ax - 1 = 0$ is in-between -1 and 0, and the other root of the equation is in-between 0 and 1.

(6) For any real number x, $(a - 2)x^2 - (a - 2)x + 2 > 0$ is always true.

(7) The equation $x^2 - ax + 2a - 3 = 0$ has at least one real number solution in the range $-2 \le x \le 4$.

(8) For a function $f(x) = x^2 - 2ax + 3$, $f(x) \ge a$ (where $-1 \le x \le 1$).

(9) For a function $f(x) = x^2 - 2ax + a$, $f(x) > 0$ in the range $0 < x < 1$.

(10) There exists the value of x such that $a(x^2 + x + 1) > x$.

(11) An inequality $x^2 - 4x \ge a^2 - 4a$ is always true in the range $-1 \le x \le 1$.

(12) For any real number x, $ax^2 + 2\sqrt{2}x + a - 1 \ge 0$ is always true.

(13) When the two roots of the equation $x^2 + 2ax + a + 6 = 0$ are α and β, find the range of a such that $0 < \alpha < 1 < \beta$.

#45 Find the value.

(1) The graph of $y = x^2 - 2(a + 1)x + a^2 + 4a$ and the line $y = mx + n$ intersect at one point, not depending on the value of the real number a. Find the value of $m + n$.

(2) A parabola $y = x^2 + 2ax + a$ and the x-axis have two intersection points A and B. Find the positive number a such that $\overline{AB} = \sqrt{3}$.

(3) When the leading coefficient of a quadratic function $y = f(x)$ is 1, the graph of the function and a line $y = a$ have the two intersection points $(1, a), (5, a)$. When the solution of the inequality $f(x) < f(-1) + 2$ is $m < x < n$, find the value of mn.

(4) 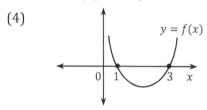 When the graph of a function $f(x) = ax^2 + bx + c$ is shown as the Figure, the inequality $bx^2 + cx + a > 0$ has solution $m < x < n$. Find the value of $m - n$.

(5) When an inequality $x + 2a \le x^2 \le 2x + b$ is always true in the range $-1 \le x \le 1$, find minimum value of $b - a$.

(6) The graph of a function $y = x^2 + 2ax + b$ intersects the lines $y = -x + 3$ and $y = 3x + 5$ at one point, respectively. Find the value of $a + b$.

(7) When the graph of a function $y = x^2 + ax - 2$ and a line $y = -2x + 3$ intersect at two points A and B, the midpoint of the segment $\overline{AB}$ is $(4, -2)$. Find the value of the constant a.

(8)

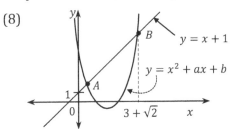

When the graph of a function $y = x^2 + ax + b$ and a line $y = x + 1$ intersect at two points A and B, the x-coordinate of the point B is $3 + \sqrt{2}$ as shown in Figure.

For rational numbers a and b, find the value of $a + b$.

(9) When the graph of a function $f(x) = x^2 - ax + 4$ and the x-axis intersect at two points A and B, length of the segment $\overline{AB}$ is $2\sqrt{2}$.

Find minimum value of $f(x)$ in the range $-\sqrt{6} \le x \le \sqrt{6}$.

(10)

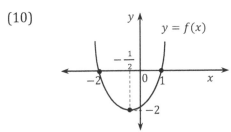

The graph of a function $y = f(x)$ is shown as the Figure. Find the sum of all three different real number solutions of the equation

$$f(f(x)) = 0.$$

(11)

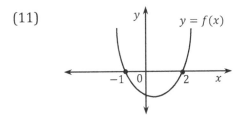

The graph of a function $y = f(x)$ is shown as the Figure. When the inequality $f(\frac{x+a}{2}) \le 0$ has solution $-3 \le x \le 3$, find the value of a constant a.

(12) When the graph of a function $y = 2x^2$ and a line $y = ax + b$ (where a and b are constants) intersect at one point, the line will be a tangent line of a circle $x^2 + (y + 1)^2 = 1$. Find the value of $a^2 + b$ when $b < 0$.

#46 For each given fractional function, draw the graph, and then state the range of $f(x)$ that satisfies the given domain.

(1) $f(x) = \frac{x-1}{x-2}$ $(3 \le x \le 4)$

(2) $f(x) = \frac{-2x+4}{x-3}$ $(1 \le x \le 5, \ x \ne 3)$

(3) $f(x) = \frac{3x}{3x-2}$ $(x \ge 0, \ x \ne \frac{2}{3})$

(4) $f(x) = \frac{2x+1}{2x-1}$ $\left(-1 \le x < \frac{1}{2} \text{ or } \frac{1}{2} < x \le 1\right)$

#47 Maximum and minimum values

(1) Given the fractional function $f(x) = 2x + \frac{1}{x}$ $\left(\frac{1}{2} \le x \le 4\right)$, draw the graph of the function and find maximum value.

(2) Given the fractional function $f(x) = x - \frac{2}{x}$ $\left(\frac{1}{2} \le x \le 4\right)$, draw the graph of the function and find minimum value.

(3) A function $y = \frac{ax+b}{x+c}$ is symmetric with respect to a point $(2, 1)$, and passes through a point $(3, 3)$. Find maximum and minimum values in the range $-1 \le x \le 1$.

#48 For each of the following fractional function, state how each graph has been translated from the graph of $y = \frac{1}{x}$.

(1) $y = \frac{1}{x-2}$

(2) $y = \frac{1}{x} - 2$

(3) $y = \frac{1}{x-1} + 2$

(4) $y = \frac{1}{x+2} - 1$

#49 For each of the following fractional function, state how each graph has been translated from the graph of $y = \frac{1}{2x}$.

(1) $y = \frac{1}{2x-2}$

(2) $y = \frac{1}{2x} - 1$

(3) $y = \frac{-2x-1}{2x+2}$

(4) $y = \frac{2x-3}{2x-4}$

#50 Find the equations of the curves obtained when the graph of $y = \sqrt{-2x}$ is translated as follows:

(1) -2 units along the y-axis

(2) 3 units along the x-axis

(3) $-\frac{1}{2}$ units along the x-axis and 4 units along the y-axis

#51 Solve the following fractional equations.

(1) $\frac{2}{x-1} - \frac{1}{x+1} = \frac{x}{x-1}$

(2) $\frac{x+3}{x+1} + \frac{x}{x-2} = -\frac{1}{x-2}$

(3) $\frac{2}{x-2} - \frac{x}{x^2-3x+2} = -1$

#52 Solve the following inequalities by using graphs.

(1) $\dfrac{2}{x-2} \le x - 1$

(2) $\dfrac{x+1}{x-1} \ge x + 1$

(3) $-1 < \dfrac{4-3x}{x-2} \le 1$

(4) $\dfrac{4}{x-2} < x + 1, \ \ x > 2$

#53 Draw graph of each radical function, and find maximum and minimum values within the given domain.

(1) $f(x) = \sqrt{x+1} - 2 \quad (0 \le x \le 2)$

(2) $f(x) = -\sqrt{x+2} + 1 \ (-1 \le x \le 2)$

(3) $f(x) = \sqrt{3-x} - 2 \quad (-1 \le x \le 2)$

(4) $f(x) = -\sqrt{3-2x} + 1 \ (-1 \le x \le 1)$

(5) $f(x) = 2 - \sqrt{x+1} \quad (0 \le x \le 3)$

#54 Draw the graph and use it to solve the given radical equation.

(1) $\sqrt{3-x} = x - 3$

(2) $\sqrt{2x+1} = -2x + 1$

#55 Solve the following radical inequalities by using graphs.

(1) $\sqrt{2x-3} > x - 3$

(2) $\sqrt{2x+1} < \dfrac{x+2}{2}$

(3) $\sqrt{x} - 1 < \sqrt{5-x}$

(4) $\sqrt{x} - 1 > \sqrt{5-x}$

(5) $\sqrt{2-x} < x < \sqrt{x+6}$

(6) $\sqrt{3-2x} > -\dfrac{2x+1}{\sqrt{2}}$

#56 Solve the following inequalities by using graphs.

(1) $(2x-3)(x^2 - 4x - 5) < 0$

(2) $(3-x)(x^2 - 2x - 15) \ge 0$

(3) $(x+1)(x^2 - 2x - 3) \le 0$

(4) $(3-x)(x^2 - 6x + 9) > 0$

(5) $x^3 - 3x^2 + 4 \le 0$

(6) $(2x-1)(1-x)(x^2 - 2x - 15) \le 0$

(7) $(x+2)(1-x)(x-3)^2 \ge 0$

(8) $(x^2 + 2x - 3)(x^2 + 2x + 6) \le 0$

(9) $(x^2 - x + 2)(x^2 - 3x - 4)(x^3 + 4x) \le 0$

(10) $(x^2 - 3x - 4)^2 > 0$

(11) $(2-x)(x-3)(x^2 - 6x + 9) \le 0$

#57 For a function $y = \sqrt{1-2x} + 3$, find the inverse function and determine the domain and range of the inverse function.

#58 Find the value.

(1) For two functions $f(x) = \dfrac{x+4}{x-3}$ and $g(x) = \sqrt{2x-4}$ with the same domain $\{x \mid x > 3\}$,

find the value of $(f \circ (g \circ f)^{-1} \circ f)(4)$.

(2) For a function $f(x) = \sqrt{2x-1}$, there is a function $g(x)$ such that $(g \circ f)(x) = \dfrac{2x-1}{x+1}$.

Find the value of $g(3)$.

#59 Minimum and maximum values

(1) When an inequality $ax \leq \dfrac{x+1}{x-2} \leq bx + 1$ is always true in the interval $3 \leq x \leq 4$,

find the maximum value of a and minimum value of b (where a and b are real numbers).

(2) For a point $P(0, 2)$ in a coordinate plane, a point Q lies on a graph of $y = \dfrac{4}{x} + 2$ in the

same plane. Find minimum value of the length of the segment $\overline{PQ}$.

#60 When the minimum value of a function $y = \sqrt{|x| + 1}$ is a point A and the two intersection

points of the graph of the function and a line $y = 2$ are B and C, find the area of the triangle

$\triangle ABC$.

#61 Find the value.

(1) When the graph of $f(x) = \dfrac{ax}{x+2}$ is symmetric with respect to the line $y = x$, find the real

number a.

(2) If the inverse functions of rational functions $f(x) = \dfrac{ax+b}{cx+d}$ $(d > 0)$ and $g(x) = \dfrac{x-1}{2x+3}$ exist,

then $(f \circ g)(x) = x$. When the axes of symmetries for a rational function $y = f(x)$ are

$x = m$ and $y = n$, find the value of $m - n$.

(3)

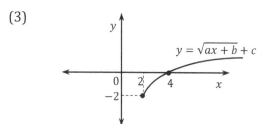

$y = \sqrt{ax + b} + c$

When the graph of a function $y = \sqrt{ax + b} + c$ is shown as the Figure,

find the value of $a + b + c$.

(4) For a function $y = \dfrac{4x+a}{x+2}$, the function has maximum value 3 and minimum value 2 in the

interval $b \leq x \leq -3$ (where a and b are real numbers). Find the value of $a + b$.

(5) The graph of a fractional function $y = \frac{-4x+3}{2x-1}$ is symmetric with respect to the line $y = ax + b$, $a \neq 0$. Find the real number $a^2 - b^2$.

(6) When the graph of a function $y = \sqrt{x+1}$ is translated with 3 units along the x-axis, -1 unit along the y-axis, we reflect the translated graph through the line $x = 1$. Then, it will be the same as the graph of $y = \sqrt{ax+b} + c$. Find the value of $a + b + c$.

#62 Find the range.

(1) When an equation $\sqrt{x+2} = x + a$ has two different real number solutions, find the range of a.

(2) When the graphs of two functions $y = \frac{x+2}{x-2}$ and $y = \sqrt{x+a}$ intersect at two different points, find the range of a.

(3) When the graph of $y = \frac{a}{x-2} + 1$ passes all 4 quadrants in a coordinate plane, find the range of the real number a.

(4) Let the axes of symmetries of a function $y = \frac{x+1}{2x-1}$ be $x = m$, $y = n$. When the graph of y and a line $y = ax + \frac{m}{n}$ intersect at the range: $a \leq \alpha$ or $a \geq \beta$, find the value of $(\alpha - \beta)^2$ (where α and β are constants).

(5) When the graphs of two functions $f(x) = \frac{1}{3}x^2 + a$ and $g(x) = \sqrt{3x - 3a}$ $(x \geq 0, \ a \geq 0)$ intersect at two different points, find the range of real number a.

(6) For two points $A = A(2, 3)$ and $B = B(3, 2)$, the segment $\overline{AB}$ intersects the graph of $y = \sqrt{ax + 2}$, $a > 0$. Find the range of a.

#63 Find the value.

(1) For a fractional equation $\frac{x^2+2x}{x-1} - \frac{16(x-1)}{x(x+2)} - 6 = 0$, find the sum of all real number solutions.

(2) For a fractional equation $\frac{x^2+x+1}{x-2} - \frac{x+2}{x-1} = \frac{3}{(x-1)(x-2)} - 2$, find the sum of all real number solutions.

(3) When the fractional equation $\frac{2x}{x^2-4} - \frac{1}{x+2} = a$ has no solution, find the value of a.

(4) For a fractional equation $x^2 + \frac{1}{x^2} = \frac{x^2+1}{x}$, find the value of the real number solution.

(5) For a fractional equation $\frac{1}{(x-1)x} + \frac{1}{x(x+1)} + \frac{1}{(x+1)(x+2)} = \frac{3}{4}$, let α and β be the two roots.

Find the value of $\frac{\beta}{\alpha} + \frac{\alpha}{\beta}$.

(6) There is a% salt solution containing 30 ounces of salt. If 50 ounces of water is evaporated, then it will be a $(a + 2)$% salt solution. Find the value of a.

(7) For a fractional equation $f(x) = \frac{-x+3}{x-2}$ and its inverse function $g(x)$, find the two different real number roots such that $f(x) + g(x) = 0$.

(8) When a fractional equation $\frac{1}{x-1} - \frac{1}{x} - \frac{1}{x+3} + \frac{1}{x+4} = 0$ has a root α,

find the value of $2\alpha + 3$.

(9) When a fractional equation $\frac{x^2+2x+a-11}{x-2} = a$ has no real number solution,

find the sum of all possible integers for a.

(10)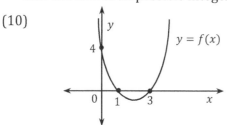

When the graph of a function $y = f(x)$ is shown as the Figure, $g(x) = \frac{1}{x} - \frac{2}{x+4}$.

Find the number of real roots such that $g(f(x)) = 0$.

(11) For a function $f(x) = x^2 - 5x + 4$, the fractional equation $\frac{f(x)}{x+a} - \frac{f(x)}{x+b} = 0$ has no solution. Find the value of $(a - b)^2$ (where a and b are real numbers).

#64 Find the value.

(1) Find the value of real number a so that the radical equation $\sqrt{1 - x^2} = x + a$ has a real number solution.

(2) When a radical equation $x^2 - 2x + \sqrt{x^2 - 2x + a} = 4$ has a root 3, find the value of the other root.

(3) For a radical equation $\sqrt{(a + 2)x + a + 3} = x$, find minimum value of real number a so that the equation has a real number root.

(4) When a radical equation $\sqrt{x + 5} = x + a$ has an extraneous root -1, find the real number root of the equation (where a is a constant).

(5) For a radical equation $\sqrt{x^2 - 3x + 1} - \sqrt{x^2 - 3x} = \frac{1}{2}$, find the sum of all roots.

(6) For a function $f(x) = \sqrt{2x + 1} + x$, find all real number solutions such that $f(f(x)) = 1$.

#65 Find the range.

(1) For a radical equation $\sqrt{x+1} = |x+a|$, find the range of a so that the product of the two real number roots of the equation is negative.

(2) For a positive real number a, find the range of a so that the radical equation $\sqrt{2a+x} + \sqrt{2a-x} = 4a$ has a real number solution.

#66 A container is filled with 100 ounces of 8% salt water solution. Suppose 6 ounces of water is evaporated per day. How many days do we need to have more than 10% salt water solution?

#67 Solve each inequality.

(1) $x^4 + x^3 \le x^2 + x$

(2) $\dfrac{2}{x-1} + \dfrac{3}{x-2} + \dfrac{2}{(x-1)(x-2)} + 3 \le 0$

(3) $\dfrac{x^2-x-30}{|x(x-1)|} \le 0$

(4) $(|x|-1)(x^2 - 2|x| - 15) \le 0$

(5) $\begin{cases} (x-1)(x-2)^2(x-4) \ge 0 \\ \dfrac{2x-3}{x+2} \le 1 \end{cases}$

#68 Find the range.

(1) For any real number x, an inequality $x^4 - x^3 - x^2 - 2x < 0$ is always true. Find the range of the real number a so that a function $f(x) = -x^2 + x + a$ has always positive values.

(2) When an inequality $(x+1)(x^2 + ax + a) < 0$ has solution $x < -1$, find the range of the real number a.

(3) For any x, $(x-1)(x-2)(x-4)^2 \le 0$. Find the range of a such that $x^2 - 4x + a \le 0$

(4) For a fractional function $f(x) = \dfrac{x}{x+2}$, $(f \circ f)(x) \ge \dfrac{1}{2}$. Find the range of x.

(5) When there are three integers for x so that a system of inequalities
$\begin{cases} x^3 + x^2 - 2x - 2 > 0 \\ x^2 - (a+1)x + a \le 0 \end{cases}$ is always true, find the range of real number a.

(6) When there are two integers for x such that $x(x-a)(x-3)^2 \le 0$, find the range of real number a.

(7) If an inequality $(|x|+1)(x+a) < 0$ is true for all x, then $x^3 - x^2 - 4x + 4 < 0$ is also true. Find the range of real number a.

(8) For two quadratic expressions $f(x)$ and $g(x)$ with leading coefficients 1, their GCF and LCM are $x+2$ and $x(x+2)(x-4)$, respectively.

Find the range of x such that $\dfrac{1}{f(x)} + \dfrac{1}{g(x)} \le 0$

(9)

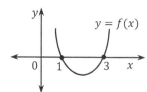

The graph of a function $y = f(x)$ is shown as the Figure.

Find the range of x such that $\dfrac{f(x-2)}{f(x)} \leq 0$.

(10) For two inequalities $x^3 - 7x^2 + 12x > 0$ and $\dfrac{x-a}{2x-3} \leq 0$, the common(intersection)

range of their solutions is $\dfrac{3}{2} < x < 3$. Find the range of the constant a.

(11) For a positive number a, there are 6 integers so that the system of inequalities
$\begin{cases} x(x + a)(x - 2a) < 0 \\ x^2 + ax - 2a^2 \leq 0 \end{cases}$ is always true. Solve the system and state the integers.

#69 Find the value.

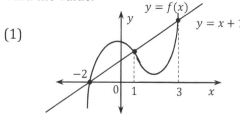

(1)

The graph of a function $y = f(x)$ and a line $y = x + 1$ intersect at $x = -2$, $x = 1$, and $x = 3$. Find maximum and minimum values of x such that $\dfrac{x}{f(2x)-1} \geq \dfrac{1}{2}$.

(2) For constants a and b, the two inequalities $\dfrac{x^2-a^2}{x^2+x+1} < 0$ and $\dfrac{1}{x+2b} < \dfrac{1}{x+1}$ have the same

solution. Find the value of $a + b$ $(a > 0)$.

(3) When the two inequalities $\sqrt{x - 2} < 4 - x$ and $\dfrac{x+a}{x+b} \leq 0$ have the same solution,

find the value of ab (where a and b are constants).

#70 For two inequalities $\dfrac{(x+1)(x-4)}{(x-1)^2} \leq 0$ and $\dfrac{x^2+3x-4}{x^3+1} \geq 0$, the common range of their solutions

is the solution of an inequality $f(x) \leq 0$. Find the expression for $f(x)$.

#71 For any two real numbers a and b, two inequalities $x(x + 1)(x + 2) > 0$ and $x^2 + ax + b \leq 0$

have the union (sum) range of their solutions, $x > -2$, and the intersection (common) range

of their solutions, $0 < x \leq 2$. Find the value of $a + b$.

#72 For any real number x $(x \neq 0)$, the inequality $x^4 - 2x^3 + ax^2 - 2x + 1 \geq 0$ is always true.

Find minimum value of a.

Chapter 6. Exponential and Logarithmic Functions

6-1 Rational Exponents and Radicals

1. Definition of $x^{\frac{m}{n}}$

For any integer power n, $x^n = \underbrace{x \cdot x \cdots\cdots x}_{n \text{ factors}}$ is x to the nth power.

Now, consider $x^{\frac{1}{n}}$ $\left(\sqrt[n]{x}\right)$, n is a positive integer.

If $(x^p)^q = x^{pq}$ is to hold, $(x^{\frac{1}{n}})^n = x^{\left(\frac{1}{n}\right) \cdot n} = x^1 = x$

Thus, $x^{\frac{1}{n}}$ must be a number whose n^{th} power is x.

> $x^3 = 8 \;\Rightarrow\; (x-2)(x^2 + 2x + 4) = 0$
> $\therefore\; x = 2$ or $x = -1 + \sqrt{3}i$ or $x = -1 - \sqrt{3}i$.
> The cube roots of 8 are $2,\; -1 + \sqrt{3}i,\; -1 - \sqrt{3}i$.
> We denote $\sqrt[3]{8} = 2$ (real number)
> When $x^3 = -8$,
> $x = -2$ or $x = 1 + \sqrt{3}i$ or $x = 1 - \sqrt{3}i$
> The cube roots of -8 are $-2,\; 1 + \sqrt{3}i,\; 1 - \sqrt{3}i$.
> The only real 3rd root is $\sqrt[3]{-8} = -2$

Note: If $n = 2$ and $x = -1$, then there is no real number a such that $a^2 = -1$.

($\because$ The square of any positive or negative number is positive.)

Therefore, $\sqrt[n]{x} = x^{\frac{1}{n}}$ cannot be defined for $x < 0$, n even.

> If $n = 2$, then the 2 of a radical $\sqrt[n]{\;}$ is usually omitted.
> That is, $\sqrt{x} = \sqrt[2]{x} = x^{\frac{1}{2}}$

(1) When n is an even positive integer and $x > 0$,

$\sqrt[n]{x} = x^{\frac{1}{n}}$ is the (unique) positive real number a such that $a^n = x$.

We call $x^{\frac{1}{n}}$ the *principal n^{th} root* of x.

If $x = 0$, then $\sqrt[n]{0} = 0^{\frac{1}{n}} = 0$

> If n is even,
> $\sqrt[n]{x} = x^{\frac{1}{n}}$ means only the positive root.
> For both real roots (positive and negative),
> $\pm\sqrt[n]{x}$ or $\pm x^{\frac{1}{n}}$

(2) When n is an odd positive integer for any real number x,

$\sqrt[n]{x} = x^{\frac{1}{n}}$ is the (unique) positive real number a such that $a^n = x$.

This is called the nth root of x.

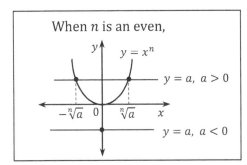

When n is an even,

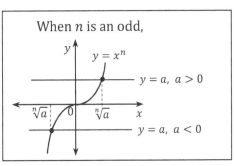

When n is an odd,

Note: ① When n is an even number (square roots, 4th roots, $\cdots$),

If $a > 0$, then there are 2 n^{th} roots, $\sqrt[n]{a}, -\sqrt[n]{a}$.

If $a < 0$, then there is no real number n^{th} root.

② When n is an odd number (cube roots, 5th roots, $\cdots$), there is only one real number n^{th} root, $\sqrt[n]{a}$.

Finally, we can easily define $x^{\frac{m}{n}}$ where m and n are integers, $n \neq 0$.

For a positive integer n and any integer , $x^{\frac{m}{n}} = (x^{\frac{1}{n}})^m$.

Example Simplify if possible.

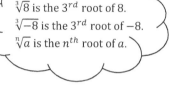

$27^{\frac{4}{6}} = 27^{\frac{2}{3}} = (\sqrt[3]{27})^2 = (3)^2 = 9$

$(-27)^{\frac{4}{6}} = (-27)^{\frac{2}{3}} = (\sqrt[3]{-27})^2 = (-\sqrt[3]{27})^2 = (-3)^2 = 9$

$(27^2)^{\frac{1}{3}} = (729)^{\frac{1}{3}} = (9^3)^{\frac{1}{3}} = 9$

$(-27)^{\frac{5}{6}} = (\sqrt[6]{-27})^5$, which is undefined, since $\sqrt[6]{-27}$ is undefined.

$(-8)^{\frac{5}{3}} = ((-8)^{\frac{1}{3}})^5 = (-2)^5 = -32$

$9^{-\frac{5}{2}} = (9^{\frac{1}{2}})^{-5} = (3)^{-5} = \frac{1}{3^5} = \frac{1}{243}$

Note: $(x^{\frac{1}{n}})^m = (x^m)^{\frac{1}{n}}$ *whenever* $x > 0$

When $x < 0$, it is true if everything is defined.

2. Properties of nth roots

For $a > 0$, $b > 0$, and positive integers m, n ($m \geq 2$, $n \geq 2$),

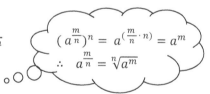

(1) $(\sqrt[n]{a})^n = a$

(2) $\sqrt[n]{a} \cdot \sqrt[n]{b} = \sqrt[n]{ab}$

(3) $\dfrac{\sqrt[n]{a}}{\sqrt[n]{b}} = \sqrt[n]{\dfrac{a}{b}}$

(4) $(\sqrt[n]{a})^m = \sqrt[n]{a^m}$

(5) $\sqrt[m]{\sqrt[n]{a}} = \sqrt[mn]{a} = \sqrt[n]{\sqrt[m]{a}}$

(6) $\sqrt[np]{a^{mp}} = \sqrt[n]{a^m}$, p is a positive integer

(7) For $a \neq 0$ and a positive integer n, $a^0 = 1$, $a^{-n} = \dfrac{1}{a^n}$

(8) For $a > 0$, integer m, positive integer n ($n \geq 2$),

$$a^{\frac{1}{n}} = \sqrt[n]{a}, \quad a^{\frac{m}{n}} = \sqrt[n]{a^m}, \quad a^{-\frac{m}{n}} = \frac{1}{a^{\frac{m}{n}}} = \frac{1}{\sqrt[n]{a^m}}$$

$a^n \cdot a^0 = a^{n+0} = a^n$		$a^n \cdot a^{-n} = a^{n-n} = a^0 = 1$
$\Rightarrow a^0 = 1$ or	0^0 is undefined.	$\Rightarrow a^{-n} = \dfrac{1}{a^n}$ or
$\dfrac{a^n}{a^n} = a^{n-n} = a^0$	If $a = 0$, then a^{-n} is undefined.	$\dfrac{1}{a^n} = 1 \div a^n = a^0 \div a^n = a^{0-n} = a^{-n}$
$\Rightarrow a^0 = 1$		$\Rightarrow a^{-n} = \dfrac{1}{a^n}$

3. Laws of Exponents

For $a > 0$, $b > 0$, and rational numbers m, n,

(1) $a^m \times a^n = a^{m+n}$

(2) $a^m \div a^n = a^{m-n}$

(3) $(a^m)^n = a^{mn}$

(4) $(ab)^n = a^n b^n$

(5) $\left(\frac{a}{b}\right)^n = \frac{a^n}{b^n}$

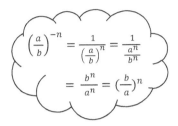

$$\left(\frac{a}{b}\right)^{-n} = \frac{1}{\left(\frac{a}{b}\right)^n} = \frac{1}{\frac{a^n}{b^n}}$$

$$= \frac{b^n}{a^n} = \left(\frac{b}{a}\right)^n$$

Note: $\quad x^a = y^b = z^c = k \implies x = k^{\frac{1}{a}}, \ y = k^{\frac{1}{b}}, \ z = k^{\frac{1}{c}}$

$$x^a = y^b \implies x = y^{\frac{b}{a}}, \ y = x^{\frac{a}{b}}, \ x^{\frac{1}{b}} = y^{\frac{1}{a}}$$

$$x^a = p, \ x^b = q \implies \left[\begin{array}{l} x^{a+b} = x^a \times x^b = pq \\ x^{a-b} = x^a \div x^b = \frac{p}{q} \end{array} \right.$$

Example Using the laws of exponents, simplify the following expressions and state the answers in terms of positive exponents.

(1) $\dfrac{6x^{-2}y^2 x^{\frac{1}{2}} y^{-1}}{2xyx^{-\frac{1}{2}}y} = \dfrac{6x^{-\frac{3}{2}}y}{2x^{\frac{1}{2}}y^2} = 3x^{-2}y^{-1} = \dfrac{3}{x^2 y}$

(2) $\left(\dfrac{-8x^{-6}y^{\frac{3}{2}}}{z^{-\frac{4}{5}}w}\right)^{-\frac{1}{3}} = \dfrac{(-8)^{-\frac{1}{3}}(x^{-6})^{-\frac{1}{3}}(y^{\frac{3}{2}})^{-\frac{1}{3}}}{(z^{-\frac{4}{5}})^{-\frac{1}{3}}(w)^{-\frac{1}{3}}} = \dfrac{(-2)^{-1}x^2 y^{-\frac{1}{2}}}{z^{\frac{4}{15}}w^{-\frac{1}{3}}} = -\dfrac{x^2 w^{\frac{1}{3}}}{2\, y^{\frac{1}{2}} z^{\frac{4}{15}}}$

(3) $\sqrt[5]{x\sqrt{x^3}} = \left(x \cdot x^{\frac{3}{2}}\right)^{\frac{1}{5}} = \left(x^{\frac{5}{2}}\right)^{\frac{1}{5}} = x^{\frac{1}{2}} = \sqrt{x}$

Example For positive real numbers x, y, rationalize the denominator of $\dfrac{\sqrt{x}-\sqrt{y}}{\sqrt{x}+\sqrt{y}}$.

Note that $(a+b)(a-b) = a^2 - b^2$ and $(\sqrt{x})^2 = x$ ($\because x$ is a positive).

Let $a = \sqrt{x}, \ b = \sqrt{y}$

Multiplying the numerator and denominator by $\sqrt{x} - \sqrt{y}$ to eliminate radicals from the denominator, we obtain

$$\frac{\sqrt{x}-\sqrt{y}}{\sqrt{x}+\sqrt{y}} = \frac{\sqrt{x}-\sqrt{y}}{\sqrt{x}+\sqrt{y}} \cdot 1 = \frac{\sqrt{x}-\sqrt{y}}{\sqrt{x}+\sqrt{y}} \cdot \left(\frac{\sqrt{x}-\sqrt{y}}{\sqrt{x}-\sqrt{y}}\right) = \frac{\left(\sqrt{x}-\sqrt{y}\right)^2}{\left(\sqrt{x}\right)^2-(\sqrt{y})^2} = \frac{\left(\sqrt{x}\right)^2 - 2\sqrt{x}\sqrt{y}+(\sqrt{y})^2}{\left(\sqrt{x}\right)^2-(\sqrt{y})^2} = \frac{x-2\sqrt{xy}+y}{x-y}$$

Example Rationalize the denominator of $\sqrt{\dfrac{7}{12}}$.

Since $12 = 4 \cdot 3 = 2^2 \cdot 3,$ multiply both the numerator and denominator by the smallest number 3 to make the denominator a perfect square.

$$\sqrt{\frac{7}{12}} = \sqrt{\frac{7}{2^2 \cdot 3}} = \sqrt{\frac{7}{2^2 \cdot 3} \cdot \frac{3}{3}} = \sqrt{\frac{21}{2^2 \cdot 3^2}} = \frac{\sqrt{21}}{\sqrt{2^2 \cdot 3^2}} = \frac{\sqrt{21}}{2 \cdot 3} = \frac{1}{6}\sqrt{21}$$

Example Rationalize the denominator of $\sqrt[3]{\dfrac{3z}{4x^4 y^2}}$.

$$\sqrt[3]{\frac{3z}{4x^4 y^2}} = \sqrt[3]{\frac{3z}{4x^4 y^2} \cdot \frac{2x^2 y}{2x^2 y}} = \sqrt[3]{\frac{6zx^2 y}{8x^6 y^3}} = \frac{\sqrt[3]{6zx^2 y}}{\sqrt[3]{8x^6 y^3}} = \frac{\sqrt[3]{6zx^2 y}}{2x^2 y}$$

6-2 Exponential Functions and Their Graphs

1. Exponential Functions

(1) Definition

For each real number x, there corresponds a unique real number a^x if $a > 0$.

Thus, the exponential function is defined as follows:

The *exponential function* with base a is the function f defined by

$$f(x) = a^x$$

where $a > 0$, $a \neq 1$, and x is any real number.

Note: ① *If $a = 1$, then $f(x) = 1^x = 1$, which is a constant function, not an exponential function.*

② *If $a < 0$, then the value of a^x may be a non-real number.*

For example, $a = -1 \Rightarrow y = (-1)^x$ has a non-real number when $= \dfrac{1}{2}$.

(2) Properties of Exponential Functions

For $a > 0, b > 0$, and x, y real,

1) $a^x \cdot a^y = a^{x+y}$
2) $(a^x)^y = a^{xy}$
3) $(ab)^x = a^x \cdot b^x$
4) $a^x > 1$ for $a > 1$, x real, and $x > 0$
5) $a^x = 1$ for $a = 1$, x real, and $x > 0$
6) $a^x < 1$ for $0 < a < 1$, x real, and $x > 0$

2. Graphs of Exponential Functions

(1) Graphs of Exponential Function $y = a^x$ $(a \neq 1)$

1) The domain of the exponential function is the set of all real values x

   ; i.e., $\{x|\ x$ is any real number$\}$, and its range is either the set of all positive real numbers if

   a is positive ; i.e., $\{y|\ y$ is any real number such that $y > 0\}$ or the set of all negative real

   numbers if a is negative ; i.e., $\{y|\ y$ is any real number such that $y < 0\}$.

2) The graph passes through points $(0, 1)$ and $(1, a)$ and

   the axis of asymptote is $y = 0$ (x-axis).

3) When $a > 1$, the graph of the function rises as x increases, called *monotone increasing*.

   When $0 < a < 1$, the graph of the function falls as x increases, called *monotone decreasing*.

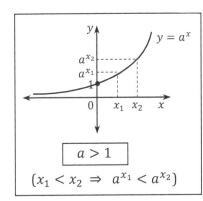
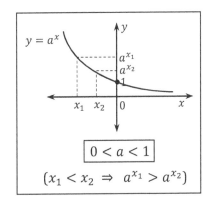

Note : For real numbers x and y such that $x < y$,

① $a^x < a^y$ for $a > 1$ ② $a^x = a^y$ for $a = 1$ ③ $a^x > a^y$ for $0 < a < 1$

4) If $a > 0$ and $b > 1$, then

   the exponential function $y = ab^x$ is an *exponential growth function*.

   If $a > 0$ and $0 < b < 1$, then

   the exponential function $y = ab^x$ is an *exponential decay function*.

Example   State whether $f(x)$ is an exponential growth or exponential decay function.

$$(1)\ f(x) = 3\left(\frac{2}{5}\right)^x \qquad (2)\ f(x) = 4\left(\frac{3}{2}\right)^x \qquad (3)\ f(x) = 10(2)^{-x}$$

(1) Since $0 < \frac{2}{5} < 1$, f is an exponential decay function.

(2) Since $\frac{3}{2} > 1$, f is an exponential growth function.

(3) $f(x) = 10(2^{-1})^x = 10\left(\frac{1}{2}\right)^x$ Since $0 < \frac{1}{2} < 1$, f is an exponential decay function.

(2) Maximum and Minimum values of Exponential Functions

For the exponential function $y = a^x$ with the domain $\{x \mid m \le x \le n\}$,

1) If $a > 0$,

then $y = a^x$ has minimum value a^m when $x = m$ and maximum value a^n when $x = n$.

2) If $0 < a < 1$

then $y = a^x$ has maximum value a^m when $x = m$ and minimum value a^n when $x = n$.

Note: *Tips for solving questions.*

① *Let $a^x = t$.*

Then, $a^{2x} = t^2$ and $a^{-x} = \dfrac{1}{t}$, $t > 0$

② *Let $a^x + a^{-x} = t$.*

Then $a^x + a^{-x} \ge 2\sqrt{a^x \cdot a^{-x}} = 2$ *(When $x = 0$, LHS = RHS)*

by the relationship between roots and coefficients.

$\therefore t \ge 2$

(3) Graphs of Exponential Function $y = a^x$ and $y = a^{-x}$ $(a \ne 1)$

There is symmetry between the graphs of $y = a^x$ and $y = a^{-x}$.

For $a^{-x} = \left(\dfrac{1}{a}\right)^x$, the graph of $y = a^{-x}$ is just like the graph of $y = a^x$ with the x-axis

reversed

in direction.

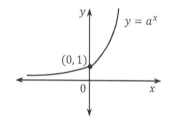
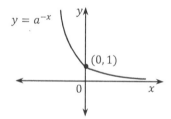

	$y = a^x$	$y = a^{-x}$
y-intercept	$(0, 1)$	$(0, 1)$
Horizontal asymptote	$y = 0$ (x-axis)	$y = 0$ (x-axis)
Domain	$(-\infty, \infty)$	$(-\infty, \infty)$
Range	$(0, \infty)$	$(0, \infty)$
Shape	Increasing	Decreasing
Reflection	Two graphs are reflections of each other through the y-axis	

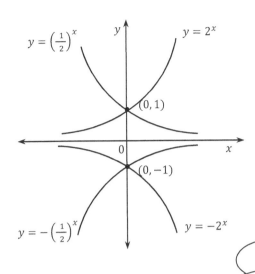

For example,

The graphs of $y = \left(\frac{1}{2}\right)^x$ and $y = 2^x$ are symmetric with respect to the y-axis.

The graphs of $y = -\left(\frac{1}{2}\right)^x$ and $y = -2^x$ are symmetric with respect to the y-axis

The graphs of $y = 2^x$ and $y = -2^x$ are symmetric with respect to the x-axis

Since the graph of $y = \left(\frac{1}{2}\right)^x$ is transformed to $y = \left(\frac{1}{2}\right)^x = (2^{-1})^x = 2^{-x}$, the graph of $y = \left(\frac{1}{2}\right)^x$ is symmetric to the graph of $y = 2^x$ with respect to the y-axis.

(4) Transformations of the Graphs of $y = a^x$ ($a > 0$, $a \neq 1$)

Translations with m units along the x-axis, n units along the y-axis	$y = a^x \Rightarrow y - n = a^{x-m}$ $\therefore y = a^{x-m} + n$
Reflections in the x-axis,	$y = a^x \Rightarrow -y = a^x \quad \therefore y = -a^x$
Reflections in the y-axis,	$y = a^x \Rightarrow y = a^{-x} \quad \therefore y = \left(\frac{1}{a}\right)^x$
Reflections in the origin $(0,0)$	$y = a^x \Rightarrow -y = a^{-x} \quad \therefore y = -\left(\frac{1}{a}\right)^x$
Reflections in the line $y = x$	$y = a^x \Rightarrow x = a^y \quad \therefore y = \log_a x$

Example Graph $y = -2\left(\frac{1}{3}\right)^{x+2} + 1$ and state the domain and range.

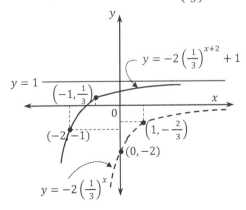

The graph of $y = -2\left(\frac{1}{3}\right)^x$ passes through $(0, -2)$ and $\left(1, -\frac{2}{3}\right)$.

Translate the graph -2 units along the x-axis and 1 unit along the y-axis. Then the graph passes through $(-2, -1)$ and $\left(-1, \frac{1}{3}\right)$.

The asymptote of the graph is the line $y = 1$.

The domain is all real numbers and the range is $y < 1$.

3. The Natural Base e (Euler Number)

For an expression $\left(1 + \frac{1}{n}\right)^n$, $\left(1 + \frac{1}{n}\right)^n$ gets closer and closer to $2.718281\cdots$ as n gets larger and larger.

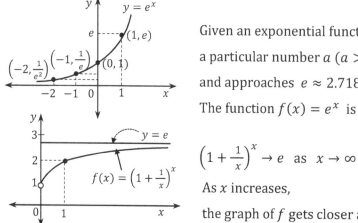

Given an exponential function $y = a^x$,

a particular number a $(a > 1)$ which is an irrational number and approaches $e \approx 2.718281$ is called the *natural base*.

The function $f(x) = e^x$ is the *natural exponential function*.

$$\left(1 + \frac{1}{x}\right)^x \to e \quad \text{as} \quad x \to \infty$$

As x increases,

the graph of f gets closer and closer to the line $y = e$.

Compound Interest

The balance A is an account earning interest compounded n times per year for t years with principal P and annual percentage rate r expressed as a decimal is given by

$$\boxed{A = P\left(1 + \frac{r}{n}\right)^{nt}}$$

As n gets closer to positive infinity, the compounded interest formula gets closer to the continuously compounded interest,

$$\boxed{A = Pe^{rt}}$$

where e is the natural base.

Example You deposit \$5,000 in an account that pays 8.5% annual interest compounded continuously. Find the balance in the account after 3 years.

Note that $P = 5000$, $r = 8.5\% = 0.085$, $n = 1$, and $t = 3$

Using the formula, $A = P\left(1 + \frac{r}{n}\right)^{nt} = 5000\left(1 + \frac{0.085}{1}\right)^{1 \cdot 3} = 5000(1.085)^3 \approx 6,386.45$

If it is compounded quarterly, then $n = 4$. $\quad \therefore A = 5000\left(1 + \frac{0.085}{4}\right)^{4 \cdot 3} = 5437.16$

Therefore, annually compounding yields \$6,386.45 − \$5437.16 = \$979.15 more than quarterly compounding. Since the question is compounded continuously, the balance is

$$A = Pe^{rt} = 5000e^{0.085(3)} = 5000e^{0.255} \approx 6,452.31$$

6-3 Logarithms

We know that $2^2 = 4$ and $2^3 = 8$.

Now, consider the value such that $2^x = 6$.

Since $2^2 < 6 < 2^3$, we expect x to be between 2 and 3.

To obtain the exact value of x, we define logarithms.

1. Definition

When $a > 1$ or $0 < a < 1$,

the inverse of the exponential function $y = a^x$ is the *logarithmic function* with base a.

If $a > 0$, $a \neq 1$, and $y = a^x$ $(x > 0)$, then the logarithm of y to the base a is a function that

yields the value x and is written $\log_a y = x$

The function $f(x) = \log_a x$ is called logarithmic function with base a.

(1) Since every function of the form $y = a^x$ defines a strictly monotone function for $a \neq 1$,

each value of y is obtained from only one x. Thus, the inverse function exists.

If $a = 1$, then $y = a^x$ is a horizontal straight line. Thus, it does not have an inverse function.

However, ① if $a > 1$, then the graph of $y = a^x$ is strictly increasing;

② if $0 < a < 1$, then the graph of $y = a^x$ is strictly decreasing.

Therefore, by the horizontal line test, $y = a^x$ has an inverse function.

For the logarithmic function with base a, $y = \log_a x$,

If $a < 0$, then $\log_{-1} 3 = x \iff (-1)^x = 3$

If $a = 0$, then $\log_0 3 = x \iff 0^x = 3$

If $a = 1$, then $\log_1 3 = x \iff 1^x = 3$

∴ There is no real number x so that the above three cases are satisfied.

Therefore, base a must be a positive such that $a \neq 1$.

(2) For the logarithmic function with base a, $y = \log_a x$,

If $x = 0$, then $\log_2 0 = x \iff 0 = 2^x$

If $x < 0$, then $\log_2(-4) = x \iff -4 = 2^x$

There is no real number x so that the above two cases are satisfied.

Therefore, x must be positive.

(3) Hence, for the logarithmic function $y = \log_a x$,

$a > 0$, $a \neq 1$, and $x > 0$ are necessary requirements.

By the definition of a logarithm, the logarithmic function $f(x) = \log_a x$ is the inverse of the exponential function $g(x) = a^x$.

That is, $f\big(g(x)\big) = \log_a a^x = x$ and $g\big(f(x)\big) = a^{\log_a x} = x$.

By reflecting the graph of $y = a^x$, we obtain the graph of $y = \log_a x$.

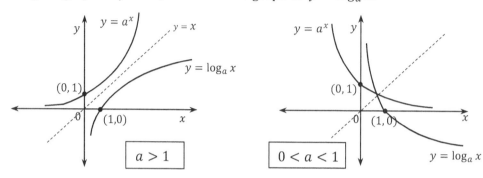

The domain of $y = \log_a x$ is the set of positive real numbers and range of values is the set of all real numbers.

Note:
(1) For a graph of $y = f(x)$, the graph of $y = f^{-1}(x)$ is obtained by reflection through the line $y = x$.
(2) For $y = \log_a x$, $a > 1$, the vertical asymptote is $x = 0$ (y-axis).
*   As x gets close to 0, the value of $\log_a x$ approaches negative infinity. Thus, $\log_a 0$ is undefined.*

Note: The logarithm of a number y to the base a is the power to which a must be raised to obtain y.

$$\log_a a^3 = 3 \qquad \log_a a^2 = 2 \qquad \log_a a^1 = 1$$

$$\log_a a^{-1} = -1 \qquad \log_a a^{-2} = -2 \qquad \log_a a^{-3} = -3$$

$$\therefore \boxed{\log_a \sqrt{a} = \log_a a^{\frac{1}{2}} = \frac{1}{2}} \quad \boxed{\log_a a^x = x} \quad \boxed{\log_a a^{x+y} = x + y} \quad \boxed{\log_a a^{\sqrt{3}} = \sqrt{3}}$$

Note: For integer values of n,

$\log_{10} 1000 = \log_{10} 10^3 = 3 \qquad\qquad \log_{10} 100 = \log_{10} 10^2 = 2$

$\log_{10} 10 = \log_{10} 10^1 = 1 \qquad\qquad \log_{10} 1 = \log_{10} 10^0 = 0$

$\log_{10} 0.1 = \log_{10} 10^{-1} = -1 \qquad\quad \log_{10} 0.01 = \log_{10} 10^{-2} = -2$

$\log_{10} 0.001 = \log_{10} 10^{-3} = -3$

Notation

① If the base is 10, then the logarithm $\log_{10} x$ is called the *common logarithm* of x.

Usually, the 10 is omitted; $\boxed{\log x = \log_{10} x}$

② If the base is e, then the logarithm $\log_e x$ is called the *natural logarithm* of x.

Usually, the e is omitted; $\boxed{\ln x = \log_e x}$

2. Properties of Logarithms

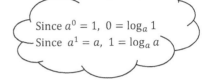

The function $f(x) = \log_a x$ is not defined for all values of x.
($\because$ Since $a > 0$, $a^y > 0$)

(1) If $a > 1$ or $0 < a < 1$, then $\log_a x = y \iff x = a^y$

(2) If $a > 1$ or $0 < a < 1$, then $\log_a x$ is defined only for $x > 0$

(3) If $a > 1$ or $0 < a < 1$, then

    ① $\log_a a = 1$ ② $\log_a 1 = 0$

Since $a^0 = 1$, $0 = \log_a 1$
Since $a^1 = a$, $1 = \log_a a$

(4) For $f(x) = a^x$ and $f^{-1}(x) = \log_a x$,

    $f : \mathbb{R} \to (0, \infty)$; $f^{-1} : (0, \infty) \to \mathbb{R}$

    where $\mathbb{R}$ is the set of all real numbers and $(0, \infty)$ is the set of all positive real numbers.

(5) If $a > 1$ or $0 < a < 1$, then

    ① $a^x = a^y \iff x = y$

    ② $\log_a x = \log_a y \iff x = y$ $(x > 0, y > 0)$

The domain of a function is the range of its inverse and vice versa.

3. Logarithms and The Arithmetic Operations

The relationships between logarithms and the operations of addition, subtraction, multiplication, division and exponentiation are as follows:

If $a > 1$ or $0 < a < 1$, then

$\log_a xyz = \log_a xy + \log_a z$
$= \log_a x + \log_a y + \log_a z$

(1) <u>Product Property</u>: $\log_a xy = \log_a x + \log_a y$ $(x > 0, y > 0)$

(2) <u>Quotient Property</u>: $\log_a \left(\dfrac{x}{y}\right) = \log_a x - \log_a y$ $(x > 0, y > 0)$

(3) <u>Power Product</u>: $\log_a x^n = n \log_a x$ $(x > 0, n$ any real number$)$

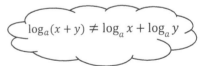

$\log_a (x + y) \neq \log_a x + \log_a y$

Note: Proofs

(1) *The logarithm of a product is the sum of the logarithms of the individual terms.*

    Let $x = a^M$, $y = a^N$

    Then, we have $xy = a^M a^N = a^{M+N}$

    By the definition of a logarithm, $M = \log_a x$, $N = \log_a y$, and $M + N = \log_a xy$

    $\therefore \log_a x + \log_a y = \log_a xy$

(3) *The logarithm of a number raised to a power is equal to the power times the logarithm of the number.*

    Let $M = \log_a x$. Then, $x = a^M$

    Raising both sides to the power n, we obtain $(a^M)^n = x^n$

    By the definition of a logarithm, $nM = \log_a x^n$

    Substituting $M = \log_a x$, we obtain $n \log_a x = \log_a x^n$.

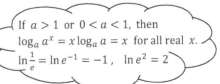

If $a > 1$ or $0 < a < 1$, then
$\log_a a^x = x \log_a a = x$ for all real x.
$\ln \dfrac{1}{e} = \ln e^{-1} = -1$, $\ln e^2 = 2$

(2) *The logarithm of a quotient is the sum of the logarithm of the numerator minus the logarithm of the denominator.*

Since $\dfrac{x}{y} = xy^{-1}$, taking the logarithm of both sides to the base a, we obtain

$$\log_a\left(\frac{x}{y}\right) = \log_a(xy^{-1}) = \log_a x + \log_a y^{-1} \quad \text{by (1)}$$
$$= \log_a x + (-1)\log_a y \quad \text{by (3)}$$
$$= \log_a x - \log_a y$$

Properties of Logarithms

❶ If $a > 1$ or $0 < a < 1$, then $\log_a\left(\dfrac{1}{x}\right) = -\log_a x$, $x > 0$

$\left(\because \log_a\left(\dfrac{1}{x}\right) = \log_a x^{-1} = (-1)\log_a x = -\log_a x\right)$

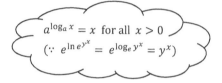

❷ $y^x = e^{x \ln y}$ $(y > 0,\ \text{all } x)$

$(\because y^x = e^{\ln y^x} = e^{x \ln y})$

❸ $\log_a b = \dfrac{\log_c b}{\log_c a}$ $(a > 0,\ a \neq 1,\ c > 0,\ c \neq 1,\ b > 0)$: Change-Of-Base Formula

$(\because$ Let $\log_a b = x$. Then, $b = a^x$

Taking the logarithm of both sides to the base c, $\log_c b = \log_c a^x = x \log_c a$

$\therefore \quad x = \dfrac{\log_c b}{\log_c a}$ Therefore, $\log_a b = \dfrac{\log_c b}{\log_c a}$)

For example, use any base a for which we can find $\log_a x$ and $\log_a b$

$$\log_2 3 = \frac{\log_{10} 3}{\log_{10} 2} = \frac{\log 3}{\log 2} \quad \text{or} \quad \log_2 3 = \frac{\log_e 3}{\log_e 2} = \frac{\ln 3}{\ln 2}$$

❹ $\log_a b = \dfrac{1}{\log_b a}$ $(a > 0,\ a \neq 1,\ b > 0,\ b \neq 1)$

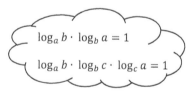

$(\because$ Let $\log_a b = x$. Then, $b = a^x$

Taking the logarithm of both sides to the base b, $\log_b b = \log_b a^x$ $\therefore\ 1 = \log_b a^x$

$\therefore\ x \log_b a = 1$; $x = \dfrac{1}{\log_b a}$ Therefore, $\log_a b = \dfrac{1}{\log_b a}$)

❺ $a^{\log_b c} = c^{\log_b a}$ $(a > 0,\ b > 0,\ b \neq 1,\ c > 0)$

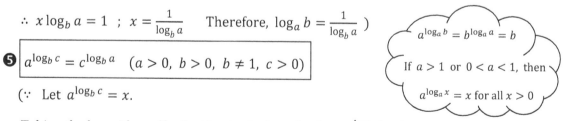

$(\because$ Let $a^{\log_b c} = x$.

Taking the logarithm of both sides to the base b, $\log_b a^{\log_b c} = \log_b x$

$\therefore\ \log_b a^{\log_b c} = (\log_b c)\log_b a = \log_b c \cdot \log_b a = \log_b a \cdot \log_b c = \log_b c^{\log_b a}$

$\therefore\ \log_b x = \log_b c^{\log_b a}$ $\therefore\ x = c^{\log_b a}$ Therefore, $a^{\log_b c} = c^{\log_b a}$)

CAUTION!

$\log_1 1 \neq 1$; $\log_1 1 \neq 0$	$a > 0,\ a \neq 1 \ \Rightarrow\ \log_a a = 1,\ \log_a 1 = 0$
$\log_a(x + y) \neq \log_a x + \log_a y$	$\log_a xy = \log_a x + \log_a y$
$\log_a(x - y) \neq \log_a x - \log_a y$	$\log_a\left(\dfrac{x}{y}\right) = \log_a x - \log_a y$
$\log_a x \cdot \log_a y \neq \log_a x + \log_a y$	
$\dfrac{\log_a x}{\log_a y} \neq \log_a x - \log_a y$	$\log_y x = \dfrac{\log_a x}{\log_a y}$
$(\log_a x)^n \neq n \log_a x$	$\log_a x^n = n \log_a x$

$$\log_{a^m} b^n = \frac{n}{m} \log_a b$$
$$(m, n:\text{ real numbers},\ m \neq 0)$$
$$\log_{a^2} b = \frac{1}{\log_b a^2} = \frac{1}{2\log_b a}$$
$$= \frac{1}{2}\frac{1}{\log_b a} = \frac{1}{2}\log_a b$$

4. The Mantissa and The Characteristic

(1) Definition

Consider the relationships between a number and its logarithm.

Since logarithms to base 10 are most common, we consider the relationship between a number and its logarithm to base 10.

For example, $10^{\frac{1}{2}} = 10^{0.5} = 3.1623 \cdots\cdots ①$

$$10^{0.5} \cdot 10^1 = 10^{1.5}$$
$$10^{0.5} \cdot 10^2 = 10^{0.5+2} = 10^{2.5}$$

Converting to the logarithmic form, $\log 3.1623 = 0.5$

Multiplying both sides of the exponential equation ① by $10, 10^2, 10^3,$ and $10^4,$ we get

$10^{1.5} = 31.623, \quad 10^{2.5} = 316.23, \quad 10^{3.5} = 3162.3, \quad 10^{4.5} = 31623,$ respectively.

Converting the above exponential equations to their logarithmic forms,

$\log 31.623 = 1.5, \quad \log 316.23 = 2.5, \quad \log 3162.3 = 3.5, \quad \log 31623 = 4.5$, respectively.

As the decimal point is moved to the right one digit, the corresponding value of the logarithm increases by 1.

Let $\log A = n + \alpha$ where n is an integer, α is a positive decimal part $(0 \leq \alpha < 1)$.

The integer n is called the *characteristic* (: the left of the decimal point) of the number and the α is called *mantissa* (: the right of the decimal point) of the number.

For example, $\log 123 = 2.0899 = \boxed{2} + \boxed{0.0899}$

Characteristic   Mantissa

$$\log 100 = \log 10^2 = 2 = \boxed{2} + \boxed{0.0}$$

(2) Properties of The Characteristics (Positive and Negative Characteristics)

1) For $\log A$, if A has n digits for integer part, then the characteristic is $n - 1$.

$$\log \underbrace{\blacksquare\,\blacksquare\,\cdots\cdots\,\blacksquare.\,\blacksquare\blacksquare\,\cdots}_{n \text{ digits}} = (n-1).\times\times\times$$

For example, $\log 31.625 = 1.5$ $\log 316.25 = 2.5$ $\log 3162.5 = 3.5$

2 digits $2 - 1 = 1$ 3 digits $3 - 1 = 2$ 4 digits $4 - 1 = 3$

Note that the negative characteristics are represented by $\overline{\blacksquare}$.

$$\log 0.00123 = -2.9101 = -2 - 0.9101 = -2 - 0.9102 + 1 - 1 = (-2 - 1) + (1 - 0.9101)$$
$$= -3 + 0.0899$$
$$= \overline{3}.0899$$

2) For $\log A$, if the first non-zero digit in the deciaml part appears n^{th} away from the decimal point of the number, then the characteristic is $\bar{n} = -n$.

That is, $\log 0.\underbrace{00\cdots\cdots 0}_{n \text{ digits}}\overset{(\neq 0)}{\blacksquare}\blacksquare\blacksquare\cdots = \bar{n}.\times\times\times$

$\overline{3}.0899$ $= -3 + 0.899$	
$\mathbb{X}$	
-3.0899 $= -3 - 0.0899$	

For example, when $\log 3.25 = 0.5119$,

$\log 32.5 = 1 + 0.5119$ $\log 325000 = 5 + 0.5119$

2 digits $2 - 1 = 1$ 6 digits $6 - 1 = 5$

$\log 0.000325 = \overline{4}.5119 = -4 + 0.5119$

4 digits

$$\log(0.325)^2 = 2\log 0.325 = 2(\overline{1}.5119) = 2(-1 + 0.5119)$$

1 digit

$$\log \sqrt[5]{32500} = \frac{1}{5}\log 32500 = \frac{1}{5}(4 + 0.5119)$$

5 digits

Note: For $A > 0$, $B > 0$,

> ① $\log A = n + \alpha$ *where n is an integer, α is a positive decimal part $(0 \leq \alpha < 1)$.*
>
> $\Leftrightarrow n \leq \log A < n + 1$
>
> $\Leftrightarrow [\log A] = n$
>
> $\Leftrightarrow A = (\blacksquare.\blacksquare\blacksquare\blacksquare\cdots\cdots) \times 10^n, \ (n \text{ integer})$
>
> $\Leftrightarrow 10^n \leq A < 10^{n+1}$

② $\log A$ and $\log B$ *have the same mantissa.*

$\Leftrightarrow \log A - \log B = (Integer)$

$\Leftrightarrow \log A - [\log A] = \log B - [\log B]$

$\Leftrightarrow \dfrac{A}{B} = 10^k,\ (k\ integer)$

(3) Antilogarithms

For a given logarithm, the *antilogarithm* is the number whose logarithm is given.

That is, if $\log x = a$, then $x = anti \log a$

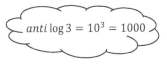
$anti \log 3 = 10^3 = 1000$

Note:

$anti \log_b c = b^c \qquad anti \log c = 10^c$

$anti \log_a(b + c) = a^{b+c} = a^b \cdot a^c = anti \log_a b \cdot anti \log_a c$

Example Given $anti \log 1.2345 = 17.2$

Find (1) $anti \log 0.2345$ (2) $anti \log(6.2345 - 10)$

Since $anti \log 1.2345 = 10^{1.2345} = 10^{0.2345} \cdot 10^1 = 17.2$,

(1) $anti \log 0.2345 = 10^{0.2345} = (17.2) \cdot 10^{-1} = 1.72$

(2) $anti \log(6.2345 - 10) = 10^{6.2345-10} = 10^{6.2345} \cdot 10^{-10} = 10^{0.2345} \cdot 10^6 \cdot 10^{-10}$

$$= 10^{0.2345} \cdot 10^{-4} = (1.72) \cdot 10^{-4} = 0.000172$$

(4) Natural Logarithms

Consider the natural number e for the base of a logarithm.

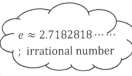
$e \approx 2.7182818\cdots$
; irrational number

The number e is natural just as the number $\pi \approx 3.14159$ is natural.

The number π is the ratio of the circumference of a circle to its diameter.

Similarly, e is defined as the limit of the term $\left(1 + \dfrac{1}{n}\right)^n$ as n increases without bounds,

and is approximately 2.718 ; $\displaystyle\lim_{n\to\infty} \left(1 + \dfrac{1}{n}\right)^n = 2.718$

Every term of the form a^x can be expressed as a term of the form e^y.

Let $a^x = e^y$.

Taking the natural logarithm of both sides, we obtain $\ln a^x = \ln e^y$; $x \ln a = y \ln e$

Since $\ln e = 1$, $y = x \ln a$

Converting a^x to an exponent with base e, $\boxed{a^x = e^{x \ln a}}$

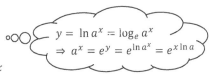
$y = \ln a^x = \log_e a^x$
$\Rightarrow a^x = e^y = e^{\ln a^x} = e^{x \ln a}$

For example, when $\ln 5 = 1.60944$, $8^x = e^{x \ln 8} = e^{1.60944x}$

6-4 Logarithmic Functions and Their Graphs

1. Logarithmic Functions and Their Graphs

For a positive real number x, a function of the form $f(x) = \log_a x$ is a logarithmic function.

(1) Graphs of Logarithmic Function $y = \log_a x$ ($a > 0$, $a \neq 1$, $x > 0$)

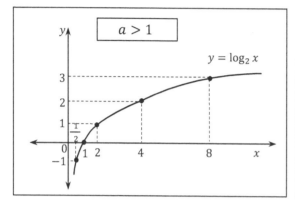

Note that $\log_{\frac{1}{2}} x \iff -\log_2 x$

$$\left(\because \log_{\frac{1}{2}} x = \frac{\log_2 x}{\log_2 \frac{1}{2}} = \frac{\log_2 x}{\log_2 2^{-1}} = \frac{\log_2 x}{-\log_2 2} = \frac{\log_2 x}{-1} = -\log_2 x \right)$$

Thus, the graph of $y = \log_2 x$ and the graph of $y = \log_{\frac{1}{2}} x$ are symmetric along the x-axis.

$y = a^x$ ($a > 0$, $a \neq 1$) $\iff x = \log_a y$

Switching x and y, $y = \log_a x$ ($a > 0$, $a \neq 1$)

Therefore, the inverse of an exponential function $y = a^x$ is $y = \log_a x$.

Since $y = a^x$ and $y = \log_a x$ are inverse functions of each other, the graph of $y = a^x$ and $y = \log_a x$ are symmetric along the line $y = x$.

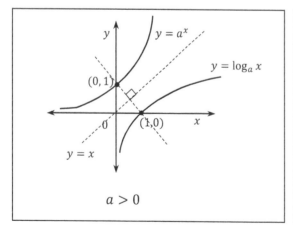

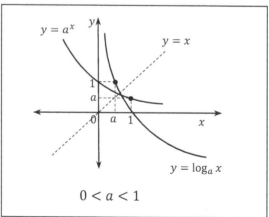

(2) Properties of Logarithmic Function $y = \log_a x$ $(a > 0,\ a \neq 1,\ x > 0)$

① The domain is $\{x | x > 0\}$: a set of all positive real numbers and

range is $\mathbb{R}$: a set of all real numbers.

② The graph passes through a point $(1, 0)$.

③ The vertical asymptote to the graph is $x = 0$ (y-axis).

④ If $a > 1$, then the function is increasing.

If $0 < a < 1$, then the function is decreasing.

(3) Transformations of the Graphs of $y = \log_a x$ $(a > 0,\ a \neq 1,\ x > 0)$

Translations with m units along the x-axis, n units along the y-axis	$y = \log_a x \Rightarrow y - n = \log_a(x - m)$ $\therefore\ y = \log_a(x - m) + n$
Reflections in the x-axis	$y = \log_a x \Rightarrow -y = \log_a x \qquad \therefore\ y = -\log_a x$
Reflections in the y-axis	$y = \log_a x \Rightarrow y = \log_a(-x) \qquad \therefore\ y = \log_a(-x)$
Reflections in the origin	$y = \log_a x \Rightarrow -y = \log_a(-x) \quad \therefore\ y = -\log_a(-x)$
Reflections in the line $y = x$	$y = \log_a x \Rightarrow x = \log_a y \qquad \therefore\ y = a^x$

Example   Using the graph of $y = \log_2 x$, sketch the graph of each function.

(1) $y = \log_2(-x)$ (2) $y = \log_2\left(\dfrac{1}{x}\right)$ (3) $y = \log_2(x - 1)$ (4) $y = \log_2 2x$

(1) : Reflection in the y-axis

(2) $y = \log_2\left(\dfrac{1}{x}\right) = \log_2 x^{-1} = -\log_2 x$: Reflection in the x-axis

(3) : Translation with 1 unit along the x-axis

(4) $y = \log_2 2x = \log_2 2 + \log_2 x = 1 + \log_2 x$

$\therefore\ y - 1 = \log_2 x$: Translation with 1 unit along the y-axis

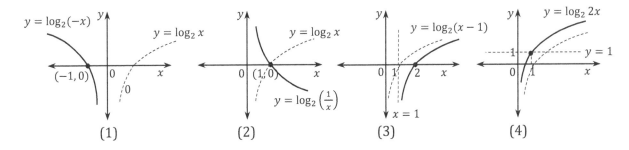

6-5 Exponential and Logarithmic Equations

1. Exponential Equations

An equation in which the variable is contained in one or more exponents (powers) is called *exponential equation*.

Solving Exponential Equations

If two powers with the same base are equal, then their exponents must be equal.

That is, for $a > 0$, $a \neq 1$, $[$ if $a^x = a^y$, then $x = y$ $]$

(1) Exponential equation with two terms:

1) If possible, convert each term so that both are expressed as a power of the same base;

$a^{f(x)} = a^{g(x)} \Rightarrow f(x) = g(x)$ or $a = 1$

For example, $2^x = 32 \Rightarrow 2^x = 2^5 \quad \therefore x = 5$

$\left(\frac{1}{9}\right)^x = \sqrt{3} \Rightarrow (3^{-2})^x = 3^{\frac{1}{2}} \Rightarrow 3^{-2x} = 3^{\frac{1}{2}} \Rightarrow -2x = \frac{1}{2} \quad \therefore x = -\frac{1}{4}$

2) If it is not possible to express each number as a rational power of the same base, take the common logarithm of each side of the equation.

For example, to solve the equation $3^{2x} = 12$ for ,

take logarithm of each side. Then $\log 3^{2x} = \log 12$

$\therefore \quad 2x \log 3 = \log 12$

Solving for x, $\quad x = \dfrac{\log 12}{2 \log 3} = \dfrac{1}{2} \left(\dfrac{\log 12}{\log 3} \right)$

> If $a^x = b$, then $\log a^x = \log b$
> $\therefore \quad x \log a = \log b$
> $\therefore \quad x = \frac{\log b}{\log a}$

3) If the exponents of positive bases are the same, compare the bases or find the value so that the exponent is zero; $a^{f(x)} = b^{f(x)} \Rightarrow a = b$ or $f(x) = 0$

For example, $(x + 1)^x = 3^x \ (x > -1) \Rightarrow$ When $x \neq 0$, $x + 1 = 3 \quad \therefore x = 2$

When $x = 0$, $1^0 = 3^0$ (True)

Therefore, $x = 0$, $x = 2$

(2) Exponential equation with three terms:

If the exponential equation contains terms $a^x, a^{2x}, \cdots\cdots$, then substitute a^x for t $(t > 0)$ and solve the equation of t.

For example, $2^{2x+1} + 3 \cdot 2^x - 2 = 0$

$\Rightarrow$ Let $2^x = t, \ t > 0$

Then, $2t^2 + 3t - 2 = 0$ $\quad \therefore (t+2)(2t-1) = 0$ $\quad \therefore t = -2$ or $t = \dfrac{1}{2}$

Since $t > 0$, $t = \dfrac{1}{2}$

$\therefore \ 2^x = \dfrac{1}{2}$ $\qquad$ Therefore, $x = -1$

Example Solve each exponential equation.

$\quad$ (1) $4^x = \dfrac{1}{16}$ $\quad$ (2) $2^{2x} - 2^x - 6 = 0$ $\quad$ (3) $10^{2x-1} + 3 = 15$ $\quad$ (4) $3^{x-1} = 2^x$

(1) $4^x = \dfrac{1}{16}$ $\Rightarrow 2^{2x} = 2^{-4}$ $\quad \therefore 2x = -4$ $\quad \therefore x = -2$

(2) $2^{2x} - 2^x - 6 = 0$ $\Rightarrow$ Let $X = 2^x$

$\qquad\qquad\qquad\qquad$ Then, $X^2 - X - 6 = 0$ $\quad \therefore (X-3)(X+2) = 0$ $\quad \therefore X = 3$ or $X = -2$

$\qquad\qquad\qquad\qquad$ Since $X > 0$, $X = 3$ $\quad \therefore 2^x = 3$ $\quad$ Therefore, $x = \log_2 3$

(3) $10^{2x-1} + 3 = 15$ $\Rightarrow 10^{2x-1} = 12$

$\qquad\qquad\qquad\qquad$ Taking common logarithm of each side, $\log 10^{2x-1} = \log 12$

$\qquad\qquad\qquad\qquad$ Since $\log 10^x = x$, $2x - 1 = \log 12$

$\qquad\qquad\qquad\qquad \therefore \ x = \dfrac{1}{2}(\log 12 + 1) = \dfrac{1}{2} + \dfrac{1}{2}\log 12 = \dfrac{1}{2} + \log\sqrt{12}$

(4) $3^{x-1} = 2^x$ $\Rightarrow \log 3^{x-1} = \log 2^x$

$\qquad\qquad\quad \Rightarrow (x-1)\log 3 = x\log 2$

$\qquad\qquad\quad \Rightarrow (\log 3 - \log 2)x = \log 3$

$\qquad\qquad\quad \Rightarrow x = \dfrac{\log 3}{\log 3 - \log 2}$

2. Logarithmic Equations

For a logarithm of b to the base a, if a or b contains unknown variable, then the equation is called a *logarithmic equation*.

Solving Logarithmic Equations

(1) $\log_a f(x) = \log_a g(x)$ $\ \Rightarrow \ f(x) = g(x)$

$\quad \log_a f(x) = b$ $\ \Rightarrow \ f(x) = a^b$

(2) Substitute $\log_a x$ for X

(3) Take common logarithm of each side.

Example   Solve each logarithmic equation.

$$(1)\ \log_2 x + \log_2(x-6) = 4 \qquad (2)\ (\log x)^2 = \log x^2 \qquad (3)\ x^{\log x} = x$$

(1) $\log_2 x + \log_2(x-6) = 4 \ \Rightarrow\ \log_2 x(x-6) = 4 \ \Rightarrow\ x(x-6) = 2^4$

$\therefore\ x^2 - 6x - 16 = (x-8)(x+2) = 0 \qquad \therefore\ x = 8 \text{ or } x = -2$

Since $x > 0$ and $x - 6 > 0$, $x > 6 \qquad \therefore\ x = -2$ is not a solution.

Therefore, $x = 8$

(2) $(\log x)^2 = \log x^2 \ \Rightarrow\ $ Let $\log x = X$.

Then, $X^2 = 2X \qquad \therefore\ X(X-2) = 0 \qquad \therefore\ X = 0$ or $X = 2$

If $X = 0$, then $\log x = 0 \qquad \therefore\ x = 1$

If $X = 2$, then $\log x = 2 \qquad \therefore\ x = 10^2 = 100$

(3) $x^{\log x} = x \ \Rightarrow\ $ Taking common logarithm of each side, $\log x^{\log x} = \log x$

$\therefore\ (\log x)(\log x) = \log x \ ;\ \log x\,(\log x - 1) = 0$

$\therefore\ \log x = 0$ or $\log x = 1 \qquad \therefore\ x = 1$ or $x = 10$

6-6   Exponential and Logarithmic Inequalities

1.   Exponential Inequalities

For $a^m > a^n$,

(1) If $a > 1$, then $a^m > a^n \iff m > n$

(2) If $0 < a < 1$, then $a^m > a^n \iff m < n$

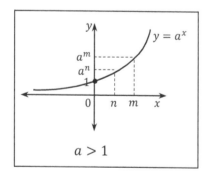

$a > 1$

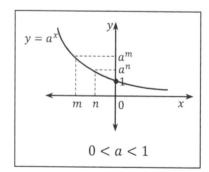

$0 < a < 1$

Example   Solve each inequality.

$$(1)\ 2(\sqrt{2})^x < \sqrt[3]{4} \qquad\qquad (2)\ x^{2x+1} > x^{x+3} \ (x > 0)$$

(1) $2(\sqrt{2})^x < \sqrt[3]{4} \ \Rightarrow\ 2 \cdot 2^{\frac{1}{2}x} < 4^{\frac{1}{3}} \qquad \therefore\ 2^{1+\frac{1}{2}x} < 2^{\frac{2}{3}}$

Since the base is greater than 1, $1 + \dfrac{1}{2}x < \dfrac{2}{3} \qquad \therefore\ x < -\dfrac{2}{3}$

(2) Case 1: If $x > 1$, then $2x + 1 > x + 3$ $\quad \therefore x > 2$

Since $x > 1$ and $x > 2$, $x > 2$ $\cdots\cdots$ ①

Case 2: If $0 < x < 1$, then $2x + 1 < x + 3$ $\quad \therefore x < 2$

Since $0 < x < 1$ and $x < 2$, $0 < x < 1$ $\cdots\cdots$ ②

By ① and ②, $0 < x < 1$ or $x > 2$

2. Logarithmic Inequalities

For $\log_a m > \log_a n$

(1) If $a > 1$, then $\log_a m > \log_a n \iff m > n$

(2) If $0 < a < 1$, then $\log_a m > \log_a n \iff m < n$

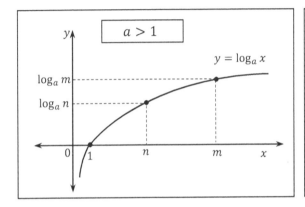
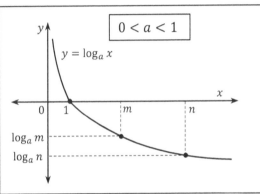

Example   Solve each logarithmic inequality.

$\quad$ (1) $\log_2(x - 3) > \log_4(x - 1)$ $\quad$ (2) $\log_{0.5}(x^2 - 23) - \log_{0.5}(x - 5) < \log_{0.5} 7$

(1) Since $x - 3 > 0$ and $x - 1 > 0$, $x > 3$ $\cdots\cdots$ ①

$\quad$ Since $\log_a b = \dfrac{\log_c b}{\log_c a}$, $\log_2(x - 3) > \dfrac{\log_2(x-1)}{\log_2 4}$

$\quad \therefore \log_2(x - 3) > \dfrac{\log_2(x-1)}{2}$ $\qquad \therefore 2\log_2(x - 3) > \log_2(x - 1)$

$\quad \therefore \log_2(x - 3)^2 > \log_2(x - 1)$ $\qquad \therefore (x - 3)^2 > x - 1$

$\quad \therefore x^2 - 7x + 10 = (x - 5)(x - 2) > 0$ $\qquad \therefore x > 5$ or $x < 2$ $\cdots\cdots$ ②

$\quad$ By ① and ②, $x > 5$

(2) Since $x^2 - 23 > 0$ and $x - 5 > 0$, $x > 5$ $\cdots\cdots$ ①

$\quad \log_{0.5}(x^2 - 23) - \log_{0.5}(x - 5) < \log_{0.5} 7$ $\quad \Rightarrow \log_{0.5}(x^2 - 23) < \log_{0.5} 7 + \log_{0.5}(x - 5)$

$\quad$ Since $0.5 < 1$, $x^2 - 23 > 7(x - 5)$

$\quad \therefore x^2 - 7x + 12 = (x - 4)(x - 3) > 0$ $\qquad \therefore x > 4$ or $x < 3$ $\cdots\cdots$ ②

$\quad$ By ① and ②, $x > 5$

3. Magnitude

Examples (1) When $0 < x < 1$, compare A and B: $A = x^{x^2}$, $B = x^{2x}$

(2) When $a > 1 > b > 0$ and $ab > 1$, compare $A, B,$ and C:

$$A = \log_{a^2} b, \ B = \log_a b^2, \ C = \log_b a^2$$

(1) Since $0 < x < 1$, $x - 2 < -1 < 0$ $\therefore x^2 - 2x = x(x - 2) < 0$

$\therefore x^2 < 2x$ $\therefore x^{x^2} > x^{2x}$ $\therefore A > B$

(2) Since $a > 1 > b > 0$, $\log a > \log 1 > \log b$ $\therefore \log a > 0$ and $0 > \log b$ ······①

Since $ab > 1$, $\log ab > \log 1$ $\therefore \log ab > 0$ ······②

$$A - B = \log_{a^2} b - \log_a b^2 = \frac{\log b}{\log a^2} - \frac{\log b^2}{\log a} = \frac{\log b}{2 \log a} - \frac{2 \log b}{\log a}$$

$$= \frac{\log b - 4 \log b}{2 \log a} = \frac{-3 \log b}{2 \log a} > 0 \quad \text{by ①} \qquad \therefore A > B$$

$$B - C = \log_a b^2 - \log_b a^2 = \frac{\log b^2}{\log a} - \frac{\log a^2}{\log b} = \frac{2 \log b}{\log a} - \frac{2 \log a}{\log b}$$

$$= \frac{2 (\log b)^2 - 2 (\log a)^2}{\log a \cdot \log b} = \frac{2 (\log b + \log a)(\log b - \log a)}{\log a \cdot \log b} > 0 \quad \text{by ① and ②}$$

$\therefore B > C$

Therefore, $A > B > C$

Exercises

#1 Simplify the following given expressions.

(1) $a^5 \times a^3$

(2) $a^2 \div a^6$

(3) $a^3 \div a^3$

(4) $(a^2)^3$

(5) $(ab)^2$

(6) $(a^2 b^5)^3$

(7) $(-3a^2 b^3)^3$

(8) $(a^3)^2 a^5$

(9) $\left(\dfrac{ab^2}{c^3}\right)^4$

(10) $\left(\dfrac{-3ab^2}{2c^3}\right)^5$

#2 Evaluate each expression.

(1) 3^{-2}

(2) 2^{-4}

(3) 4^0

(4) 5^{-3}

(5) $(2^3)^0$

(6) $2^{-1} \div 2^{-3}$

(7) $2^2 \div 2^{-3}$

(8) $3^0 \div 3^4$

(9) $2^{-10} \div 2^{-2} \times 2^5$

(10) $8^3 \times 16^{-2} \div 32$

(11) $(3 \times 2^{-3})^{-2}$

(12) $(-2)^2 (2^{-3} \times 3^2)^{-1}$

(13) $(-10)^0$

#3 Evaluate each expression.

(1) $\sqrt[3]{64}$

(2) $\sqrt[3]{-64}$

(3) $\sqrt[4]{81}$

(4) $\sqrt[5]{-32}$

(5) $\sqrt[4]{2} \cdot \sqrt[4]{8}$

(6) $\dfrac{\sqrt[3]{2}}{\sqrt[3]{16}}$

(7) $\dfrac{\sqrt[4]{64}}{\sqrt[4]{4}}$

(8) $\sqrt[3]{8^2}$

(9) $\sqrt[3]{\left(\dfrac{8}{27}\right)^2}$

(10) $(-\sqrt[3]{3})^9$

(11) $\sqrt[4]{(-2)^4}$

(12) $\sqrt[3]{0.0001} \cdot \sqrt[3]{10}$

(13) $\sqrt[3]{\sqrt[4]{8}} \cdot \sqrt{\sqrt[4]{64}}$

(14) $\sqrt[9]{3^3} \cdot \sqrt[12]{3^8}$

(15) $\dfrac{\sqrt[4]{400}}{\sqrt{10}}$

#4 Evaluate following expressions.

(1) $\sqrt{16}$

(2) The square roots of 16

(3) $\sqrt[4]{16}$

(4) $\sqrt[4]{-16}$

(5) The 4th root of 16

(6) Cube root of 8

(7) Cube root of -8

(8) $\sqrt[3]{-8}$

(9) The 5th root of 32

(10) The 5th root of -32

#5 Evaluate each expression.

(1) $16^{\frac{1}{2}}$

(2) $4^{\frac{3}{2}}$

(3) $32^{\frac{4}{5}}$

(4) $8^{-\frac{2}{3}}$

(5) $64^{-\frac{3}{2}}$

(6) $(-64)^{\frac{2}{3}}$

(7) $\left\{\left(\frac{9}{16}\right)^{-\frac{3}{4}}\right\}^{\frac{2}{3}}$

(8) $\sqrt[3]{16} \div \sqrt[3]{2^{10}}$

(9) $16^{\frac{1}{3}} \div 24^{\frac{2}{3}} \div 18^{\frac{1}{3}}$

(10) $2^{\frac{1}{3}} \times 3^{\frac{1}{2}} \div 6^{\frac{1}{6}} \div \left(\frac{3}{2}\right)^{\frac{1}{3}}$

(11) $\sqrt{2^3 \cdot 3} \div \sqrt[4]{2^2 \cdot 3^5} \times \sqrt[6]{2 \cdot 3^{10}}$

#6 (Magnitude) Compare the following numbers.

(1) $\sqrt[3]{3}$, $\sqrt[4]{9}$, $\sqrt[5]{27}$

(2) $\left(\frac{1}{2}\right)^{\frac{1}{3}}$, $\left(\frac{1}{2}\right)^{-2}$, $\sqrt{\frac{1}{2}}$, $\frac{1}{4}$

(3) $\sqrt[6]{16}$, $\sqrt[12]{81}$, $\sqrt[9]{64}$

(4) $\sqrt{2}$, $\sqrt[3]{3}$, $\sqrt[4]{5}$, $\sqrt[6]{6}$

(5) $\sqrt[3]{3}$, $\sqrt[4]{5}$, $\sqrt{\sqrt[3]{9}}$

(6) $\left(\sqrt{2\sqrt[3]{3}}\right)^2$, $\left(\sqrt[3]{3\sqrt{2}}\right)^2$, $\left(\sqrt[3]{2\sqrt{3}}\right)^2$

(7) $\sqrt{3\sqrt{3}}$, $\sqrt{3^{\sqrt{3}}}$, $(\sqrt{3})^{\sqrt{3}}$

(8) For positive integers a and b such that $1 < a^{b-3} < b^{a-5}$,

compare the numbers A, B, C: $A = a^{\frac{1}{a-5}} \cdot b^{\frac{1}{b-3}}$, $B = a^{-\frac{1}{a-5}} \cdot b^{\frac{1}{b-3}}$, $C = a^{\frac{1}{a-5}} \cdot b^{-\frac{1}{b-3}}$

#7 Simplify each expression.

(1) $\sqrt[3]{\dfrac{\sqrt[4]{a}}{\sqrt{a}}} \div \sqrt{\dfrac{\sqrt[6]{a}}{\sqrt[3]{a}}}$

(2) $\sqrt{\dfrac{\sqrt[3]{a}}{\sqrt[4]{a}}} \times \sqrt[4]{\dfrac{\sqrt{a}}{\sqrt[3]{a}}} \times \sqrt[3]{\dfrac{a}{\sqrt[4]{a}}}$

(3) $\sqrt{\dfrac{a}{\sqrt{a}}} \cdot \sqrt[3]{a^2}$

(4) $\sqrt{a\sqrt{a\sqrt{a}}}$

(5) $\sqrt{a\sqrt{a\sqrt{a\sqrt{a}}}}$

(6) $\left(a^{\frac{1}{4}} + b^{\frac{1}{4}}\right)\left(a^{\frac{1}{4}} - b^{\frac{1}{4}}\right)\left(a^{\frac{1}{2}} + b^{\frac{1}{2}}\right)$

(7) $(a - b) \div \left(a^{\frac{1}{3}} - b^{\frac{1}{3}}\right)$

(8) $(a - a^{-1}) \div \left(a^{\frac{1}{2}} - a^{-\frac{1}{2}}\right)$

#8 Find the value.

(1) When $x^{\frac{1}{2}} + x^{-\frac{1}{2}} = 3$, ① $x + x^{-1}$ ② $x^2 + x^{-2}$ ③ $x^{\frac{3}{2}} + x^{-\frac{3}{2}}$

(2) When $x + \dfrac{1}{x} = 7$, $\sqrt{x} + \dfrac{1}{\sqrt{x}}$

(3) When $e^{2x} = 5$, ① $\left(\dfrac{1}{e^2}\right)^{-3x}$ ② $\dfrac{e^x + e^{-x}}{e^x - e^{-x}}$ ③ $\dfrac{e^{3x} + e^{-3x}}{e^x + e^{-x}}$

(4) When α and β are two roots of an equation $x^2 - 4x + 1 = 0$, $\dfrac{4^{\alpha} \cdot 8^{\beta}}{(2 \cdot 4^{\alpha})^{\beta}}$

(5) When $\sqrt[3]{x} + \dfrac{1}{\sqrt[3]{x}} = 2$ for positive x, $\sqrt[3]{x^4} + \dfrac{1}{\sqrt[3]{x^4}}$

(6) When $a^{\frac{1}{2}} - a^{-\frac{1}{2}} = 3$ $(a > 0)$, $\dfrac{a^{\frac{3}{2}} - a^{-\frac{3}{2}} + 5}{a + a^{-1} + 3}$

(7) When $a^{2x} = 7 + 4\sqrt{3}$ $(a > 0, \ a \neq 1)$, $\dfrac{a^x - a^{-x}}{a^{2x} - a^{-2x}}$

(8) When $a^x = b^y = 2^z$ and $\dfrac{1}{x} + \dfrac{1}{y} - \dfrac{4}{z} = 0$ $(xyz \neq 0, \ a > 0, \ b > 0)$, $\dfrac{1}{ab}$

(9) When $a^x = 3$, $(ab)^y = 3^4$, $(abc)^z = 3^5$ ($a, b,$ and c are positive numbers), $3^{\frac{1}{x} + \frac{4}{y} - \frac{5}{z}}$

(10) When $a = 2^m 3^n$, $\sqrt{\dfrac{a}{2}} \times \sqrt[3]{\dfrac{a}{4}}$

(11) When $f(x) = \dfrac{1 + x + x^2 + \cdots\cdots + x^{10}}{x^{-2} + x^{-3} + \cdots\cdots + x^{-12}}$, $f(\sqrt[24]{2})$

(12) When $2^{3x} = 5$ (x is a real number), $\dfrac{2^{5x} + 2^{-4x}}{2^{2x} + 2^{-x}} + \dfrac{2^{2x} - 2^{-4x}}{2^{-4x} + 2^{-x}}$

(13) When $x^{\frac{1}{2}} = e^{\frac{1}{2}} + e^{-\frac{1}{2}}$ $(e > 1)$, $\dfrac{x - 2 - \sqrt{x^2 - 4x}}{x - 2 + \sqrt{x^2 - 4x}}$

#9 Find the value.

(1) When a point $P(a, b)$ lies on the line that has x-intercept 2 and y-intercept 1, find the minimum value of $\sqrt{2^a} + 2^b$.

(2) When α and β are two roots of an equation $x^2 + 2ax + 4 = 0$, $\dfrac{\alpha^{-1} - \beta^{-1}}{\alpha^{-2} - \beta^{-2}} = \dfrac{3}{2}$ $(\alpha > 0, \ \beta > 0)$. Find the value of the constant a.

(3) For real numbers $x, y,$ and z, $2^x = 10$, $20^y = 10$, and $a^z = 10$ $(a > 0)$. When $\dfrac{1}{x} - \dfrac{1}{y} + \dfrac{1}{z} = 3$, find the value of a.

(4) For real numbers x and y, $3^x = 5$ and $5^{\frac{y}{2}} = 27$. Find the value of xy.

(5) For real numbers $a, b,$ and c such that $abc = 12$, $2^a = 3^2$ and $3^b = 5^3$. Find the value of 5^c.

(6) When $a = 2^{\frac{2}{3}}$ and $b = 3^{\frac{1}{4}}$, find the value of mn such that $a^m b^n = 36$.

(7) Find the value of the constant a such that $4^a = \left(\sqrt{2 + \sqrt{3}} - \sqrt{2 - \sqrt{3}} \right)^{\frac{1}{3}}$.

(8) A function $y = 3^{-x^2 + 4x - 3}$ has the maximum value when $x = a$. Find the value of a.

(9) When a function $y = a^{x^2 - 4x + 6}$ has the maximum value $\dfrac{1}{100}$,

   find the value of the constant a $(a > 0)$.

(10) A function $y = (2^x - 1)^2 + (2^{-x} - 1)^2$ has the minimum value when $x = a$.

   Find the value of a.

#10 Compare the graph of each of the following with the graph of $f(x) = 2^x$.

   Identify the domain and range of each function.

   (1) $g(x) = 2^{x+1}$ $\qquad\qquad$ (2) $h(x) = 2^x - 3$ $\qquad\qquad$ (3) $k(x) = -2^x$

#11 For each function, find the maximum and minimum values within the given domain.

   (1) $y = 3 \cdot 2^x$ $(-1 \le x \le 1)$ $\qquad\qquad$ (4) $y = 4^x + 2^{x+1} - 2$ $(-1 \le x \le 1)$

   (2) $y = 2^{1+2x}$ $(-1 \le x \le 1)$ $\qquad\qquad$ (5) $y = -4^x + 2^{x+3}$ $(2 \le x \le 3)$

   (3) $y = 3^x \cdot 2^{-x}$ $(0 \le x \le 2)$ $\qquad\qquad$ (6) $y = \left(\frac{1}{4}\right)^x - 2\left(\frac{1}{2}\right)^x - 1$ $(-2 \le x \le 3)$

#12 Find the minimum value for each function.

   (1) $y = 2^{2+x} + 2^{2-x}$ $\qquad\qquad\qquad\qquad$ (2) $y = 4^x + 4^{-x} - 2(2^x + 2^{-x}) + 7$

#13 Solve for x.

   (1) $\left(\frac{2}{3}\right)^{x^2 - 2} = \left(\frac{3}{2}\right)^{-2x-1}$ $\qquad\qquad\qquad$ (3) $\left(\frac{1}{4}\right)^x + \frac{1}{4}\left(\frac{1}{2}\right)^x > 4\left(\frac{1}{2}\right)^x + 1$

   (2) $x^{x^2} > x^{2x+15}$ $(x > 0)$

#14 (Transformations)

   (1) If the graph of a function $y = 2^{2x}$ is translated with m units along the x-axis and n units

   along the y-axis, then the graph will be the same as the graph of a function $y = \dfrac{2^{2x} - 3}{16}$.

   Find the value of $m + n$

(2) If the graph of a function $f(x) = 2^x$ is translated with m units along the x-axis and n units along the y-axis, then the graph will be the same as the graph of a function $y = g(x)$.

By the translation, a point $P(1, f(1))$ will be moved to a point $P'(4, g(4))$.

When the graph of $y = g(x)$ passes through a point $(0, 1)$, find the value of mn.

#15 Solve exponential equation.

(1) When an equation $4^x = 2^{x+2} + a$ of x has two different real number solutions, find the range of the constant a.

(2) When an equation $8^x - 4^{x+1} - 2^{x+3} + 8 = 0$ has two different real roots, find the sum of the roots.

(3) Find the sum of all roots of an equation $2^{2x} - 2^{x+2} + 8 = 0$.

(4) Find the product of all roots of an equation $(x-2)^{x^2+2x-5} = (x-2)^{4x+3}$, $x > 2$

#16 Solve each inequality.

(1) $\dfrac{1}{2^x - 1} - \dfrac{2}{4^x + 2^x + 1} \leq \dfrac{5}{8^x - 1}$

(2) $\left(\dfrac{1}{2}x - 1\right)^{x^2 - 8x} < \left(\dfrac{1}{2}x - 1\right)^{4x - 27}$, $x > 2$

#17 Sketch the graph of a function.

(1)

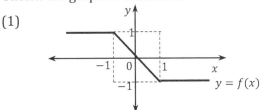
$y = f(x)$

When the graph of a function $y = f(x)$ is shown as the Figure, sketch the graph of a function $y = 2^{1-f(x)}$.

(2) The three functions $f(x)$, $g(x)$, and $h(x)$ satisfy the following conditions:

① $f(x) = f(-x)$,

② $g(-x) = -g(x)$, and

③ $h(x) = f(x) + g(x)$.

When $h(x) = 3^x$, sketch the graph of a function $y = f(x)$.

#18 Find the range.

(1) Find the range of x so that $\log_{x-1}(-3x^2 + 22x - 24)$ is defined for all real number x.

(2) Find the range of a so that $\log_{2a-1}(ax^2 + 3ax + 1 + a)$ is defined for all real number x.

#19 Find the value of x for each expression.

(1) $\log_3 81 = x$

(2) $\log_8 0.25 = x$

(3) $\log_x 16 = 2$

(4) $\log_{2\sqrt{2}} 64 = x$

(5) $\log_x 16 = \dfrac{2}{3}$

(6) $\log_9 x = 1.5$

(7) $\log_4(\log_{81} x) = -1$

#20 Express each expression.

(1) When $\log_{10} 2 = a$ and $\log_{10} 3 = b$, express the following expressions with a and b.

① $\log_{10} 6$ ② $\log_{10} 5$ ③ $\log_{10} 800$ ④ $\log_{10}\left(\dfrac{3}{5}\right)^{-10}$ ⑤ $\log_{10} \sqrt[4]{500}$

(2) When $45^a = 27$ and $5^b = 81$, express $\dfrac{3}{a} - \dfrac{4}{b}$ with a and b.

(3) When $\log_2 3 = a$ and $\log_3 5 = b$, express $\log_{30} 20$ with a and b.

(4) When $10^a = x$, $10^b = y$, and $10^c = z$, express $\log_{xy}\sqrt{y^2 z}$ with a, b, and c.

(5) When $\log_2 10 = \dfrac{1}{a}$ and $\log_3 10 = \dfrac{1}{b}$, express $\log_{\sqrt{48}}\sqrt[3]{24}$ with a and b.

(6) When $x^a = y^b = 6$ for positive numbers x and y, express $\log_{xy} x^3$ with a and b. $(xy \neq 1)$

#21 Simplify the expression.

(1) $10^{\log 5}$

(2) $\log_2(32)^x$

(3) $\log_3\left(\dfrac{1}{9}\right)$

(4) $\log_9 3$

(5) $\log_5\left(\dfrac{\sqrt[3]{5}}{25}\right)$

(6) $\log_2 3 \cdot \log_3 2$

(7) $\log_2 3 \cdot \log_3 4 \cdot \log_4 2$

(8) $\log_2 6 - \log_4 9$

(9) $2\log_2 \sqrt{3} + 4\log_2 3$

(10) $2\log\dfrac{3}{2} - \log\dfrac{5}{3} + 3\log 2 + \dfrac{1}{2}\log 25$

(11) $5^{2\log_5 4 + \log_5 3 - 3\log_5 2}$

#22 Find the value.

(1) When α and β are two different roots of an equation $x^2 - 3x + 3 = 0$,

find the value of $\log_4(\alpha + \beta^{-1}) + \log_4(\alpha^{-1} + \beta) + \log_4 \alpha\beta$.

(2) When $\log_2 a$ and $\log_2 b$ are two different roots of an equation $x^2 - 3x - 4 = 0$

$(a > 0,\ a \neq 1,\ b > 0,\ b \neq 1)$, find the value of $\log_{a^2} 2 + \log_b \sqrt{2}$.

(3) For any rational numbers a and b, $a\log_{10} 20 + \dfrac{b}{\log_8 100} + 5 = 0$. Find the value of $a + b$.

(4) For any real numbers a, b, and c $(a > 1,\ b > 1,\ c > 1)$, $\log_a c : \log_b c = 1 : 2$

Find the value of $\log_a b + \log_b a$.

(5) When $a \log_2 3 = 5$ and $\log_2 b = 1 - \log_2(\log_3 2)$ for any real numbers a and b,

find the value of ab.

#23 Compare the numbers.

(1) $\log_2 3$, $\log_3 2$

(2) $\log(\log 2) + \log(\log 3)$, $2\log(\log \sqrt{6})$

(3) $\log \dfrac{5}{2}$, $\dfrac{\log 2 + \log 3}{2}$

(4) $\log_a b$, $\log_b a$, $\log_a \left(\dfrac{a}{b}\right)$, $\log_b \left(\dfrac{b}{a}\right)$ when $1 < a < b < a^2$

#24 When $\log 2.34 = 0.3692$, determine each logarithm.

(1) $\log 23.4$ (4) $\log(0.234)^2$

(2) $\log 234000$ (5) $\log \sqrt[3]{2340}$

(3) $\log 0.00234$

#25 When $\log 34.5 = 1.5378$, find the value of x.

(1) $\log x = 4.5378$ (2) $\log x = \overline{4}.5378$

#26 When $\log 2 = 0.3010$ and $\log 3 = 0.4771$, determine how many digits are in the integer part of the number or when the first non-zero digit appears in the decimal part of the number.

(1) 6^{100} (4) 5^{-50}

(2) 6^{25} (5) $2^{100} \div 3^{200}$

(3) $(5^5)^5$

#27 (Mantissa) Find the value.

(1) When $\log x$ and $\log x^2$ have the same mantissa, find the value of x. $(10 \le x < 100)$

(2) When $\log x$ and $\log x^3$ have the same mantissa, find the value of x. $(100 \le x < 1000)$

(3) When $\log x$ and $\log \dfrac{1}{x}$ have the same mantissa, find the value of x. $(10 < x < 1000)$

(4) When the sum of mantissas of $\log x$ and $\log \sqrt{x}$ is 1, $(100 \le x < 1000)$,

find the mantissa of $\log x$.

(5) When $\log x = n + \alpha$, (n is an integer, $0 \le \alpha < 1$),

find the value of $\log x$ $(x > 0)$ such that $n^2 + \alpha^2 = \dfrac{37}{9}$

(6) For positive real numbers x and y, the following conditions are satisfied:

① $\log x$ and $\log y$ have the same characteristic.

② $\log x$ and $\log \dfrac{1}{y}$ have the same mantissa.

③ $\log x^2 y^3 = 17.05$

Find the value of $x - y$.

#28 When $10^{0.3456} = 2.2162$, determine the *anti*log of each of the following number and express it in decimal form.

(1) *anti*log 0.3456

(2) *anti*log 2.3456

(3) *anti*log$(0.3456 - 2)$

(4) *anti*log(-2.6544)

#29 Find the value.

(1) Let the integer part and decimal part of the number $\log_3 18$ be a and b, respectively. Find the value of $2^a + 3^b$

(2) Given $\log 2 = 0.3010$, $\log 3 = 0.4771$, and $\log 7 = 0.8451$, find the sum of the two numbers in the highest place value and the ones place value of the number 6^{100}.

(3) For any positive numbers x and y such that $\log x = 3 + \alpha$ $(0 < \alpha < \frac{1}{4})$ and $\log y = 1 + \beta$ $\left(\frac{1}{2} < \beta < 1\right)$, $\dfrac{x^2}{y}$ has n digits in the integer part. Find the value of n.

#30 The characteristic and mantissa of the number $\log 2013$ are m and α, respectively.

The characteristic and mantissa of the number $\log \dfrac{1}{2013}$ are n and β, respectively.

For a triangle $\triangle OAB$ (where $O = O(0,0)$, $A = A(m, \alpha)$, $B = B(n, \beta)$ in a coordinate plane), find the coordinate of the centroid of the triangle $\triangle OAB$.

#31 Find the inverse of each function.

(1) $y = 2 \cdot 3^{x-2}$

(2) $y = 1 + \log_3(x + 2)$

(3) $y = \log_{10}(x + \sqrt{x^2 - 1})$, $x \geq 1$

(4) $y = \dfrac{1}{2^x + 1}$

#32 Solve each exponential equation for x.

(1) $4^{2x-1} = 7^{x+2}$

(2) $3 \cdot 5^{2x} = 2 \cdot 4^{3x}$

(3) $\dfrac{2^x + 2^{-x}}{3} = 4$

#33 Solve for x.

(1) $\left[\log_3(x-1)\right]^2 - 2\log_3(x-1) = 15$

(2) $2\log_3(x+1) - \log_3(x+4) = 2\log_3 2$

(3) $x^{\log x} - 1000\left(\dfrac{1}{x}\right)^2 = 0, \ \ x > 1$

(4) $\log_{\frac{1}{3}}(x-3) + \log_{\frac{1}{3}}(x-5) > -1$

(5) $(\log_2 x)(\log_2 4x) \le 24$

#34 Find the value.

(1) Find the value of a such that $\log_2(\log_3(\log_4 a)) = 0$

(2) When an equation $(\log x)(\log 2x) = \log 3x$ has two different real roots α and β,

find the value of $\alpha\beta$.

(3) For any real numbers x and y $(x > y)$ such that $\begin{cases} \log_2 xy = \dfrac{7}{2} \\ \log_2 x \cdot \log_2 y = \dfrac{3}{2} \end{cases}$, find the value of $\dfrac{x}{y}$.

(4) For a function $f(x) = |\log_2 x - 1|$, $f(a) = f(b)$ where a, b are two different real

numbers. Find the value of ab.

(5) When the solutions of a system of equations $\begin{cases} \dfrac{1}{\log_x 2} + \dfrac{3}{\log_y 8} = 3 \\ \log_2 3x + \log_{\sqrt{2}} y = \log_2 48 \end{cases}$ are $x = \alpha$ and

$y = \beta$, find the value of $\alpha + \beta$.

(6) When an equation $\left(\log_2 \dfrac{x}{2}\right)^2 - 30\log_8 x + 31 = 0$ has two roots α and β,

find the value of $\alpha\beta$.

(7) When an equation $a^{2x} - 2a^x = 3$ $(a > 0, \ a \ne 1)$ has a solution $x = \dfrac{1}{5}$,

find the value of the constant a.

#35 Maximum and minimum values

(1) For a function $y = \log_2(x+1), \ 0 \le x \le 3$, find the maximum and minimum values.

(2) Find the minimum value of $\log_4(x^2 - 4x + 8)$.

(3) For a function $y = \log_2(-x^2 + 4x + 6), \ 0 \le x \le 5$,

find the maximum and minimum values of y.

(4) For two positive numbers x and y such that $x + y = 6$,

find the maximum or minimum value of $\log_{\frac{1}{3}} x + \log_{\frac{1}{3}} y$.

(5) When the minimum value of $y = 2^{a+x} + 2^{a-x}$ is 4, find the value of a.

(6) Find the minimum value of $4 \log_a b + 9 \log_b a$ $(a > 1, \ b > 1)$.

(7) Find the maximum value of $y = \log 5x \cdot \log \frac{20}{x}$ $\left(\frac{1}{5} < x < 20 \right)$.

(8) Find the minimum value of $y = 2(\log_2 2x)^2 + \log_2 (2x)^2 + 6 \log_2 x + 12$.

(9) Find the maximum and minimum values of $y = (\log_2 x)^2 - \log_2 x^4 + 7, \ 4 \le x \le 8$.

(10) Find the maximum value of $y = 100x^4 \div x^{\log x}$.

(11) Find the maximum and minimum values of $y = 100x^{6 - \log x}, \ \frac{1}{10} \le x \le 1000$.

(12) When the minimum value of $y = (\log_3 x)^2 - 3 \log_3 x^2 + a, \ 1 \le x \le 9$ is 0,

find the maximum value of y.

(13) When $x > 1$ and $y > 1$, find the minimum value of $\dfrac{\log_x 2 + \log_y 2}{\log_{xy} 2}$.

(14) When the difference of the maximum and minimum values of x

such that $|a - \log_2 x| \le 1$ is 15, find the value of a.

(15) When $x \ge 0, \ y \ge 0$, and $xy = 1000$,

find the maximum and minimum values of $\log x \cdot \log y$.

#36 Find the range.

(1) For any real number x, an inequality $(1 - \log_2 a)x^2 + 2(1 - \log_2 a)x + \log_2 a > 0$ is

always true. Find the range of a positive number a.

(2) When the roots of an equation $(\log_2 x)^2 - \log_2 x^3 + a = 0$ are in-between $\dfrac{1}{2}$ and 8,

find the range of real number a.

(3) Find the range of x such that $\begin{cases} \log_3 |x - 1| < 3 \\ \log_2 x + \log_2 (x + 2) \ge 3 \end{cases}$

Chapter 1. The Number System

#1 (1) 2   (2) $-2a$ (3) $2a$ (4) 4   (5) $2b - 2c$ (6) $a - b$ (7) $a + 2b$

#2 (1) -7 (2) -6

#3 (1) $\pm\sqrt{2}$ (2) ± 0.1 (3) $\pm\dfrac{4}{3}$ (4) 0

#4 (1) 3   (2) 5   (3) 8   (4) -6 (5) $-\dfrac{2}{3}$ (6) $2^{\frac{2}{3}}$ (7) -5 (8) 3   (9) -10 (10) $-10 + 5^{\frac{3}{2}}$

   (11) 4   (12) $2^{\frac{1}{3}}$ (13) $2^{\frac{1}{6}}5^{-\frac{1}{6}} = \dfrac{2^{\frac{1}{6}}}{5^{\frac{1}{6}}}$

#5 (1) 3   (2) $\dfrac{\sqrt{2}}{4}$ (3) $\dfrac{2\sqrt{6}}{3}$ (4) $2^{\frac{2}{3}} \cdot 3^{\frac{1}{2}} \cdot 5^{\frac{5}{6}}$ (5) $\dfrac{3\sqrt{6}}{5}$ (6) $\sqrt{3}$ (7) $3\sqrt[3]{2}$ (8) $\dfrac{3\sqrt{2}+2}{7}$ (9) 1

#6 (1) $\dfrac{3\sqrt{2}}{2}$ (2) $\dfrac{3\sqrt{5}}{10}$ (3) $\dfrac{\sqrt{6}}{2}$ (4) $-\dfrac{7}{2}\sqrt{3}$ (5) $\sqrt{15} - 1$ (6) $\dfrac{5\sqrt[3]{4}}{2}$ (7) $\dfrac{5\sqrt{6}}{6} - \sqrt{5}$

   (8) $4(2 - \sqrt{3})$ (9) $5\sqrt{6}$

#7 (1) $d = 2\sqrt{2}$, $m = (2, 3)$ (2) $d = \sqrt{89}$, $m = (\frac{1}{2}, -1)$ (3) $d = \sqrt{74}$, $m = \left(\frac{1}{2}, \frac{1}{2}\right)$

   (4) $d = \sqrt{22}$, $m = \left(\frac{3-\sqrt{2}}{2}, \frac{3+\sqrt{2}}{2}\right)$ (5) $d = \sqrt{\frac{73}{4}}$, $m = \left(-2, -\frac{1}{4}\right)$

#8 (1) -3 (2) 2

#9 (1) $a = -1$, $b = 1$ (2) $a = \dfrac{22}{3}$, $b = -\dfrac{4}{3}$ (3) $a = 1$, $b = -3$

   (4) $a = \dfrac{3}{5}$, $b = \dfrac{9}{5}$ (5) $a = 2$, $b = 1$

#10 (1) $2 + 3i$ (2) $4\sqrt{5}\,i$ (3) $3 + 4i$ (4) $8 + 5i$ (5) $-2\sqrt{5}$ (6) $\dfrac{2}{3}$ (7) -75

   (8) $9 - 19i$ (9) $\dfrac{-4}{5} + \dfrac{7}{5}i$ (10) $-\dfrac{\sqrt{6}}{3}i$ (11) $\dfrac{\sqrt{15}}{5}$ (12) $\dfrac{-1}{5} - \dfrac{2\sqrt{6}}{5}i$ (13) -9 (14) $i + 1$

#11 (1) $-2 - 2\sqrt{3}i$ (2) 8   (3) $-\dfrac{1}{2} + \dfrac{\sqrt{3}}{2}i$

#12 $a = \sqrt{3}$ #13 $a = -\dfrac{8}{3}$ or $a = -2$ #14 $x = 3$

#15 $2d - c$ #16 $-2, -1, 0, 1, 2, 3, 4, 5$

#17 (1) 2   (2) 5   (3) $2\sqrt{2}$ #18 (1) 2 (2) 1

#19 $a = -1$, $b = 4$ #20 $\dfrac{b}{a} + \dfrac{a}{b} = 6$

Chapter 2. Polynomials

#1 (1) $5x - 3y$ (2) $-2x - 7y$ (3) $5x^3 + 2x^2 - 3x + 11$ (4) $3x^3 - 7x^2 + 3x - 2$

(5) $2x^3 - x^2 - 16x + 15$ (6) $x^2 - 4y^2 - 4y - 1$ (7) $x^4 + 2x^3 + 3x^2 + 2x - 3$

(8) $x^4 - 6x^3 + x^2 + 24x - 20$

#2 $\frac{3}{2}ab$ **#3** $\frac{2}{3}$ **#4** -1 **#5** $\frac{5}{4}$

#6 (1) $x = 1,\ y = 1,\ z = -5$ (2) $a = 3,\ b = 2,\ c = -4$

#7 (1) $(x+1)(x-1)(x^2-5)$ (2) $(x^2+x+4)(x^2-x+4)$

(3) $(x^2+2x+4)(x^2-2x+4)$ (4) $(x^2+xy+y^2)(x^2-xy+y^2)$

(5) $(x^2-2x-4)(x^2-2x+2)$ (6) $(x^2-x+6)(x^2-x-5)$

(7) $(x+5)(x-3)(x^2+2x+4)$ (8) $(x-2)(x+y-2)$

(9) $(x+z)(x-z)(x-y)$ (10) $(y-z)(x-y)(x-z)$

(11) $(xy-1)(xy-3z)$ (12) $(3x-y-2)^2$ (13) $(-2x+1)^3$

#8 (1) ① 5 ② $\frac{5}{-2}$ ③ 1 ④ 9

(2) 12 (3) 54 (4) 12 (5) $\frac{1}{4}$ (6) -6

(7) $12\sqrt{3}$ (8) 140 (9) $\frac{\sqrt{3}}{9}$ (10) -9 (11) -12

#9 (1) 36 (2) $\frac{38}{13}$ (3) $\pm\sqrt{13}$ (4) 96 (5) -1 (6) $3+2\sqrt{5}$ (7) $-4-\sqrt{3}i$ (8) -3

#10 $(x-2y-3)(x-y+1)$

#11 (1) $\frac{25}{4}$ (2) 4 (3) $\frac{9}{2}$ (4) $\frac{4}{25}$ (5) $\frac{4}{5}$ or $-\frac{4}{5}$ (6) -20 or 20 (7) 8 or -8

(8) 8 or -4 (9) 7 or -17 (10) $\frac{1}{16}x^2$ (11) 31 or -29 (12) $\frac{1}{4}$

(13) $-12\frac{1}{4}$ (14) $6\frac{1}{4}$ (15) 9

#12 (1) -3 (2) -2 (3) -2 (4) -2 (5) $\frac{3}{2}$

#13 (1) $3\frac{1}{3}$ (2) -4 (3) -6 (4) -1

#14 (1) $(x+2)(2x^2-x+3)$ (2) $(x+1)^2(x-2)$

    (3) $(x-1)(x+1)(2x-1)(3x+1)$ (4) $(x+2)(x-1)(2x-3)$

#15 (1) 1 and -3 (2) $\frac{1}{2}, -1+\sqrt{3}$, and $-1-\sqrt{3}$ (3) $1, \frac{1}{2}, \frac{-1+\sqrt{7}\,i}{4}$, and $\frac{-1-\sqrt{7}\,i}{4}$

#16 (1) $2x^3-5x^2+5x-2$ (2) x^2-1 (3) x^3+4x^2+5x+2

#17 (1) $x^2+\frac{5}{2}x-\frac{1}{4}, -\frac{21}{4}$ (2) $x^2+\frac{1}{2}x-\frac{9}{4}, \frac{7}{4}$

#18 (1) $-3x+4$ (2) -4 (3) -4 (4) x^2-x+2 (5) -8 (6) $-2x^2+10x-7$

#19 $a=\frac{5}{2}, \ b=1$

#20 (1) Isosceles triangle   (2) Right triangle

#21 (1) 9800   (2) 1880   (3) -200 (4) 6560   (5) 21   (6) $\frac{101}{200}$ (7) 2^{16} (8) 1

    (9) 34   (10) 8700   (11) 900   (12) 1600   (13) 1001   (14) $\frac{3^{16}-1}{2}$ (15) $-\frac{3^2\cdot 97}{10^4}$

#22 3   #23 (1) $-\frac{x}{y}$ (2) $\frac{1}{x}$ (3) $\frac{y}{x}$ (4) $\frac{3x+5}{2x+3}$

#24 (1) $A=(x-1)^2$, $B=(x-1)(x-3)$ or $A=(x-1)(x-3)$, $B=(x-1)^2$

    (2) $A=(x-1)(x+3)$, $B=(x-1)(x-2)$ or $A=(x-1)(x-2)$, $B=(x-1)(x+3)$

    (3) $A=(x+1)(x-2)$, $B=(x+1)(x-4)$ or $A=(x+1)(x-4)$, $B=(x+1)(x-2)$

#25 -2 #26 11   #27 10

#28 (1) $\sqrt{5}+\sqrt{2}$ (2) $\frac{\sqrt{6}-\sqrt{2}}{2}$ (3) $\sqrt{2}$ (4) $2\sqrt{3}$ (5) $\frac{x-\sqrt{x^2-4}}{2}$ (6) $\frac{\sqrt{2}-\sqrt{6}}{2}$

#29 (1) $-1+\frac{\sqrt{3}}{3}$ (2) $10\sqrt{2}$ (3) $\frac{\sqrt{6}-\sqrt{2}}{2}$ (4) $-2\sqrt{2}-2$

#30 $\frac{10}{x(x+10)}$ #31 7

#32 (1) $\frac{6-\sqrt{2}}{2}$ (2) $\sqrt{2}-1$ (3) 3

#33 $\frac{3}{10}$ #34 $2a$ #35 $\sqrt{x}$

Chapter 3. Equations and Inequalities

#1 (1) $x = 3$ (2) $x = 5$ (3) $x = \frac{1}{6}$ (4) $x = 1$ (5) $x = -\frac{7}{25}$ (6) $x = -2$ (7) $x = 3$ (8) $x = \frac{6}{11}$

#2 (1) 2 (2) 0 (3) 6 (4) $\frac{7}{8}$ (5) $-\frac{2}{5}$ (6) $-3\frac{2}{5}$ (7) $-\frac{1}{36}$ (8) 4

#3 (1) $\frac{1}{6}$ (2) -4 and 6 (3) $\frac{7}{2}$

#4 (1) $3, -1$ (2) $1, \frac{5}{2}$ (3) $0, 2$ (4) $-2, 1$ (5) $-6, 1$ (6) $1, -\frac{1}{2}$

#5 (1) ± 2 (2) $\pm \frac{\sqrt{5}}{3}$ (3) $1 \pm \sqrt{5}$ (4) $\frac{-5 \pm \sqrt{3}}{2}$ (5) $2 \pm \frac{1}{2}$

#6 (1) $\frac{3}{2} \pm \frac{\sqrt{21}}{2}$ (2) $1, -\frac{7}{2}$ (3) $-\frac{3}{2} \pm \frac{\sqrt{29}}{2}$ (4) $1, \frac{1}{3}$

#7 (1) 0 (2) $\pm 2\sqrt{5}$ (3) ± 1 (4) $\frac{9}{8}$ (5) $6\frac{1}{8}$ (6) 2 (7) $-1 \pm \frac{4\sqrt{6}}{3}$

#8 (1) $\frac{53}{16}$ (2) $2\frac{1}{3}$ (3) 4

#9 (1) $a < -2$ (2) $a < \frac{3}{4}$ (3) $a = \frac{1}{8}$ (4) $a > \frac{1}{12}$ (5) $a = \pm\sqrt{5}$ (6) $a = -6$ (7) $a = -3$

   (8) $a = -2$ (9) $a = 8$

#10 (1) $1 \pm \sqrt{5}$ (2) $\frac{-5 \pm \sqrt{37}}{6}$ (3) $\frac{1 \pm \sqrt{6}}{5}$ (4) $-1, \frac{5}{2}$ (5) $3 \pm \sqrt{5}$ (6) $\frac{3 \pm \sqrt{3}}{2}$ (7) $\frac{1 \pm \sqrt{21}}{2}$

   (8) $\frac{-7 \pm \sqrt{17}}{2}$ (9) $\frac{8 \pm \sqrt{19}}{5}$ (10) $-3 \pm \frac{\sqrt{10}}{2}$

#11 (1) 0 (2) -2 (3) $\frac{9}{2}$ (4) $\frac{13}{30}$ (5) $\frac{17}{15}$ (6) $\frac{2}{7}$

#12 (1) 2 different real number solutions. (2) No real number solution

   (3) A double solution (4) 2 different real number solutions

   (5) 2 different real number solutions (6) No real number solution

#13 (1) $a > 6$ (2) $a < -3$ or $-3 < a < \frac{3}{2}$ (3) $a = \frac{-4 \pm 2\sqrt{31}}{9}$ (4) $a = 2 + 2\sqrt{3}$ (5) $a = -\frac{1}{2}$

   (6) $a < 1$ (7) $0 < a < \frac{9}{8}$ (8) $0 < a < \frac{1}{3}$ (9) $a < 0$ (10) $a < -3$ or $-3 < a \le \frac{3}{2}$

   (11) $a = 2$ (12) $\frac{1}{8} < a < \frac{3}{2}$ (13) $a = -1$ or $a = -13$

#14 (1) 1) $-\dfrac{5}{3}$　　2) $\dfrac{37}{9}$　　3) $\pm\dfrac{7}{3}$　　4) $\begin{cases} -\dfrac{35}{9} & \text{when } \alpha - \beta = \dfrac{7}{3} \\ \dfrac{35}{9} & \text{when } \alpha - \beta = -\dfrac{7}{3} \end{cases}$　　5) $\dfrac{5}{2}$

(2) 1) $-3k$　　2) $2k^2 - 4k - 1$　　3) $5k^2 + 8k + 2$　　4) $k^2 + 16k + 4$　　5) $\dfrac{5k^2 + 8k + 2}{2k^2 - 4k - 1}$

(3) 46　(4) 9　(5) -16　(6) 5　(7) 3　(8) $-\dfrac{1}{2}$　(9) -16　(10) 0　(11) 6　(12) 0

#15 (1) $x = -3$ or $x = 2$　　　(2) No solution　　　(3) $x = \dfrac{3 \pm \sqrt{41}}{-4}$

(4) $x = -\dfrac{4}{3}$ or $x = -2$　　(5) No solution

#16 (1) When $k = 6$, $x = -3$ (A double root)　or　When $k = 2$, $x = -1$ (A double root)

(2) $x = -\dfrac{1}{2}$ (A double root)

(3) $x = -\dfrac{4}{3}$ (A double root)

#17 5

#18 (1) $x^2 + 5x + 6 = 0$　　　(2) $x^2 + \dfrac{2}{3}x + \dfrac{1}{3} = 0$　　　(3) $x^2 - \sqrt{6}x + 1 = 0$

#19 (1) $x = -1 + \sqrt{2}$ or $x = -1$　　(2) $x = 2$ or $x = 4$

#20 $\sqrt{3}$

#21 (1) $x = 1$, $x = i$, $x = -i$　　　　　(2) $x = 1$, $x = \dfrac{3 + \sqrt{7}\,i}{4}$, $x = \dfrac{3 - \sqrt{7}\,i}{4}$

(3) $x = 5$, $x = \dfrac{-1 + \sqrt{3}\,i}{2}$, $x = \dfrac{-1 - \sqrt{3}\,i}{2}$　　　(4) $x = 1$, $x = -1$, $x = \dfrac{-1 + \sqrt{5}}{2}$, $x = \dfrac{-1 - \sqrt{5}}{2}$

(5) $x = 2$, $x = \dfrac{1}{2}$, $x = \dfrac{-1 + \sqrt{3}\,i}{2}$, $x = \dfrac{-1 - \sqrt{3}\,i}{2}$

(6) $x = \dfrac{-5 + \sqrt{13}}{2}$, $x = \dfrac{-5 - \sqrt{13}}{2}$, $x = \dfrac{-5 + \sqrt{3}\,i}{2}$, $x = \dfrac{-5 - \sqrt{3}\,i}{2}$

(7) $x = -1$, $x = 1 + i$, $x = 1 - i$

#22 (1) 1　　(2) 18

#23 (1) 0　　　(2) -1　　　(3) -1　　　(4) -2　　　(5) 1

#24 (1) -7　　　(2) -5　　　(3) -6

#25 $a > 3$

#26 (1) 4   (2) 2   (3) 3   (4) -5

#27 $(x, y) = (-6, 5)$

#28 (1) 8   (2) -1 (3) -7 (4) -2 (5) $-\frac{3}{2}$ (6) $-\frac{3}{2}$ (7) -9 (8) $\frac{5}{11}$

#29 3

#30 (1) $(x, y) = (1, 0)$, $(x, y) = (0, 1)$, $(x, y) = (1, -1)$, $(x, y) = (-1, 1)$

   (2) $(x, y) = (1, 3)$, $(x, y) = (3, 1)$, $(x, y) = (-1, -3)$, $(x, y) = (-3, -1)$

#31 (1) ① $-1 \le 2x + 1 \le 3$ ② $-5 \le -3x - 2 \le 1$ ③ $-3\frac{1}{4} \le \frac{1}{4}x - 3 \le -2\frac{3}{4}$

   (2) ① $-2 < y < \frac{2}{3}$ ② $2 < y < \frac{10}{3}$ ③ $-4 \le x \le -1$

#32 (1) 6   (2) -6 (3) $\frac{3}{4}$ (4) $-\frac{4}{3}$ (5) -2 (6) $-\frac{1}{4}$ (7) -3 (8) $-\frac{8}{3}$

#33 (1) $x < -\frac{5}{3}$ (2) $x < -19$

#34 (1) $x = 2$

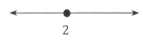

   (2) All real numbers except -3.

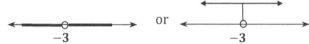   or 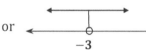

   (3) No solution

   (4) $-1 \le x \le 2$

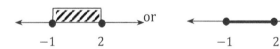   or 

   (5) $-\frac{3}{2} < x < \frac{1}{2}$ or $\frac{1}{2} < x < \frac{5}{2}$.

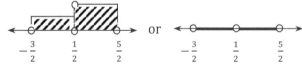

   (6) $1 < x < 2$ or $-4 < x < -3$.

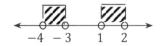

(7) $-\dfrac{5}{3} \le x \le 1$

(8) $-\dfrac{3}{2} < x < \dfrac{5}{2}$

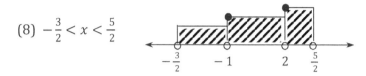

(9) $x > \dfrac{7}{4}$

(10) $\dfrac{8}{5} \le x \le 10$

(11) $x < -3$ or $x > 4$

(12) $-1 < x < 7$

(13) $x > 3$

(14) $x < -\dfrac{9}{2}$ or $x > \dfrac{11}{2}$

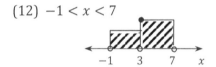

#35 Maximum value is $\dfrac{4}{3}$ and minimum value of is 1.

#36 (1) $a = 2$, $b = -1$ (2) $a = 9$

#37 (1) 5 (2) $-\dfrac{1}{5}$ (3) 1 (4) 19 (5) $\dfrac{5}{3}$ (6) -2

#38 (1) $k > 1$ (2) $k \ge -5$ (3) $5 < k \le 6$ (4) $k \le 0$

#39 (1) 7 (2) 0

#40 -14

#41 (1) $a = -2$, $b = 10$ (2) 13   (3) -4

#42 (1) $a > 2$ or $a < 0$ (2) $a > \frac{1}{2}$ (3) $1 \le a < 4$ (4) $\frac{1}{2} < a < 1$ (5) $a > 4$

     (6) $a > \frac{3}{2}$ (7) $2 \le a < 3$

#43 (1) $0 \le x < 3$ (2) $-1 \le x < 1$

#44 (1) $a + b = 1 \Rightarrow b = 1 - a$

$$a^2 + b^2 - ab = (a + b)^2 - 2ab - ab = (a + b)^2 - 3ab$$

$$= 1 - 3a(1 - a) = 3a^2 - 3a + 1 = 3(a^2 - a) + 1$$

$$= 3\left(\left(a - \frac{1}{2}\right)^2 - \left(\frac{1}{2}\right)^2\right) + 1 = 3\left(a - \frac{1}{2}\right)^2 + \frac{1}{4}$$

Since $3\left(a - \frac{1}{2}\right)^2 \ge 0$, $3\left(a - \frac{1}{2}\right)^2 + \frac{1}{4} > 0$

$\therefore \ a^2 + b^2 - ab > 0$

Therefore, $a^2 + b^2 > ab$

(2) Since $a + b = 1$, $a = 1 - b$

Show that : $1 > ab$

$$1 - ab = 1 - (1 - b)b = b^2 - b + 1 = \left(b - \frac{1}{2}\right)^2 - \left(\frac{1}{2}\right)^2 + 1 = \left(b - \frac{1}{2}\right)^2 + \frac{3}{4}$$

Since $\left(b - \frac{1}{2}\right)^2 \ge 0$, $\left(b - \frac{1}{2}\right)^2 + \frac{3}{4} > 0$

$\therefore \ 1 - ab > 0$

$\therefore \ 1 > ab$

Therefore, $a + b > ab$

(3) Since $a > 0$, $b > 0$, and $c > 0$,

    ① $\frac{a+b}{2} \ge \sqrt{ab}$ ② $\frac{b+c}{2} \ge \sqrt{bc}$ and ③ $\frac{c+a}{2} \ge \sqrt{ca}$

   $\therefore \ \left(\frac{a+b}{2}\right)\left(\frac{b+c}{2}\right)\left(\frac{c+a}{2}\right) \ge \sqrt{a^2 b^2 c^2}$

   $\therefore \ \frac{(a+b)(b+c)(c+a)}{8} \ge abc$

Therefore, $(a + b)(b + c)(c + a) \ge 8abc$

(4) $a^3 + b^3 - ab(a + b) = a^3 + b^3 - a^2b - ab^2$

$$= a^2(a - b) + b^2(b - a)$$
$$= (a - b)(a^2 - b^2)$$
$$= (a - b)(a + b)(a - b)$$
$$= (a - b)^2(a + b)$$

Since $(a - b)^2 \geq 0$ and $a + b > 0$, $(a - b)^2(a + b) \geq 0$.

$\therefore$ $a^3 + b^3 - ab(a + b) \geq 0$

Therefore, $a^3 + b^3 \geq ab(a + b)$

When $a = b$, LHS (Left Hand Side) $= RHS$ (Right Hand Side)

#45 (1) $a^2 + b^2 - ab = \left(a - \frac{1}{2}b\right)^2 - \left(\frac{1}{2}b\right)^2 + b^2 = \left(a - \frac{1}{2}b\right)^2 + \frac{3}{4}b^2$

Since $\left(a - \frac{1}{2}b\right)^2 \geq 0$ and $\frac{3}{4}b^2 \geq 0$, $\left(a - \frac{1}{2}b\right)^2 + \frac{3}{4}b^2 \geq 0$

$\therefore$ $a^2 + b^2 - ab \geq 0$

$\therefore$ $a^2 + b^2 \geq ab$

When $\left(a - \frac{1}{2}b\right)^2 = 0$ and $\frac{3}{4}b^2 = 0$, the equality becomes true.

$\therefore$ $LHS = RHS$, when $a = \frac{1}{2}b$ and $b = 0$

That is, $LHS = RHS$ when $a = b = 0$

(2) Note that: $a > 0$, $b > 0$ $\Rightarrow$ $\frac{2ab}{a+b} \leq \sqrt{ab} \leq \frac{a+b}{2}$

Since $\frac{a+\frac{9}{a}}{2} \geq \sqrt{a \cdot \frac{9}{a}}$, $\frac{a+\frac{9}{a}}{2} \geq 3$

$\therefore$ $a + \frac{9}{a} \geq 6$

Multiplying both sides by a, $a^2 - 6a + 9 \geq 0$ $\therefore$ $(a - 3)^2 \geq 0$

When $a = 3$, $LHS = RHS$

(3) $\left(a + \frac{1}{b}\right)\left(b + \frac{1}{a}\right) = ab + 2 + \frac{1}{ab}$

Since $a > 0$ and $b > 0$, $\frac{ab+\frac{1}{ab}}{2} \geq \sqrt{ab \cdot \frac{1}{ab}}$

$\therefore$ $ab + \frac{1}{ab} \geq 2$ $\therefore$ $ab + \frac{1}{ab} + 2 \geq 4$

$\therefore$ $\left(a + \frac{1}{b}\right)\left(b + \frac{1}{a}\right) \geq 4$

When $ab = 1$, $ab + \frac{1}{ab} + 2 = 4$; that is, $LHS = RHS$ when $ab = 1$

#46 (1) 4   (2) 36   (3) $2\sqrt{3}$ (4) 20   (5) 27   (6) $\dfrac{9}{4}$ (7) 3   (8) 9   (9) 0

#47 (1) $\dfrac{1}{5}$ (2) 4   (3) $\sqrt{15}$ (4) 3

Chapter 4. Elements of Coordinate Geometry and Transformations

#1 (1) $4 + \sqrt{3}$　(2) $\sqrt{74}$　(3) $2\sqrt{10}$　(4) 5

#2 (1) $P\left(-\frac{3}{2}, 0\right)$　(2) $P\left(\frac{1}{2}, \frac{5}{2}\right)$　(3) $(2, 4)$　(4) $(-2, 11)$　(5) $\left(\frac{3}{2}, -\frac{1}{2}\right)$

#3 (1) $(-1, 2)$　(2) $(7, -6)$　(3) $\left(\frac{-1}{2}, \frac{3}{2}\right)$

#4 (1) $\pm 2\sqrt{3} - 1$　(2) $37 + \sqrt{73}$　(3) 6　(4) 1　(5) 5　(6) $\frac{1 + \sqrt{5}}{2}$　(7) 18　(8) 16

#5 $\dfrac{75\sqrt{3}}{4}$

#6 (1) $2x - y - 3 = 0$　(2) $y - 5 = 0$　(3) $2x + 3y - 10 = 0$　(4) $y = -2x - 6$

(5) $3x - y - 1 = 0$　(6) $2x + y - 2 = 0$　(7) $x - 2 = 0$　(8) $2x + y + 1 = 0$

(9) $4x - 7y + 20 = 0$　(10) $3x + 5y - 14 = 0$　(11) $8x - y + 13 = 0$　(12) $x - y + 3 = 0$

(13) $8x - 3y - 12 = 0$　(14) $x + 1 = 0$　(15) $y + 4 = 0$

#7 (1) $2x - y - 1 = 0$　(2) $3x + y - 9 = 0$　(3) $x - 2 = 0$　(4) $y - 3 = 0$　(5) $3x + 4y - 18 = 0$

(6) $3x + 2y - 12 = 0$　(7) $3x - y - 3 = 0$　(8) $y - 3 = 0$　(9) $x - 2 = 0$

#8 (1) -1　(2) $\frac{3}{4}$　(3) -3　(4) $\frac{1}{2}$　(5) 4　(6) -3　(7) $\frac{2}{5}$　(8) 2　(9) 9

#9 (1) $\frac{4}{3}$　(2) -1　(3) $\frac{4}{3}$　　**#10** (1) $\frac{7}{3}$　(2) 2　(3) 6　(4) $-\frac{9}{4}$

#11 (1) $-\frac{3}{2}$　(2) 1　(3) $-\frac{1}{2}$　(4) $\frac{21}{16}$

#12 (1) $5x + 7 = 0$　(2) $3x - 5 = 0$　(3) $3x + 2y + 8 = 0$

#13 24　　**#14** (1) $-\frac{19}{5}$　(2) 0

#15 $l_1 /\!/ l_2 \Rightarrow a = 2$, $l_1 \perp l_2 \Rightarrow a = \frac{2}{3}$　　**#16** $a = 4$　　**#17** (1) $\frac{10\sqrt{13}}{13}$　(2) $\frac{\sqrt{10}}{5}$

#18 $k = -\frac{37}{3}$ or $k = \frac{41}{3}$　　**#19** $3x - 2y + 2\sqrt{13} = 0$ or $3x - 2y - 2\sqrt{13} = 0$

#20 21 (unit^2)　　**#21** $2\sqrt{2}$　　**#22** $\frac{3\sqrt{2}}{2}$ when $a = -\frac{1}{2}$　　**#23** $\sqrt{85}$

#24 (1) $(x - 1)^2 + (y - 3)^2 = 10$　(2) $(x - 1)^2 + (y + 3)^2 = 16$

(3) $(x + 1)^2 + (y - 2)^2 = 4$ (4) $x^2 + (y - 1)^2 = 5$

#25 $x = 4$ and $y = -2$ **#26** $4\sqrt{15}$ **#27** Minimum value is 3 and maximum value is 7.

#28 12

#29 If $D > 0$, then the circle and the line intersect at 2 different points.

If $D = 0$, then the circle and the line intersect at only one point.

If $D < 0$, then the circle and the line do not intersect at any point.

#30 $a > -\dfrac{3}{4}$ **#31** 6

#32 x-intercept (when $y = 0$) is $x = \dfrac{9\sqrt{5}}{5}$ and y-intercept (when $x = 0$) is $y = \dfrac{9}{2}$

#33 6   **#34** Maximum value of $\overline{AB}$ is 6 and minimum value of $\overline{AB}$ is $3\sqrt{2}$.

#35 $6\dfrac{4}{5}$ **#36** $a = 4$ **#37** $x = 1$ and $y = 1 + \sqrt{2}$ **#38** $\dfrac{4\sqrt{5}}{5}$

#39 $y = -\dfrac{1}{3}x + \dfrac{5}{3}$ **#40** (1) $2\sqrt{3}$ (2) $\sqrt{15}$ **#41** $a = 3 + 2\sqrt{3}$ and $b = 0$

#42 (1) 10   (2) -2 (3) 2   **#43** $x^2 + y^2 = 5$

#44 (1) $a = 2$ (2) $a = 1 - \sqrt{15}$ or $a = 1 + \sqrt{15}$ (3) $a = 3 + 2\sqrt{5}$ (4) $a = 3$

#45 $a = -2$ and $b = 1$ **#46** (1) -1 (2) $-4 + 2\sqrt{13}$ or $-4 - 2\sqrt{13}$ (3) 1

#47 $\sqrt{61}$ **#48** $\left(-\dfrac{6}{5}, \dfrac{23}{5}\right)$ **#49** $P\left(\dfrac{1}{12}, \dfrac{5}{6}\right)$ **#50** $\sqrt{2}$ **#51** $\left(x + \dfrac{4}{5}\right)^2 + \left(y - \dfrac{2}{5}\right)^2 = 4$

#52 (1)

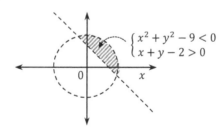

$$\begin{cases} x^2 + y^2 - 9 < 0 \\ x + y - 2 > 0 \end{cases}$$

(2) $|x - y| \le 2$

$\Rightarrow \;\; -2 \le x - y \le 2$

$\Rightarrow \;\; -2 - x \le -y \le 2 - x$

$\Rightarrow \;\; -(-2 - x) \ge y \ge -(2 - x)$

$\Rightarrow \;\; x - 2 \le y \le x + 2$

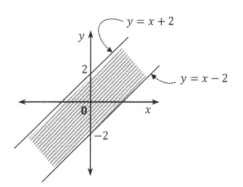

(3)

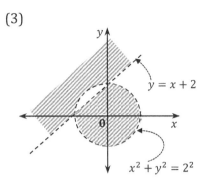

(4)

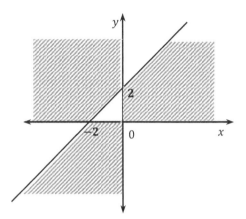

#53 $-2\sqrt{3} < a \leq -2$ or $2 \leq a < 2\sqrt{3}$

#54 16 **#55** (1) $1 < a < 2$ (2) $-\dfrac{3}{2} < a < -\dfrac{1}{2}$

#56 (1) 3 (2) $\dfrac{20}{9}$ (3) 9 (4) 8 **#57** $\dfrac{1}{5}$

#58 Maximum value of is 2 and minimum value of is 1.

Chapter 5. Functions

#1 (1) 4   (2) 3   (3) 6   (4) $\dfrac{15}{4}$ **#2** $\{-3, -1, 1, 3\}$

#3 (1) $\{-4, 0, 2, 4\}$ (2) $\{-4, 0, 4\}$

   (3) The set of $x's$ in the real numbers such that x is not equal to 3.   (4) $[-3, 3]$

#4 (1) 5   (2) $\dfrac{1}{2}$ **#5** $x = \sqrt{2}$ or $x = -\sqrt{2}$

#6 Odd function

#7

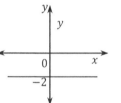

(1) Even

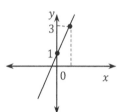

(2) Neither

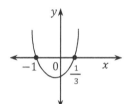

(3) Neither

$$f(x) = 3x^2 + 2x - 1$$
$$= (3x - 1)(x + 1)$$

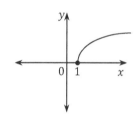

(4) Neither

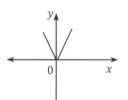

(3) Even

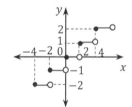

(6) Neither

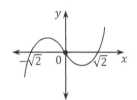

(7) Odd

$$f(-x) = (-x)^3 - 2(-x)$$
$$= -x^3 + 2x = -(x^3 - 2x) = -f(x)$$

#8 2   **#9** (1) $1 \le a \le 3$ (2) $0 < a < 1$

#10 (1) -12 (2) $\dfrac{1}{4}$ (3) -4 (4) $-\dfrac{1}{3}$ (5) 1   (6) $\dfrac{1}{8}$ (7) 3   (8) $\dfrac{3}{7}$ (9) -3

   (10) -12 (11) -9 (12) 21   (13) -8

#11 (1) -1 (2) -2 (3) $-\dfrac{5}{2}$ (4) $\dfrac{5}{2}$ (5) -3

#12 (1) 10   (2) -2 (3) 2

#13 (1) x-intercept (when $y = 0$) is 3 and y-intercept (when $x = 0$) is 3

(2) x-intercept is $-\frac{3}{2}$; $\left(-\frac{3}{2}, 0\right)$ and y-intercept is -6 ; $(0, -6)$

(3) x-intercept is -12 and y-intercept is 6

#14 (1) $2x - y - 3 = 0$ (2) $y - 5 = 0$ (3) $y = -\frac{2}{3}x + \frac{10}{3}$ (4) $2x + y + 6 = 0$

(5) $3x - y - 1 = 0$ (6) $2x + y - 2 = 0$ (7) $x - 2 = 0$ (8) $2x + y + 1 = 0$

(9) $4x - 7y + 20 = 0$ (10) $3x + 5y - 14 = 0$ (11) $8x - y + 13 = 0$ (12) $x - y + 3 = 0$

(13) $8x - 3y - 12 = 0$ (14) $x + 1 = 0$ (15) $y + 4 = 0$

#15 (1) $2x - y - 1 = 0$ (2) $3x + y - 9 = 0$ (3) $x - 2 = 0$ (4) $y - 3 = 0$ (5) $3x + 4y - 18 = 0$

(6) $3x + 2y - 12 = 0$ (7) $3x - y - 3 = 0$ (8) $y - 3 = 0$ (9) $x - 2 = 0$

#16 (1) -1 (2) $\frac{3}{4}$ (3) -3 (4) $\frac{1}{2}$ (5) 4 (6) -3 (7) $\frac{2}{5}$ (8) 2 (9) 9

#17 (1) $\frac{4}{3}$ (2) -1 (3) $\frac{4}{3}$ 　 **#18** (1) $\frac{7}{3}$ (2) 2 (3) 6 (4) $-\frac{9}{4}$

#19 (1) $-\frac{3}{2}$ (2) 1 (3) $-\frac{1}{2}$ (4) $\frac{21}{16}$

#20 (1) $5x + 7 = 0$ (2) $3x - 5 = 0$ (3) $3x + 2y + 8 = 0$

#21 (1) -1 (2) -4 (3) 6 　 **#22** (1) 6 (2) 12 (3) 6 (4) 24

#23 1 　　　 **#24** (1) $4x + 3$ (2) $2x - 5$ (3) $8x + 7$ (4) $2x - 8$

#25 (1) $\frac{x}{2} + \frac{3}{2}$ (2) $2x - 3$

#26 (1) $\frac{1}{2}$ (2) -2 (3) 0 (4) 3

#27

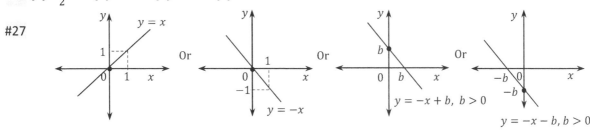

#28 (1) Inverse   (2) Inverse   (3) Neither   (4) Direct   (5) Direct   (6) Direct

#29 g and h

#30 (1)

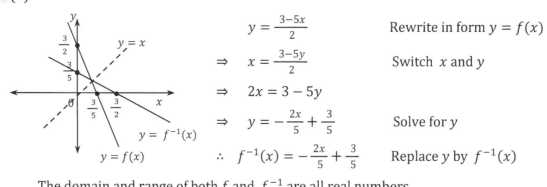

$$y = \frac{3-5x}{2} \qquad \text{Rewrite in form } y = f(x)$$

$$\Rightarrow \quad x = \frac{3-5y}{2} \qquad \text{Switch } x \text{ and } y$$

$$\Rightarrow \quad 2x = 3 - 5y$$

$$\Rightarrow \quad y = -\frac{2x}{5} + \frac{3}{5} \qquad \text{Solve for } y$$

$$\therefore \quad f^{-1}(x) = -\frac{2x}{5} + \frac{3}{5} \qquad \text{Replace } y \text{ by } f^{-1}(x)$$

The domain and range of both f and f^{-1} are all real numbers.

(2)

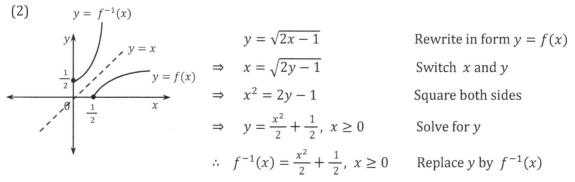

$$y = \sqrt{2x - 1} \qquad \text{Rewrite in form } y = f(x)$$

$$\Rightarrow \quad x = \sqrt{2y - 1} \qquad \text{Switch } x \text{ and } y$$

$$\Rightarrow \quad x^2 = 2y - 1 \qquad \text{Square both sides}$$

$$\Rightarrow \quad y = \frac{x^2}{2} + \frac{1}{2}, \ x \geq 0 \qquad \text{Solve for } y$$

$$\therefore \quad f^{-1}(x) = \frac{x^2}{2} + \frac{1}{2}, \ x \geq 0 \qquad \text{Replace } y \text{ by } f^{-1}(x)$$

The domain of f is the interval $\left[\frac{1}{2}, \infty\right)$ and the range of f is the interval $[0, \infty)$.

The domain of f^{-1} is the interval $[0, \infty)$ and the range of f^{-1} is the interval $\left[\frac{1}{2}, \infty\right)$.

(3)

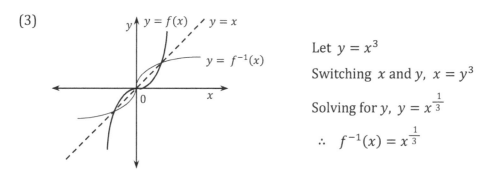

Let $y = x^3$

Switching x and y, $x = y^3$

Solving for y, $y = x^{\frac{1}{3}}$

$$\therefore \quad f^{-1}(x) = x^{\frac{1}{3}}$$

(4) Let $y = \frac{1}{x-1}$

Switching x and y, $x = \frac{1}{y-1}$

Solving for y, $y - 1 = \frac{1}{x}$ $\qquad \therefore \quad y = \frac{1}{x} + 1$

$$\therefore \quad f^{-1}(x) = \frac{1}{x} + 1$$

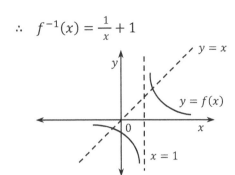 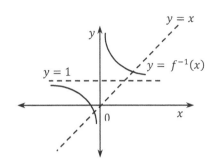

#31 (1) $\frac{5}{3}$ (2) -2 (3) -3 (4) -16 (5) $-\sqrt{5}$ (6) $\frac{5}{2}$ (7) 2   (8) 2   (9) 5

#32 (1) $a > 2$ or $a < -2$ (2) $0 \le a < \frac{1}{2}$ (3) $-2 \le a \le 2$

#33 (1) $3\sqrt{2}$ (2) $2\sqrt{2}$ **#34** $\frac{1}{2}g(x)$

#35 (1) 10   (2) -2

     (3)   i) When $x \ge 0$, unlimited numbers of intersection points
        ii) When $x \le 0$, only one point $x = 0$.

#36 (1) True   (2) False   (3) True

#37 (1) Max: 1, Min: -3 (2) 13   (3) $\frac{1}{2}$ (4) 39   (5) Max: 0, Min: -8 (6) Max: 1, Min: 0

     (7) $f(2)$ (8) 9   (9) Max: 2, Min: -2 (10) $-\frac{1}{3}$

#38 (1) $a \ge -\frac{1}{5}$ (2) $-3 < a < -2$ (3) $a > 2$ (4) $-3 < a < 1$ (5) $a > 7$ (6) $-15 < a < -3$

     (7) $1 < a < \frac{5}{4}$ (8) $-2 \le a \le 0$ (9) $0 < a < \frac{-1+\sqrt{5}}{4}$ (10) $-6 + 2\sqrt{10} < a < \frac{1}{2}$

#39 (1) $\{0, 1\}$ (2) $\{0, 1\}$ (3) $\{0, -1\}$ **#40** 7 integers

#41 (1) $x < -1$ or $x > -1$ (2) All real numbers   (3) $x = 5$ (4) No solution

#42 (1) $a = -1$, $b = -2$ (2) $a = 6$, $b = 8$ (3) $a = 4$, $b = -4$ (4) $a = -2$, $b = -4$

#43 (1) $1 - 2\sqrt{2} < a < 1 + 2\sqrt{2}$ (2) $a > 1$ or $a < -\frac{1}{3}$ (3) $0 < a \le 3$

#44 (1) $-5 < a < 7$ (2) $-2\sqrt{3} < a < 2\sqrt{3}$ (3) $a > 2\sqrt{2}$ or $a < -2\sqrt{2}$ (4) $a \le -3$

     (5) $-1 < a < 1$ (6) $2 \le a < 10$ (7) $a \le 2$, $a \ge 6$ (8) $-4 \le a \le \frac{4}{3}$ (9) $0 \le a < 1$

     (10) $a > -1$ (11) $1 \le a \le 3$ (12) $a \ge 2$ (13) $-6 < a < -\frac{7}{3}$

#45 (1) -2 (2) $\frac{3}{2}$ (3) -9 (4) $-\frac{5}{4}$ (5) $\frac{25}{8}$ (6) $\frac{25}{4}$ (7) -10 (8) 3 (9) -2

(10) $-\frac{3}{2}$ (11) 1 (12) 70

#46 (1) $\frac{3}{2} \le f(x) \le 2$ (2) $f(x) \ge -1$ or $f(x) \le -3$ (3) $f(x) \le 0$ or $f(x) > 1$

(4) $f(x) \le \frac{1}{3}$ or $f(x) \ge 3$

#47 (1)

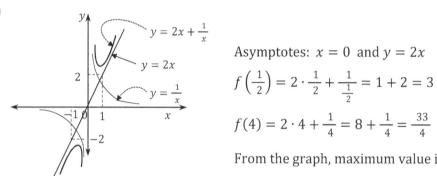

Asymptotes: $x = 0$ and $y = 2x$

$f\left(\frac{1}{2}\right) = 2 \cdot \frac{1}{2} + \frac{1}{\frac{1}{2}} = 1 + 2 = 3$

$f(4) = 2 \cdot 4 + \frac{1}{4} = 8 + \frac{1}{4} = \frac{33}{4}$

From the graph, maximum value is $f(4) = \frac{33}{4}$

(2)

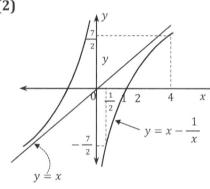

Asymptotes: $x = 0$ and $y = x$

$f\left(\frac{1}{2}\right) = \frac{1}{2} - \frac{2}{\frac{1}{2}} = \frac{1}{2} - 4 = -\frac{7}{2}$

$f(4) = 4 - \frac{2}{4} = 4 - \frac{1}{2} = \frac{7}{2}$

From the graph, minimum value is $f\left(\frac{1}{2}\right) = -\frac{7}{2}$

(3) Since $y = \frac{ax+b}{x+c}$ is symmetric with respect to a point $(2, 1)$, the axes of symmetries are

$x = 2$ and $y = 1$. $\therefore y = \frac{k}{x-2} + 1 = \frac{k+(x-2)}{x-2}$

Since the function y passes through a point $(3, 3)$,

$3 = \frac{k+(3-2)}{3-2} = k + 1$

$\therefore k = 2$ $\therefore y = \frac{2}{x-2} + 1$

Since $f(-1) = \frac{2}{-1-2} + 1 = \frac{1}{3}$ and $f(1) = \frac{2}{1-2} + 1 = -1$,

maximum value is $f(-1) = \frac{1}{3}$ and minimum value is $f(1) = -1$

#48 (1) Translation: 2 units along the x-axis   (2) Translation: -2 units along the y-axis

(3) Translation: 1 unit along the x-axis and 2 units along the y-axis

(4) Translation: -2 units along the x-axis and -1 unit along the y-axis

#49 (1) Translation: 1 unit along the x-axis

(2) Translation: -1 unit along the y-axis

(3) Translation: -1 unit along the x-axis and -1 unit along the y-axis

(4) Translation: 2 units along the x-axis and 1 unit along the y-axis

#50 (1) $y = \sqrt{-2x} - 2$ (2) $y = \sqrt{-2(x-3)}$ (3) $y = \sqrt{-2\left(x + \frac{1}{2}\right)} + 4$

#51 (1) $x = \sqrt{3}$ or $x = -\sqrt{3}$ (2) $x = 1$ or $x = -\frac{5}{2}$ (3) $x = 0$

#52 (1) $0 \le x < 2$, $x \ge 3$ (2) $x \le -1$, $1 < x \le 2$ (3) $1 < x \le \frac{3}{2}$ (4) $x > 3$

#53 (1) Max: $\sqrt{3} - 2$ Min: -1 (2) Max: 0 Min: -1 (3) Max: 0 Min: -1

(4) Max: 0 Min: $-\sqrt{5} + 1$ (5) Max: 1 Min: 0

#54 (1) $x = 3$ (2) $x = 0$

#55 (1) $\frac{3}{2} \le x < 2$, $2 < x < 6$ (2) $-\frac{1}{2} \le x < 0$, $x > 4$ (3) $0 \le x < 4$ (4) $4 < x \le 5$

(5) $1 < x \le 2$ (6) $-\frac{5}{2} < x \le \frac{3}{2}$

#56 (1) $x < -1$, $\frac{3}{2} < x < 5$ (2) $x \le -3$, $3 \le x \le 5$ (3) $x \le 3$ (4) $x < 3$ (5) $x \le -1$

(6) $x \le -3$, $\frac{1}{2} \le x \le 1$, $x \ge 5$ (7) $-2 \le x \le -1$, $x = 3$ (8) $-3 \le x \le 1$

(9) $x \le -1$, $0 \le x \le 4$ (10) $x < -1$, $-1 < x < 4$, $x > 4$ (11) $x \le 2$ or $x \ge 3$

#57 $y = \frac{1}{2} - \frac{(x-3)^2}{2}$ The domain is $x \ge 3$ and the range is $\le \frac{1}{2}$.

#58 (1) 34 (2) $\frac{3}{2}$

#59 (1) max maximum value of a is $\frac{5}{8}$, minimum value of b is 1 (2) $2\sqrt{2}$

#60 3 (unit^2) **#61** (1) -2 (2) 2 (3) -4 (4) 6 (5) $-\frac{5}{4}$ (6) -2

#62 (1) $2 \le a < \frac{9}{4}$ (2) $a \ge 2$ (3) $a > 2$ (4) 192 (5) $0 \le a < \frac{3}{4}$ (6) $\frac{2}{3} \le a \le \frac{7}{2}$

#63 (1) 2   (2) -2 (3) $a = 0$ or $a = -\frac{1}{4}$ (4) $x = 1$ (5) $-\frac{13}{6}$ (6) $a = 10$

(7) $x = \frac{-1+\sqrt{13}}{2}$ or $x = \frac{-1-\sqrt{13}}{2}$ (8) 0   (9) 59   (10) two   (11) 9

#64 (1) $-1 \le a \le \sqrt{2}$ (2) $x = -1$ (3) -3 (4) $x = 4$ (5) 3   (6) $x = 1-\sqrt{2}$, $x = 5-\sqrt{10}$

#65 (1) $-1 < a < 1$ (2) $\frac{1}{4} \le a \le \frac{1}{2}$ **#66**   At least 4 days

#67 (1) $0 \le x \le 1$ or $x = -1$ (2) $\frac{1}{3} \le x < 1$ or $1 < x < 2$

(3) $-5 \le x < 0, \ 0 < x < 1, \ 1 < x \le 6$ (4) $1 \le x \le 5$ or $-5 \le x \le -1$

(5) $-2 < x \le 1$ or $4 \le x \le 5$ or $x = 2$

#68 (1) $a \ge 2$ (2) $0 \le a \le 4$ (3) $a \le 0$ (4) $-4 \le x < -2$ or $-2 < x < -\frac{4}{3}$

(5) $4 \le a < 5$ (6) $-1 < a < 1$ (7) $a \ge 2$ (8) $-2 < x < 0$ or $2 \le x < 4$

(9) $1 < x < 3$ or $3 < x \le 5$ (10) $3 \le a < 4$

(11) $3 \le a < 4$ and the integers are $1, \ 2, \ 3, -6, -5, -4$.

#69 (1) Maximum value of x is $\frac{3}{2}$ and minimum value of x is -1.   (2) $\frac{1}{2}$ (3) 6

#70 $f(x) = \frac{a(x-4)}{x-1}, \ a > 0$ **#71** -3 **#72** 2

Chapter 6. Exponential and Logarithmic Functions

#1 (1) a^8 (2) $\dfrac{1}{a^4}$ (3) 1   (4) a^6 (5) a^2b^2 (6) a^6b^{15} (7) $-27a^6b^9$ (8) a^{11}

(9) $\dfrac{a^4b^8}{c^{12}}$ (10) $\dfrac{-243a^5b^{10}}{32c^{15}}$

#2 (1) $\dfrac{1}{9}$ (2) $\dfrac{1}{16}$ (3) 0   (4) $\dfrac{1}{125}$ (5) 1   (6) 4   (7) 32   (8) $\dfrac{1}{81}$ (9) $\dfrac{1}{8}$ (10) $\dfrac{1}{16}$

(11) $\dfrac{64}{9}$ (12) $\dfrac{32}{9}$ (13) 1

#3 (1) 4   (2) -4 (3) 3   (4) -2 (5) 2   (6) $\dfrac{1}{2}$ (7) 2   (8) 4   (9) $\dfrac{4}{9}$

(10) -27 (11) 2   (12) 0.1   (13) 2   (14) 3   (15) $\sqrt{2}$

#4 (1) 4   (2) ± 4 (3) 2   (4) Undefined   (5) ± 2 (6) 2   (7) -2 (8) -2 (9) 2   (10) -2

#5 (1) 4   (2) 8   (3) 16   (4) $\dfrac{1}{4}$ (5) $\dfrac{1}{8^3} = \dfrac{1}{512}$ (6) 16   (7) $\dfrac{4}{3}$ (8) $\dfrac{1}{4}$ (9) $\dfrac{1}{6\sqrt[3]{3}}$

(10) $\sqrt{2}$ (11) $2^{\frac{7}{6}} \times 3^{\frac{11}{12}}$

#6 (1) $\sqrt[3]{3} < \sqrt[4]{9} < \sqrt[5]{27}$ (2) $\dfrac{1}{4} < \sqrt{\dfrac{1}{2}} < \left(\dfrac{1}{2}\right)^{\frac{1}{3}} < \left(\dfrac{1}{2}\right)^{-2}$ (3) $\sqrt[12]{81} < \sqrt[6]{16} = \sqrt[9]{64}$

(4) $\sqrt[6]{6} < \sqrt{2} < \sqrt[3]{3} < \sqrt[4]{5}$ (5) $\sqrt[3]{3} = \sqrt{\sqrt[3]{9}} < \sqrt[4]{5}$ (6) $\left(\sqrt[3]{2\sqrt{3}}\right)^2 < \left(\sqrt[3]{3\sqrt{2}}\right)^2 < \left(\sqrt{2\sqrt[3]{3}}\right)^2$

(7) $\sqrt{3\sqrt{3}} < \sqrt{3^{\sqrt{3}}} = (\sqrt{3})^{\sqrt{3}}$ (8) $A > B > C$

#7 (1) 1   (2) $\sqrt[3]{a}$ (3) $a^{\frac{7}{12}}$ (4) $a^{\frac{7}{8}}$ (5) $a^{\frac{15}{16}}$ (6) $a - b$ (7) $a^{\frac{2}{3}} + a^{\frac{1}{3}}b^{\frac{1}{3}} + b^{\frac{2}{3}}$ (8) $a^{\frac{1}{2}} + a^{-\frac{1}{2}}$

#8 (1) ① 7   ② 47   ③ 18   (2) 3   (3) ① 125   ② $\dfrac{3}{2}$ ③ $\dfrac{21}{5}$ (4) 2^6 (5) 2   (6) $\dfrac{41}{14}$ (7) $\dfrac{1}{4}$

(8) $\dfrac{1}{16}$ (9) $\dfrac{a}{c}$ (10) $2^{\frac{5m-7}{6}} \cdot 3^{\frac{5n}{6}}$ (11) $\sqrt{2}$ (12) $8\dfrac{1}{5}$ (13) $\dfrac{1}{e^2}$

#9 (1) $2\sqrt{2}$ (2) $-\dfrac{4}{3}$ (3) 10^4 (4) 6   (5) 4   (6) 24   (7) $\dfrac{1}{12}$ (8) 2   (9) $\dfrac{1}{10}$ (10) 0

#10 (1) The domain is $(-\infty, \infty)$ and the range is $(0, \infty)$.

(2) The domain is $(-\infty, \infty)$ and the range is $(-3, \infty)$.

(3) The domain is $(-\infty, \infty)$ and the range is $(-\infty, 0)$.

#11 (1) Max: 6 Min: $\dfrac{3}{2}$ (2) Max: 8 Min: $\dfrac{1}{2}$ (3) Max: $\dfrac{9}{4}$ Min: 1   (4) Max: 6 Min: $-\dfrac{3}{4}$

(5) Max: 16 Min: 0   (6) Max: 7 Min: -2

#12 (1) 8   (2) 5

#13 (1) $x = 3$ or $x = -1$ (2) $x > 5$ or $0 < x < 1$ (3) $x < -2$

#14 (1) $\dfrac{29}{16}$ (2) $\dfrac{21}{8}$ **#15** (1) $-4 < a < 0$ (2) 2   (3) 3   (4) 12

#16 (1) $0 < x \le 1$ (2) $2 < x < 3$ or $4 < x < 9$

#17 (1) 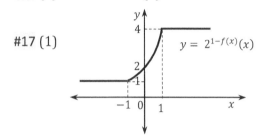   (2)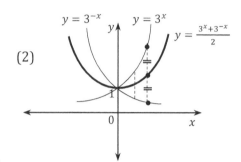

#18 (1) $\dfrac{4}{3} < x < 2$ or $2 < x < 6$ (2) $\dfrac{1}{2} < x < \dfrac{4}{5}$

#19 (1) 4   (2) $-\dfrac{2}{3}$ (3) 4   (4) 4   (5) 64   (6) 27   (7) 3

#20 (1) ① $a + b$ ② $1 - a$ ③ $3a + 2$ ④ $-10(b - 1 + a)$ ⑤ $\dfrac{3}{4} - \dfrac{a}{4}$

　　　(2) 2   (3) $\dfrac{2 + ab}{1 + a + ab}$ (4) $\dfrac{2b + c}{2(a + b)}$ (5) $\dfrac{2(3a + b)}{3(4a + b)}$ (6) $\dfrac{3b}{a + b}$

#21 (1) 5   (2) $5x$ (3) -2 (4) $\dfrac{1}{2}$ (5) $-\dfrac{5}{3}$ (6) 1   (7) 1   (8) 1   (9) $5 \log_2 3$

　　　(10) $\log 54 = \log 2 + 3 \log 3$ (11) 6

#22 (1) 2   (2) $-\dfrac{3}{8}$ (3) $-\dfrac{5}{3}$ (4) $\dfrac{5}{2}$ (5) 10

#23 (1) $\log_2 3 > \log_3 2$ (2) $\log(\log 2) + \log(\log 3) < 2\log(\log\sqrt{6})$ (3) $\log\dfrac{5}{2} > \dfrac{\log 2 + \log 3}{2}$

　　　(4) $\log_a\left(\dfrac{a}{b}\right) < \log_b\left(\dfrac{b}{a}\right) < \log_b a < \log_a b$

#24 (1) 1.3692   (2) 5.3692   (3) -2.6308 (4) -1.2616 (5) 1.1231

#25 (1) 34500   (2) 0.000345

#26 (1) 78 digits in the integer part   (2) 20 digits in the integer part

　　　(3) 18 digits in the integer part

　　　(4) The 1st non-zero number appears 35th away from the decimal point.

　　　(5) The 1st non-zero number appears 66th away from the decimal point.

#27 (1) 10 (2) 10^2 or $10^{\frac{5}{2}}$ (3) $10^{\frac{3}{2}}$, 10^2, or $10^{\frac{5}{2}}$ (4) $\frac{2}{3}$ (5) $\frac{7}{3}$ or $-\frac{5}{3}$ (6) 0

#28 (1) 2.2162 (2) 221.62 (3) 0.022162 (4) 0.0022162

#29 (1) 6 (2) 12 (3) 5 **#30** $\left(-\frac{1}{3}, \frac{1}{3}\right)$

#31 (1) $y = \log_3\left(\frac{9}{2}x\right), \ x > 0$ (2) $y = 3^{x-1} - 2$ for real number x

     (3) $y = \frac{1}{2}(10^x + 10^{-x}), \ x \geq 0$ (4) $y = \log_2\left(\frac{1-x}{x}\right)$

#32 (1) $x = \dfrac{\ln 4 + 2\ln 7}{2\ln 4 - \ln 7}$ (2) $x = \dfrac{\ln\frac{2}{3}}{\ln\left(\frac{25}{64}\right)}$ (3) $x = \dfrac{\ln(6+\sqrt{35})}{\ln 2}$ or $x = \dfrac{\ln(6-\sqrt{35})}{\ln 2}$

#33 (1) $x = 244$ (2) $x = 5$ (3) $x = 10$ (4) $5 < x < 6$ (5) $2^{-6} \leq x \leq 2^4$

#34 (1) $a = 4^3$ (2) $\alpha\beta = 5$ (3) $4\sqrt{2}$ (4) 4 (5) 6 (6) 2^{12} (7) 3^5

#35 (1) Max: 2 Min: 0 (2) Min: 1 (3) Max: $1 + \log_2 5$ Min: 0 (4) Max: -2 (5) $a = 1$

     (6) Min: 12 (7) Max: 1 (8) Min: -2 (9) Max: 4 Min: 3 (10) Max: 10^6

     (11) Max: 10^{11} Min: 10^{-5} (12) Max: 8 (13) Min: 4

     (14) $a = 1 + \log_2 5$ (15) Max: $\frac{9}{4}$ Min: 2

#36 (1) $\sqrt{2} < a \leq 2$ (2) $0 < a \leq \frac{9}{4}$ (3) $2 \leq x < 28$

Index

A

B

C

D

CPSIA information can be obtained
at www.ICGtesting.com
Printed in the USA
LVHW021406210723
752924LV00010B/112